Pixel Detectors

Particle Acceleration and Detection

springer.com

The series *Particle Acceleration and Detection* is devoted to monograph texts dealing with all aspects of particle acceleration and detection research and advanced teaching. The scope also includes topics such as beam physics and instrumentation as well as applications. Presentations should strongly emphasise the underlying physical and engineering sciences. Of particular interest are

- contributions which relate fundamental research to new applications beyond the immediate realm of the original field of research

- contributions which connect fundamental research in the aforementioned fields to fundamental research in related physical or engineering sciences

- concise accounts of newly emerging important topics that are embedded in a broader framework in order to provide quick but readable access of very new material to a larger audience

The books forming this collection will be of importance for graduate students and active researchers alike.

Series Editors:

Professor Alexander Chao
SLAC
2575 Sand Hill Road
Menlo Park, CA 94025
USA

Professor Christian W. Fabjan
CERN
PPE Division
1211 Genève 23
Switzerland

Professor Rolf-Dieter Heuer
DESY
Gebäude 1d/25
22603 Hamburg
Germany

Professor Takahiko Kondo
KEK
Building No. 3, Room 319
1-1 Oho, 1-2 1-2 Tsukuba
1-3 1-3 Ibaraki 305
Japan

Professor Franceso Ruggiero
CERN
SL Division
1211 Genève 23
Switzerland

Leonardo Rossi Peter Fischer
Tilman Rohe   Norbert Wermes

Pixel Detectors

From Fundamentals to Applications

With 177 Figures

Springer

Dr. Leonardo Rossi
Instituto Nazionale di Fisica Nucleare
Dipartimento di Fisica
Via Dodecaneso 33
16146 Genova
Italy
E-mail: leonardo.rossi@ge.infn.it

Dr. Tilman Rohe
Paul Scherrer Institut
5232 Villigen
Switzerland

Dr. Peter Fischer
Institut für Technische Informatik
Universität Mannheim
B6, 26, 68131 Mannheim
Germany

Dr. Norbert Wermes
Physikalisches Institut
Universität Bonn
Nussallee 12, 53115 Bonn
Germany

Library of Congress Control Number: 2005930884

ISSN 1611-1052

ISBN-10 3-540-28332-3 Springer Berlin Heidelberg New York
ISBN-13 978-3-540-28332-4 Springer Berlin Heidelberg New York

This work is subject to copyright. All rights are reserved, whether the whole or part of the material is concerned, specifically the rights of translation, reprinting, reuse of illustrations, recitation, broadcasting, reproduction on microfilm or in any other way, and storage in data banks. Duplication of this publication or parts thereof is permitted only under the provisions of the German Copyright Law of September 9, 1965, in its current version, and permission for use must always be obtained from Springer. Violations are liable for prosecution under the German Copyright Law.

Springer is a part of Springer Science+Business Media
springer.com
© Springer-Verlag Berlin Heidelberg 2006
Printed in The Netherlands

The use of general descriptive names, registered names, trademarks, etc. in this publication does not imply, even in the absence of a specific statement, that such names are exempt from the relevant protective laws and regulations and therefore free for general use.

Typesetting: by the authors and TechBooks using a Springer LATEX macro package

Cover design: *design & production* GmbH, Heidelberg

Printed on acid-free paper   SPIN: 10843298   54/TechBooks   5 4 3 2 1 0

Preface

Progress in science requires the continuous refinement of experimental methods. Particle physics is just a particular example where sophisticated experimental techniques are of utmost importance. Collisions at high-energy accelerators produce many particles whose direction, energy, and identity are measured with large and complex detectors. One of the most demanding requirements is the simultaneous detection of hundreds of particle tracks with micrometer spatial and nanosecond timing precision in the harsh radiation environment close to the point of their production.

Addressing and mastering such requirements has become possible by what one might want to call the "silicon age" of the human development. Since the discovery of the transistor effect, for which Shockley, Bardeen, and Brattain were awarded the Nobel Prize in 1956, the number of applications using selectively doped silicon has not ceased to increase. Miniaturization of electronics devices has steadily progressed over the past decades. Already in 1965 Moore had predicted the doubling of the number of transistors per integrated circuit every ≈ 2 years. Surprisingly enough, this exponential increase still holds after 40 years and it is today difficult to find a large-scale industrial product or research project where microelectronics does not play a significant role.

The development of highly segmented detectors with many million pixels, each operating as an independent intelligent sensing element, also has only become possible from the advancements in microelectronics. Hand in hand with the continuous reduction of the structure sizes, the manufacturing costs were reduced and access to the design tools was made possible. This allowed the particle physics community to adapt the microelectronics technology to the needs of the experiments. Silicon microstrip and pixel detectors and also the associated integrated readout electronics rapidly became standard building blocks for detector systems and broke the way to the precise measurement of short-lived particle decay vertices. Pixel detectors are the most recent step in this evolution, finding important applications in particle physics experiments, most notably in the detectors which are currently being built for the Large Hadron Collider. The mastering of the pixel technology has also led to first applications in other fields and more are certain to follow soon.

It is therefore time to share the know-how that has been accumulated over the last years in this field. This book may help pixel detector designers for other and future applications of the pixel technology, and it shall also contribute to informing the scientific community about the present status and possible further developments of these powerful and versatile detection devices.

We must point out that the field is in a fast evolution at the time of writing this book. Some very recent results could not be included and some statements may already be outdated at the time of printing. The many references given should help the reader to find the most recent results in the various fields.

The text is organized as follows: after Chap. 1, which introduces pixel detectors in general terms and gives some historical perspective to this development, the sensor is discussed in detail in Chap. 2, the electronics in Chap. 3, and the way to connect them in Chap. 4. Chapter 5 presents applications both in particle physics and in other fields. Chapter 6, finally, attempts a look into the future and indicates some promising developments.

This book would not have been possible without the help and encouragement of many people. Chris Fabjan suggested to address the issue of pixel detectors and has continuously stimulated the authors. Murdock Gilchriese and Peter Weilhammer did contribute to the early stage of the book. Ladislav Andricek, Michael Moll, Rainer Richter, and Renate Wunstorf helped with many useful discussions on pixel sensors, Ivan Peric did comment on the electronics part, Oswin Ehrmann on the bump-bonding processes, and Michael Overdick on imaging applications of pixel detectors. We thank Federico Antinori, Roland Horisberger, and Simon Kwan for carefully reading the description of, respectively, the ALICE, CMS, and BTeV pixel detector description. We also thank Rinaldo Cenni and Laura Opisso for their help with Latex when putting the pieces together and Rosanna Puppo and Bert Wiegers for making many drawings. Finally, we thank our families for their patience and understanding during the writing of this book.

Geneva,
May 2005

Leonardo Rossi
Peter Fischer
Tilman Rohe
Norbert Wermes

Contents

1 **Introduction** 1
  1.1 Generalities on Pixel Detectors 1
    1.1.1 Motivations for Pixel Detectors in Particle Physics.... 2
    1.1.2 Working Principle and Operating Characteristics
          of Segmented Silicon Detectors 5
    1.1.3 Hybrid Pixel Detectors 9
    1.1.4 Monolithic Pixel Detectors 12
  1.2 Evolution of Pixel Detectors in Particle Physics 13
    1.2.1 The First Pixel Detectors and Their Use in Experiments 16
    1.2.2 Other Applications 23

2 **The Sensor** 25
  2.1 Introduction 25
  2.2 Device Physics and Fundamental Sensor Properties 26
    2.2.1 Carrier Concentration 26
    2.2.2 Charge Generation and Recombination in Silicon 29
    2.2.3 Transport of Charge Carriers 37
    2.2.4 The pn-Junction 40
    2.2.5 Surface Barrier 44
    2.2.6 Metal Oxide Semiconductor Structure............... 46
    2.2.7 Punch Through 48
  2.3 Pixel Sensors and Their Properties 50
    2.3.1 Different Types of Silicon Sensors 50
    2.3.2 Leakage Current and Maximum Operation Voltage ... 51
    2.3.3 Full Depletion Voltage and Substrate Doping 53
    2.3.4 Pixel Capacitance 57
    2.3.5 Charge Motion and Signal Formation 59
    2.3.6 Spatial Resolution 61
    2.3.7 Radiation Hardness................................ 67
  2.4 Radiation-Induced Effects on Silicon 68
    2.4.1 Bulk Damage 68
    2.4.2 Surface Effects 80
  2.5 Sensor Concepts... 81
    2.5.1 Overview of Sensor Types 82

2.5.2 p^+ in n: Low-Cost Solution for Applications
  in a Low- or Medium-Radiation Environment 85
 2.5.3 n^+ in n: Solution for Applications
  in a High-Radiation Environment.................. 97
2.6 Processing of Silicon Wafers 110
 2.6.1 Production and Cleaning of Silicon 111
 2.6.2 Thermal Oxidation 113
 2.6.3 Layer Deposition 113
 2.6.4 Photolithographic Steps 114
 2.6.5 Etching 115
 2.6.6 Doping... 116
 2.6.7 Metallization 118
 2.6.8 Example of a Process Sequence 118
2.7 Detector Materials Other Than Silicon 121
 2.7.1 Gallium Arsenide................................ 122
 2.7.2 CdTe and CdZnTe.............................. 124
 2.7.3 Diamond 125

3 The Front-End Electronics 129
3.1 Introduction 129
 3.1.1 Generic Pixel Chip 130
 3.1.2 Simple Sensor Model............................. 133
 3.1.3 Generic PUC 135
 3.1.4 Module Controller Chips 143
3.2 Design Aspects... 144
 3.2.1 Typical Specifications 145
 3.2.2 Radiation-Tolerant Design 154
 3.2.3 Cross Talk...................................... 157
 3.2.4 Testability and Ease of Operation 160
3.3 Analog Signal Processing 161
 3.3.1 Charge Amplification 162
 3.3.2 Feedback and Leakage Compensation 163
 3.3.3 Hit Discrimination............................... 168
 3.3.4 Threshold Trim 170
 3.3.5 Noise in a Simple FET Amplifier 171
 3.3.6 Noise in Charge Amplifier/Shaper Combination 178
 3.3.7 FET Preamplifier 183
 3.3.8 Bipolar Amplifier................................ 184
 3.3.9 Summary....................................... 186
3.4 Readout Architectures 187
 3.4.1 Chips Without Data Buffering 188
 3.4.2 Chips with Zero Suppression and Data Buffering 188
 3.4.3 Counting Chips 197

4 Integration and System Aspects ... 201
- 4.1 Introduction ... 201
- 4.2 Modules ... 202
- 4.3 Bump Bonding ... 203
  - 4.3.1 Solder Bumping and Bonding Process ... 204
  - 4.3.2 Indium Bump-Bonding Process ... 206
  - 4.3.3 Gold-Stud Bump-Bonding Process ... 210
  - 4.3.4 Quality Control of Bump-Bonded Assemblies ... 212
  - 4.3.5 Rework of Bump-Bonded Assemblies ... 213
  - 4.3.6 Thinning of Electronics Wafers ... 214
- 4.4 "Dressing" the Modules ... 214
  - 4.4.1 Flex Hybrid ... 215
  - 4.4.2 Multichip Module Deposited ... 216
- 4.5 Support Mechanics and Cooling ... 218
  - 4.5.1 Mechanical Supports ... 219
  - 4.5.2 Cooling ... 220
- 4.6 Power and Signal Interconnect in High-Luminosity Colliders ... 222
- 4.7 Operation in High-Radiation Environment ... 223

5 Pixel Detector Applications ... 225
- 5.1 Pixel Detectors for High-Luminosity Collider Experiments ... 225
- 5.2 Large Systems in Construction ... 231
  - 5.2.1 ATLAS ... 232
  - 5.2.2 CMS ... 237
  - 5.2.3 ALICE ... 241
  - 5.2.4 BTeV ... 243
- 5.3 Pixel Detectors for Imaging Applications ... 246
- 5.4 Pixel Imaging Systems in Operation ... 250
  - 5.4.1 Counting Pixels for Radiography ... 251
  - 5.4.2 Pixel Detectors for Protein Crystallography with Synchrotron Radiation ... 254
  - 5.4.3 Autoradiography with Pixel Detectors ... 256

6 Trends and New Developments for Pixel Detectors ... 261
- 6.1 Introduction ... 261
- 6.2 Limitations and Prospects of the Hybrid Pixel Technology ... 262
  - 6.2.1 MCM-D Integration ... 263
  - 6.2.2 Interleaved Hybrid Pixels ... 265
  - 6.2.3 Active Edge Three-Dimensional Silicon Pixel Detectors 267
- 6.3 Monolithic and Semimonolithic Pixel Detectors ... 269
  - 6.3.1 Monolithic Pixels on Bulk Silicon ... 271
  - 6.3.2 Monolithic CMOS Pixels ... 272

   6.3.3 Monolithic SOI Pixels................................ 276
   6.3.4 Amorphous Silicon above CMOS Pixel Electronics.... 277
   6.3.5 DEPFET Pixels..................................... 278

Glossary 283

References 289

Index 301

1 Introduction

Abstract. Pixel detectors have originally been developed for particle physics applications. This chapter describes why this kind of detector was needed and why it could be realized only recently. The evolution of pixel detectors is described in some detail with examples of applications in experiments.

1.1 Generalities on Pixel Detectors

The notion of pixel (short for "picture element") has been introduced in image processing to describe the smallest discernable element in a given process or device. A pixel detector is therefore a device able to detect an image and the size of the pixel corresponds to the granularity of the image. The omnipresent digital cameras are a typical example of pixel detectors. In this case, photons of different energies are integrated in the sensing elements (pixel) during a short exposure time and generate an intensity distribution which is the image.

In depth description and discussion of all the variety of pixel detectors and the diversity of their underlying techniques would be beyond the scope of this book. This book will deal only with the pixel detectors which are fast (i.e. able to take millions of pictures per second) and able to detect high-energy particles and electromagnetic radiation. Charged coupled devices will not be discussed in this book; the interested reader is referred to [1,2]. In particular, detectors are discussed where the image is generated in a semiconductor and is processed electronically and where data are readout in parallel. The basic building block of such a pixel detector is sketched in Fig. 1.1, which shows how a small-volume sensor ($\approx 5 \times 10^{-3}$ mm^3) is individually connected to its own electronics. This is the so-called *hybrid pixel detector* (hybrid because electronics and sensors are fabricated separately and then mated) and will be the main subject of this book. Planar integration technology allows one to put together several thousands of those building blocks in a matrix covering few square centimeters. Matrices can then be put together to cover larger surfaces.

The images (or patterns) considered in this book are not generated by visible light, but by relativistic charged particles or photons in the kiloelectronvolt to megaelectronvolt energy range. The charges generated by ionizing radiation are transformed into images through dedicated electronics circuits.

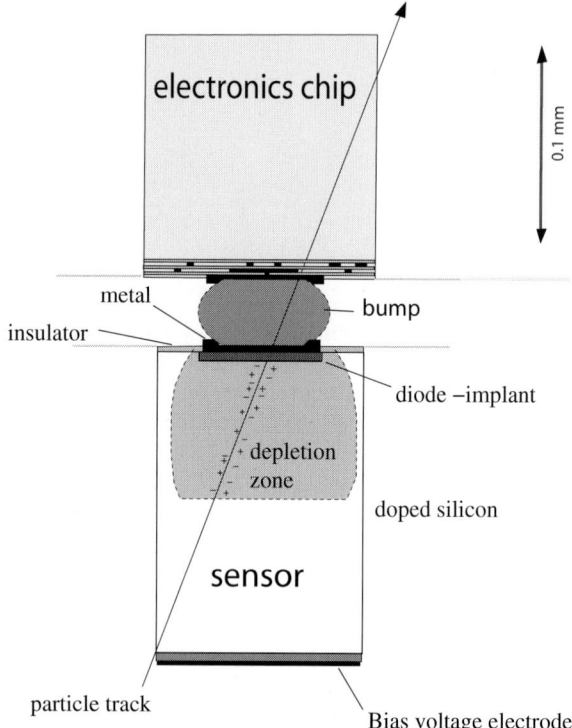

Fig. 1.1. Schematic view of one pixel cell, the basic building block of a hybrid pixel detector. The ionizing particle crosses the sensor and generates charges that, moving in the depletion region under the action of an electric field, produce signals. These are amplified, and hit pixels are identified and stored by the electronics. The thickness of the sensitive part of the detector – the depletion zone – depends on the bias voltage and on the sensor parameters, as explained in Sect. 1.2

This kind of device has been developed for the needs of particle physics, but, as it will become evident in Chap. 5, it can be used in many other fields. Particle physics applications demand high speed, good time resolution, and the ability to select hit patterns, while applications in other fields emphasize more high sensitivity and stability.

1.1.1 Motivations for Pixel Detectors in Particle Physics

The development of pixel detectors in particle physics has been primarily triggered by two specific requirements, which both became recently important and, in most applications, have to be simultaneously met:

(a) The possibility of studying short-lived particles

(b) The capability of coping with the increasing interaction rates and energies (and therefore the number of particles produced per collision) of modern particle accelerators

Scientists have been confronted with the following problem. High-energy accelerators generate elementary particle collisions at a rate of 10–100 MHz, with 10–100 particles emerging from every collision. Some rare, but interesting, particles live about 1 ps (10^{-12} s) and then decay into a few daughter particles. The topology of such a decay is sketched in Fig. 1.2.

Fig. 1.2. Topology of a short-lived particle decay, with ordinary particles emerging from the same collision. The collision vertex (V) and the decay vertex (D) are indicated. They are few millimeters apart

The tracks emerging from this decay must be measured as close as possible to the interaction point, with an accuracy of $\approx 0.1\, c\tau$ (where τ is the particle proper lifetime and c is the speed of light).[1] This gives a required measurement accuracy of $\lesssim 0.03$ mm for a lifetime in the order of a picosecond.

Accuracy is not the only important parameter, as many other particles may pass close to the decay point and this may confuse the picture. This makes it hard or impossible to study the decay even if the detector has the necessary accuracy unless enough sensing elements are available. Pixel detectors have not only enough space and time resolution, but also, and this is peculiar to them, high enough granularity to cope with the problem just described. This statement can be illustrated by the following example.

One can imagine a pixel detector as a thin layer of silicon patterned with contiguous squares of 0.1×0.1 mm^2 and assume each square to act as an

---

[1] Tracks emerging from the decay of a relativistic particle, once extrapolated back to the primary interaction vertex, miss this vertex on average by a distance $c\tau$. This distance is known as impact parameter. In order to discriminate the tracks coming from the rare heavy quark decays from other tracks emerging directly from the interaction, the impact parameter must be measured with an accuracy sensibly smaller than $c\tau$: 10% $c\tau$ is commonly assumed

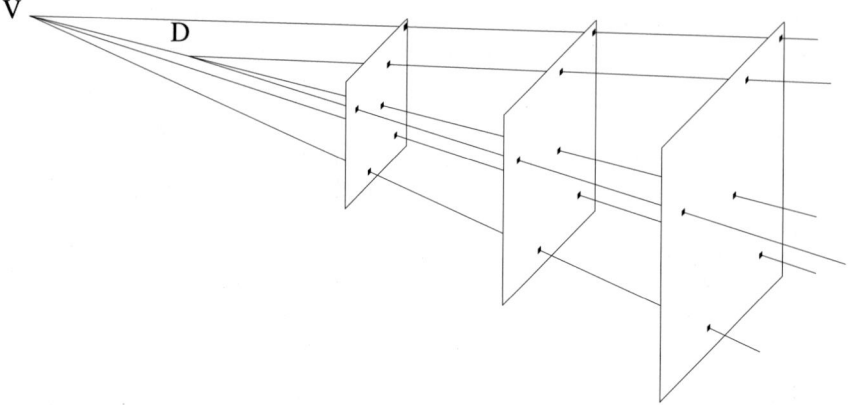

Fig. 1.3. Same decay topology as shown in Fig. 1.2. The tracks are measured by three pixel detectors with 100-μm pixel pitch. The hit pixels (i.e. the pattern "seen" by the detectors) are highlighted

independent sensing element able to detect the passage of a particle. If every sensing element is 100 % efficient and there are no spurious noise hits, Fig. 1.2 appears to a telescope made of three such detectors as shown in Fig. 1.3.

Now imagine to pattern the three silicon layers with sensing elements with the same time resolution and the same spatial accuracy but along one coordinate only. By doing so the number of readout channels is largely reduced and the problem of connecting the sensing elements to their readout channels is greatly simplified. The hit pattern is, however, much more difficult to interpret, as is apparent from Fig. 1.4. This figure represents, again in the hypothesis of 100 % efficiency and no noise, the decay topology as seen using microstrip detectors [3], which is the first kind of detector developed in particle physics to attack problems generated by the requirements (a) and (b). In this case, the same space resolution is obtained using two planes of microstrip detectors rotated by 90°, but, together with the N true strip coincidences due to the N tracks, one must also take into account $(N^2 - N)$ ghosts strip coincidences which can severely hamper the track reconstruction capability.

In short, the particle physics requirements (a) and (b) are satisfied by a detector with a high granularity able to detect multiple tracks with good space and time resolution. Moreover, its electronics should be capable of selecting the interesting patterns, which may be as few as one in millions. To do this the readout electronics should be designed to temporarily store the hit pattern belonging to an individual event which is judged possibly interesting on the basis of its topological or dynamical variables. These variables, derived from the event itself, are digitized and then used in a combinatorial circuit whose output will eventually select (or "trigger") the events of interest. Successive and increasingly selective cuts are applied to the events until their

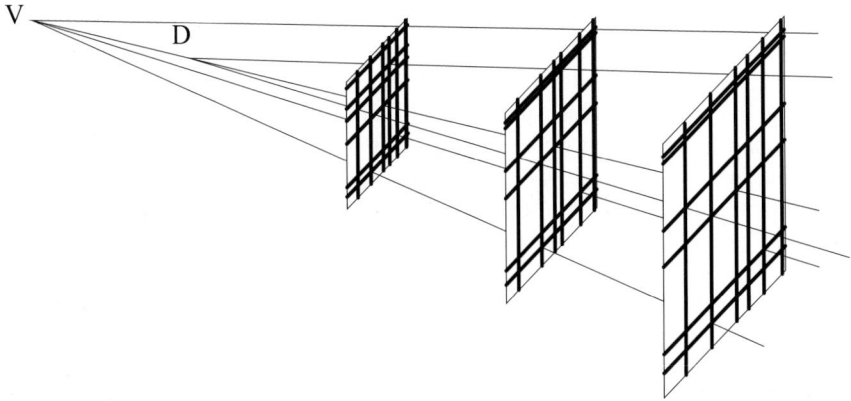

Fig. 1.4. Same decay topology as shown in Fig. 1.2. The tracks are measured by three double-sided microstrip detectors with 100-μm strip pitch. The hit strips (i.e. the pattern "seen" by the detectors) are highlighted

number is sufficiently reduced to transfer all those surviving to a computer for a complete analysis.

The typical particle physics application therefore requires that one hit pattern (image) is one event. The information is not uniformly distributed in all hit patterns, but concentrated in some rare patterns that one has to hunt for with sophisticated algorithms and appropriate readout architectures.

Other applications (like X-ray radiography) require instead that one image is made of many events (the individual X-ray conversions in the detecting medium). The information is, in this case, uniformly distributed on all events and only after summing up many of them one will obtain the image with the required quality. The readout electronics should therefore not select events, but accumulate them for a preset period of time under known and stable conditions.

Today, pixel detectors with cell sizes in the order of $100 \times 100\,\mu m^2$ can cover areas as large as a few square meters with 10^8 pixels.

1.1.2 Working Principle and Operating Characteristics of Segmented Silicon Detectors

In the previous section the pixel detector has been introduced in quite general terms. In this section some details on the operation of segmented silicon detectors are given in order to provide the minimal tools for the discussion of hybrid pixel detectors and their applications. A more detailed and complete account on pixel sensors will be presented in Chap. 2.

The pixel detector and its working principle can be best described by starting from a short description of the microstrip detector and then indicate how the latter evolved into the former. The advantage of this approach is that it gives some perspective of the technological and scientific progress on segmented semiconductor detectors and it allows one to refer to excellent books [4] on microstrip detectors, where the interested reader can find more detailed information. This approach will also naturally lead to the description of the hybrid pixel detector, i.e. a kind of detector where sensor and readout electronics are built on substrates with different and optimized characteristics. This is a mature concept, validated by a detailed R&D program initiated by the RD19 collaboration at CERN [5]. All the existing or planned applications of pixel detectors in particle physics (see Sect. 2.1 and Chap. 5) are based on the hybrid design.

Single-sided microstrip detectors [6] are a special case of semiconductor detectors in which one electrode is segmented in thin parallel strips. Ion implantation and photolithographic techniques are used to selectively dope the surfaces of the semiconductor wafer of typically 300-µm thickness and to deposit the metallization patterns necessary to extract the signals. This kind of technique, pioneered by Kemmer [3], is derived from the standard processing used in microelectronics and therefore profits from the large investments and the high-quality standards of the integrated circuit industry. A sketch of a generic microstrip detector is shown in Fig. 1.5.

It is also possible to pattern the rear contact with thin segments tilted with respect to the strips implanted on the front side. Since the pn-junction is only on the front side, special care has to be taken to prevent the strips on the rear side to be shortened by a charge accumulation layer. This more complex

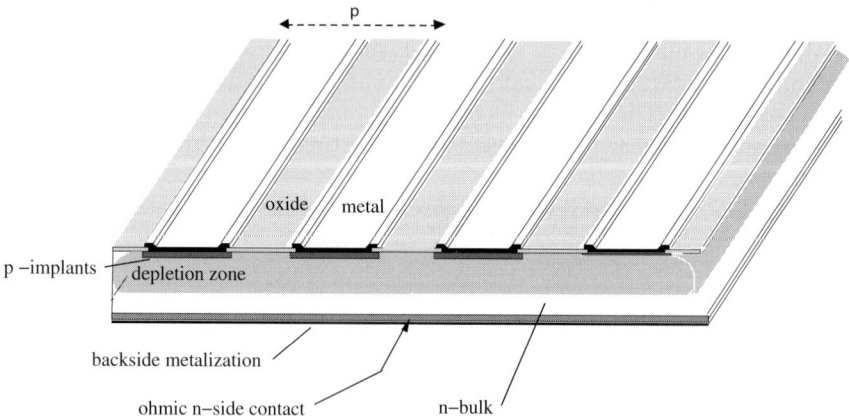

Fig. 1.5. Sketch of a single-sided microstrip detector. In this example strips are p^+ implants on n-type silicon. They are repeated at a pitch p. In between the strips, metallizations are regions with a silicon dioxide layer. The detector is operated by applying a voltage between the backside metallization and the strips. In this example the detector is only partially depleted. Increasing the voltage would extend the depletion zones toward the backside and toward the neighboring strips

process allows one to produce double-sided microstrip detectors and then to measure, with the same substrate, two coordinates, as shown in Fig. 1.4.

To describe the operation of a silicon microstrip detector one may, for instance, consider a 300-μm-thick n-doped (i.e. doped with the addition of a pentavalent impurity, like phosphorus) silicon wafer and assume that each strip is a p-implant (i.e. doped with a trivalent impurity, like boron). The doping must be such as to largely overcome the intrinsic carrier densities in silicon at room temperature ($\approx 10^{-10}$ cm^{-3}) and will therefore determine the abundance of free electrons (holes) in the n-zone (p-zone).

The resistivity ρ of doped silicon depends only on the dopant density N and on the majority carrier mobility μ according to

$$\rho = \frac{1}{eN\mu}, \qquad (1.1)$$

where e is the elementary charge. The interface region between the n-doped and the p-doped regions will be emptied of free charges through the following mechanism. The majority carriers in each region will diffuse through the junction and recombine with the opposite sign charge carriers. This will generate an electric field due to the excess charge from the immobile doping atoms, which counterbalances the diffusion and establishes an equilibrium. This equilibrium, characterized by the absence of charges which can move freely, extends to some thickness W (depletion zone), which depends on the dopant concentration N of the lower doped bulk material and on the voltage V across the junction according to

$$W = \sqrt{2\varepsilon_0\varepsilon_{\text{Si}}(V/eN)} = \sqrt{2\varepsilon_0\varepsilon_{\text{Si}}(V\mu\rho)}. \qquad (1.2)$$

Charges are built up on both sides of the junction and therefore the depletion zone can be seen as a charged capacitor of value C per unit area:

$$C = \varepsilon_0\varepsilon_{\text{Si}}/W = \sqrt{e\varepsilon_0\varepsilon_{\text{Si}}N/2V}. \qquad (1.3)$$

Increasing the reverse bias voltage V increases the thickness of the depletion zone and reduces the capacitance of the sensing element, and both these effects increase the signal-to-noise ratio (S/N). Fully depleted detectors[2] (i.e. those with the depletion zone extending to the whole thickness of the silicon layer) will give the best S/N. The nomogram shown in Fig. 1.6 correlates most of the parameters which have been discussed so far.

A 300-μm-thick n-type silicon substrate with p-implants is a set of pn-junctions (microstrips) which act as independent diodes. If these diodes are reversely polarized, e.g. applying a positive voltage on the n-side and connecting each p-implant to ground through its readout amplifier, very little current flows through them. The majority carriers experience a barrier due to

---

[2] Assuming a silicon resistivity of 4.5 kΩ cm, a 300-μm-thick n-type silicon detector fully depletes at 65 V bias voltage.

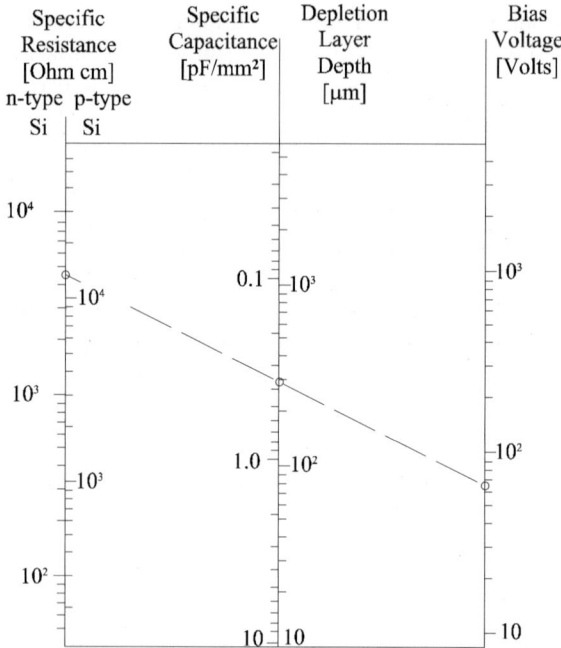

Fig. 1.6. Nomogram showing relations between the most important parameters for silicon junction detectors operation. Any straight line correlates the values of the different parameters (Elaborated from [7])

the voltage applied externally. The minority carriers (holes in this example) are constantly removed out of the depleted region by the field in the junction, thus generating a small current, known as dark current. As the carriers are thermally generated, this current depends on temperature and is also known as thermal background current.

Particles crossing the silicon detector, or photons absorbed in it, generate charged carriers (on average 1 electron–hole [e–h] pair per 3.6 eV of energy deposited). If these carriers are generated in the depletion zone, they lead to a current signal much larger than the thermal background current and which is therefore detectable. The depletion zone hence constitutes the active volume of the detector. In the undepleted regions, on the contrary, there is too low an electric field to collect charges in a short time and too many majority carriers which facilitate charge recombination.

To understand how this current is generated and how it can be detected, the case of a relativistic particle crossing the 300-μm-thick silicon detector is considered. The particle looses energy through many collisions with the electrons of the crystal and generates ≈80 e–h pairs per micrometer of path in a few micrometer wide cylinder around its trajectory. These charges drift under the action of the external electric field at a speed which depends on the electric field but saturates at values $\approx 10^7$ cm/s for fields close to 10^4 V/cm. The charges are therefore collected in less than 10 ns, resulting in a current of about 0.5 µA. During the drift the charges do not exactly follow the electric field lines, but diffuse as a consequence of the random thermal motion in the crystal lattice. Spread of the arrival position of the charge due to this effect can be described as a Gaussian distribution with standard deviation

$$\sigma = \sqrt{2Dt} \, , \tag{1.4}$$

which results in a spread of a few micrometers at the collecting electrode, assuming a typical electron diffusion constant of 35 cm²/s and a transit time of the carriers of 10 ns. The diffusion constant is higher for electrons than for holes, as it scales with the mobility.

Intense magnetic fields (B) of up to 4 T are often used in particle physics experiments to allow measuring the momenta of the charged particles through the deflection of their trajectories due to the Lorentz force. The magnetic field acts on all charged particles and therefore also on the electrons and holes drifting inside the silicon, which deviate from the electric field lines by the Lorentz angle θ_L of

$$\tan \theta_\mathrm{L} = \mu_\mathrm{H} B_\perp \approx \mu B_\perp \, , \tag{1.5}$$

where $B_\perp$ is the magnetic field component perpendicular to the electric field, μ_H is the Hall mobility, and μ is the carrier mobility (see Chap. 2.1 for details). Typical Lorentz angles range from a few to 20°.

1.1.3 Hybrid Pixel Detectors

The fabrication of a pixel sensor is very similar to the fabrication of a microstrip sensor. In the pixel case the implants have a higher segmentation, i.e. as if every microstrip diode would be further subdivided along its length. This simple change of the sensor design has many consequences at the system level and offers a variety of applications, as it will be shown later in this book.

If the area of each pixel is larger than a few square millimeters and the number of channels in the order of 10^2, the signals can be routed to the periphery of the sensor and connected to the electronics there. In this case one speaks of pad detectors and the connection to the electronics can still be done by ultrasonic wire bonding, a low-cost mature technique widely used in the semiconductor industry. One speaks of pixel detectors when the area of the sensing element is below 1 mm² and their number is in the order of 10^3–10^4. In this case a two-dimensional connectivity to the electronics as dense as the pixels themselves is necessary. The two-dimensional high-density connectivity is the key characteristics of the hybrid pixel detector and has three main consequences that are illustrated in Fig. 1.7:

(a) The connectivity between the sensor and the mating readout chip must be vertical, i.e. the connections must run out of the sensor plane.

1 Introduction

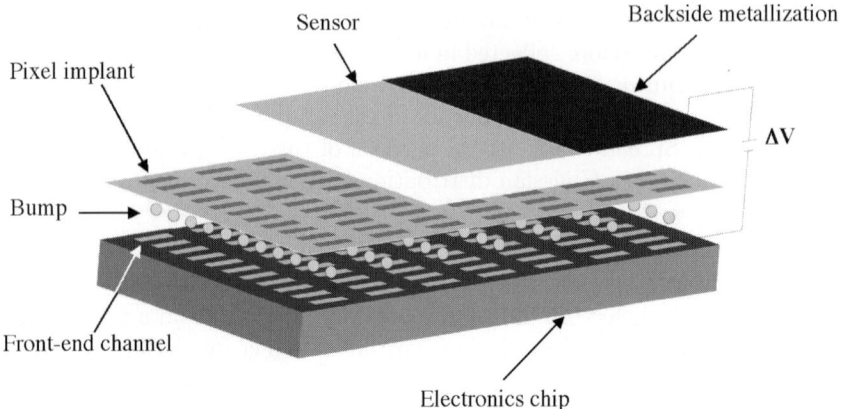

Fig. 1.7. Sketch of a "blown-up" hybrid pixel detector

(b) There must be exact matching between the size of the pixel and the size of the front-end electronics channel.
(c) The electronics chip must be very close (10–20 µm) to the sensor.

As shown in Sect. 1.1.2, the operation of the hybrid pixel sensor obeys (1.3), (1.4), and (1.5), but constraints which result from the topology of the assembly must be considered too. To deplete the sensor a sufficiently high bias voltage must be applied on the backside plane while all the pixels are grounded. This is properly done through the virtual ground of the preamplifier once the electronics is connected[3] and indicates the importance of the hybridization process for the pixel detectors. A pixel sensor can hardly be tested before being connected to the electronics with a sophisticated interconnection technique.

Any electronics chip must have some ancillary logic to extract the signal from the front-end channels, organize the information, and transmit it out. This logic cannot be distributed to all pixel cells, but has to be concentrated and is normally placed close to one edge of the chip. This means that the chip, on which there will be traffic of digital signals, is bound to extend outside the sensitive area of the sensor and overlap its cutting edge.

Since the chip is very close ($\approx$10 µm) to the sensor, designers must pay special attention to avoid the following:

(i) Large static voltage (i.e. bias voltage) on the front side or on the edge of the sensors that may give rise to destructive sparks. This implies that the guard ring structure which helps to confine the high-voltage region should be on the backside of the sensor.

---

[3] Pixel grounding can also be done via a specially designed punch-through electrode to be connected with probe needles (see Chap. 2)

(ii) Large high-frequency signals on the electronics that may induce detectable signals on the pixel metallization. This implies using low swing logic signals (e.g. LVDS) and minimizing the coupling capacitance between the sensor and the digital busses.

It should be stressed that the pixel area covered by the electronics and that covered by the sensor must be about the same. The complexity of the two parts, and therefore the production yield, is instead quite different. Front-end chips have a few million transistors per square centimeter; a typical chip size cannot, today, sensibly exceed 1 cm^2 if a high enough yield (>50%) is desired. Similar yields are obtained for sensors of several tens of square centimeter. All these considerations imply that multiple square centimeter electronics chips should be mounted on a sensor which is considerably larger ($\approx$10 cm^2). This and other hybridization issues will be discussed in Chap. 4.

Other peculiar characteristics of the pixel detectors are related to the small dimensions of the sensing elements. Each pixel covers, in fact, a very small area ($\approx 10^{-4}$ cm^2) over a thin ($\approx$300 μm) layer of silicon. It therefore exhibits a very low capacitance ($\approx$0.2–0.4 pF), which is dominated by the coupling to the neighboring pixels rather than to the backside plane. The direct interpixel coupling has to be kept to a minimum with proper sensor design (see Chap. 2) to avoid cross talk between pixels.

The low capacitance is one of the key advantages of pixel detectors since it allows fast signal shaping with very low noise, as will be shown in detail in Chap. 3. It is common to obtain noise figures of $\approx$200 e$^-$ for electronics operating at 40 MHz and therefore an S/N exceeding 100 for fully depleted 300-μm-thick sensors. This is a very comfortable situation as it allows operation in absence of spurious noise hits. A detection threshold set at, e.g., $10\sigma_{\text{noise}}$ gives, in fact, full efficiency and very low probability that a noise fluctuation exceeds the threshold. This may be looked at as a very idealized situation as other sources of hits could be conceived (e.g. electronics pickup, cross talk, low-energy photons), but measurements [8] prove that a spurious hit probability of $<10^{-8}$ per pixel can be reached under experimental conditions.

Another way of taking advantage of the excellent S/N ratio is to consider that the detector is robust enough to tolerate even a considerable signal loss. This extends the application of the hybrid pixel detector in two directions:

- To sensors which have a poor charge collection or a limited active thickness (e.g. diamond, GaAs, Cd(Zn)Te) or
- To crystalline silicon sensors damaged by high irradiation flux

In the latter case the collected charge is diminished through two effects which are both illustrated in Chap. 2: the trapping of drifting carriers due to radiation-induced defects in the crystal lattice and the reduction of the depletion depth due to the increase of the space charge [9].

Finally, smallness of the pixel means smallness of the reverse current flowing through it at depletion (typically $0.1\,\mu\text{A}/\text{cm}^2$). This reduces the parallel noise and allows operation even after considerable irradiation. After 10^{15} particles per square centimeter the reverse current density increases to $\approx 30\,\mu\text{A}/\text{cm}^2$, rendering large sensing elements difficult to operate.

In summary, the hybrid pixel detector is the ideal detector to work in the very hostile environment which exists close to the interaction region of a particle accelerator because:

- It is radiation hard (i.e. it survives high integral flux of particles).
- It provides nonambiguous three-dimensional measurements with good time resolution (i.e. it operates in high instantaneous particle flux).
- It provides the space resolution which is needed to measure short-lived particles.
- It can extract the rare patterns the physicist is looking after (i.e. it memorizes hit patterns and selectively reads out those interesting).

Hybrid pixel detectors have been shown to work in particle physics experiments [10, 11]. This success has triggered the design and the construction of detectors approaching few square meters of sensitive area and 100 millions of channels [12–14] to be operated in intense particle fluxes. Freedom in the choice of the sensitive material has also favored the application of hybrid pixel detectors in other fields, like medical diagnostics [15, 16].

1.1.4 Monolithic Pixel Detectors

Since silicon is the material most commonly used for pixel detectors, several groups have looked into the possibility of building both electronics and sensor in the same technological process. This avoids the high-density interconnection technique and the many related manipulations, and it allows one to further reduce the capacitance of each pixel and obtain a very low-noise performance. It therefore opens the possibility of a more robust and less expensive (but less versatile) detector, as the different ideas for monolithic pixel detectors, all detailed in Chap. 6, are based on some compromise between the sensor and the readout functions.

In some cases (like, for instance, in the DEPFET design [17]) the sensor part is driving the development. Simple electronics circuits (like the first stage of the amplifier of each pixel and some addressing scheme) are integrated on high-resistivity silicon.[4] The signal generation is optimal as large thicknesses can be depleted and high electric fields provide fast and efficient charge collection. The on-chip electronics treatment of the signal is minimal since all design technologies have been developed for low-resistivity silicon

---

[4]Unlike sensors, the commercial integrated circuits are built on low-resistivity ($\rho = 1\text{--}10\,\Omega\,\text{cm}$) silicon, that can be produced in large quantities at low cost

and considerable design effort has to be spent even to realize simple circuits with acceptable yield.

In other cases (like, for instance, in the MAPS design [18]) the electronics part is driving the development. The detector is realized on a thin layer of low-resistivity p-doped silicon, which is optimal for complex electronics design but does not allow having large depletion volumes and fast charge collection. The pn-junction is realized between the n-well and the p-type epitaxy, but, because of the low resistivity, the depletion is partial even on the very thin ($\approx 10\,\mu$m) epitaxy layer and the collected charge is small ($\approx 1,000\,e^-$). The charge collection from the epitaxial layer to the n-well/p-epi diode happens through drift and diffusion of the carriers and takes about 100 ns, i.e. 10 times longer than in the approaches based on high-resistivity silicon.

Another solution under study [19] can be defined as quasi-monolithic, as a further deposition step on top of a standard ASIC wafer is foreseen. In this last step, plasma-enhanced chemical vapor deposition of hydrogenated amorphous silicon produces a thin sensor of limited charge collection properties. This design, too, is driven from the electronics development. Amorphous silicon as a sensor material promises to be radiation hard. The deposition technique can be used with other sensor materials, like polycrystalline mercuric iodide, which may be interesting for their high stopping power for X-rays.

It is possible that one of the monolithic approaches could take the lead in a not too distant future and may replace the hybrid design.

1.2 Evolution of Pixel Detectors in Particle Physics

The idea of a fast semiconductor detector array able to view close details of particle collisions existed for 40 years [20] and had begun while looking for an alternative to the widely used bubble chamber detector, which suffered from data rate limitations. The integrated circuit technology accessible to particle physicists was, however, not yet mature enough in the 1960s to realize an electronics bubble chamber.

New accelerators, with increased particle energies and intensities, and the discovery of charmed particles did push again toward semiconductor detector arrays in the late 1970s. The first trials were done with multiple layers of closely spaced silicon surface barrier diodes [21,22]. With each diode one was able to measure the total charge left by the particles emerging from the interaction and therefore deduce, with some approximation, the number of particles traversing each diode. A charmed particle decay happens after few millimeters of flight path and increases, on average, this number. The detection and localization of the charm decay point could then be done through a charge measurement.

The real breakthrough was the advent of the microstrip detector, which allowed the measurements of the track parameters with the precision required to study charm and the other short-lived particles discovered soon after.

The development of the microstrip detectors was rapid as this device could be readout with existing miniaturized, but discrete, electronics. The use of integrated electronics became possible a few years later.

Pixel detectors, on the contrary, needed more sophisticated technologies even to start-up and stayed longer in the conceptual phase.

The difference in the evolution of the microstrip and the pixel detectors is illustrated by Figs. 1.8 and 1.9. These pictures are separated by 16 years and represent the first generation of microstrip detectors and the second generation of pixel detectors.

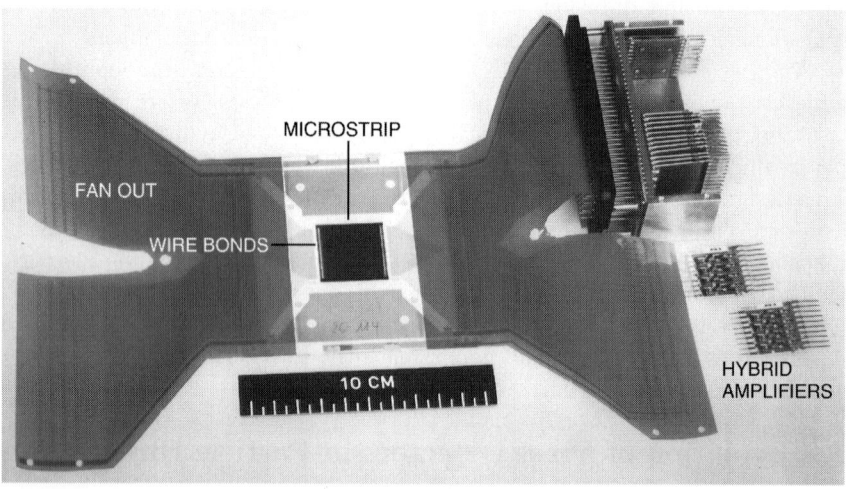

Fig. 1.8. A microstrip detector built in the year 1984 for the CERN experiment WA82. The microstrip sensor has 512 channels individually connected, through ultrasonic wire bonding followed by a ceramic/kapton fan-out, to hybrid amplifiers

The microstrip detector has a sensitive area of about 2.5×2.5 cm^2 and 512 channels at 50-µm pitch. The signals, extracted through a two-level fan-out system on ceramics and kapton, reach the preamplifiers through a massive multipin connector. The electronics is realized on ceramics using thick film technology; each circuit measures 2.5×2.5 cm^2 and contains four amplifiers.

The pixel detector has a sensitive area of 1.6×6.4 cm^2 and 46,080 rectangular pixels at pitches of 50 and 400 µm respectively. The signals are amplified and processed by 16 front-end chips bump-bonded to the sensor tile, each chip serving 2,880 pixels. A kapton circuit fan-out covers the sensor backside and is almost the only part visible in the picture of Fig. 1.9. It distributes properly decoupled low voltages to the front-end chips and connects them to the module controller chip. This integrated circuit, visible in the center of the kapton fan-out, organizes the operation of the front-end chips, collects their data and transmits them, once properly formatted, to a remote receiver.

1.2 Evolution of Pixel Detectors in Particle Physics

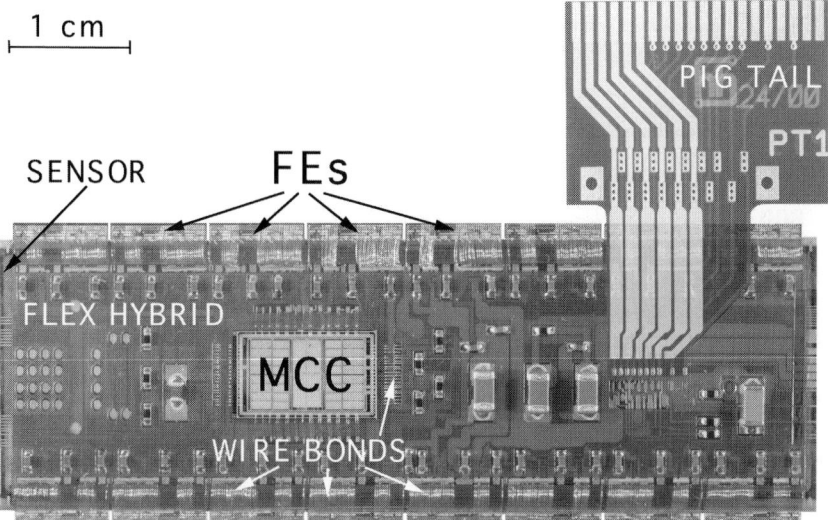

Fig. 1.9. A pixel detector prototype built in the year 2000 for the CERN experiment ATLAS. The 46,080-channel sensor is barely visible as it is covered by the flexible kapton circuit which distributes power and control signals to the front-end electronics. Connection between the front-end chips (FEs) and the kapton fan-out is made through 25-μm Al wire ultrasonically bonded. Wire bonding is also used to connect the module controller chip (MCC) to the kapton circuit. Only a few signal and power lines are necessary to operate this detector; they are traced out via the "pigtail" extension

The most noticeable difference between Fig. 1.8 and Fig. 1.9 is the level of integration of the electronics and the related connectivity problems. The pixel detector shown in Fig. 1.9 has ≈100 times more channels than does the microstrip detector shown in Fig. 1.8; it has more functionality built-in and still weights only 1% of its forefather. One can say that the evolution has eliminated most of the inactive material that was necessary before to connect hybrid electronics to multichannel sensors. The miniaturization of electronics which has contributed to widely spread consumer electronics products, like notebooks and digital cameras, has also made the development of pixel detectors possible.

Clearly the same trend has happened in the development of microstrip detectors and nowadays the bulky material that one can see in Fig. 1.8 does no longer exist as integrated circuits are used in microstrip detectors, too.

It is interesting to note that, during this same time, an electronics bubble chamber based on microstrip detectors and integrated readout chips was built and used for the study of heavy quarks in the CERN experiment WA92 [23]. Figure 1.10 shows a picture of a hadronic interaction in a thin copper target as seen by the closely spaced set of high-resolution microstrip detectors used

16   1 Introduction

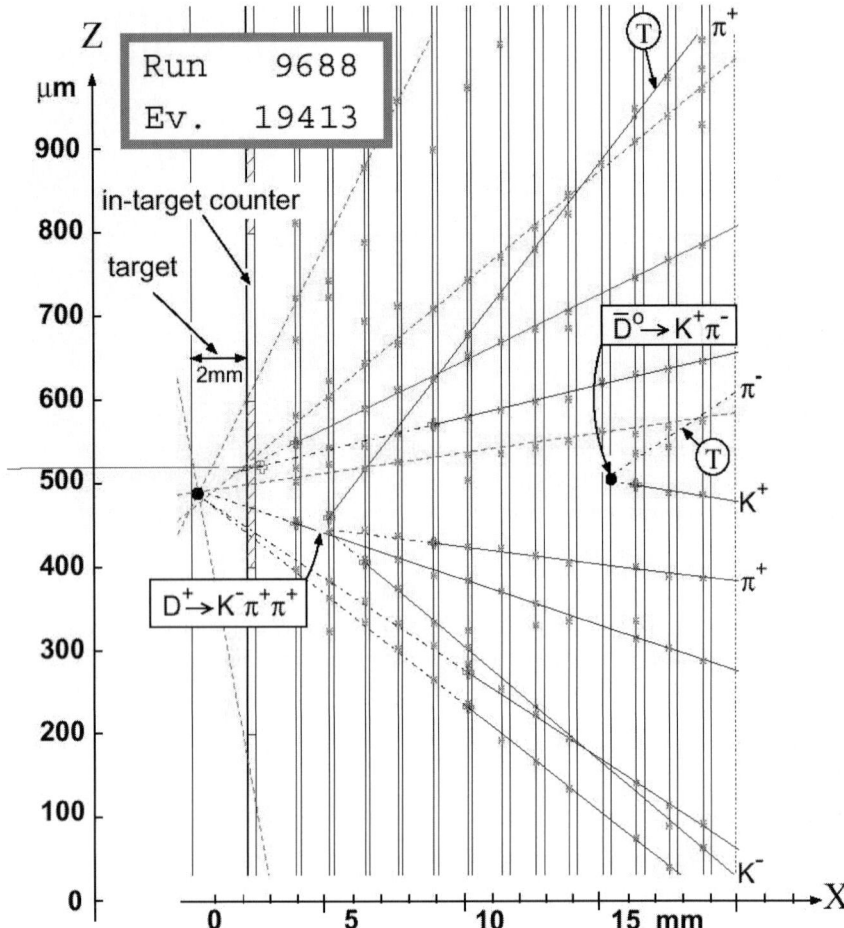

Fig. 1.10. Display of a double-charm event as recorded in the "electronics bubble chamber" of the WA92 experiment at CERN. The tracks reconstructed by the spectrometer (not shown) are superimposed to the hits (indicated as *asterisks*) left, by the same tracks, in the 1.2-mm spaced 10-µm pitch microstrip detectors. Tracks indicated with "T" have impact parameter larger than 100 µm

by the CERN experiment WA92. A charm and an anticharm particle, which have traveled few millimeters before decaying, are clearly visible.

1.2.1 The First Pixel Detectors and Their Use in Experiments

In this section the historical development of pixel detectors for particle physics applications is briefly described. More details can be found in the excellent review by Heijne [8].

The birth of pixel detectors can be traced back to the 1984 IEEE Nuclear Science Symposium where Gaalema pointed out [24] that an integrated circuit for focal plane imaging sensors, developed by Hughes Aircraft Co., could be connected, through bump bonding, to a semiconductor diode array to detect and localize X-rays.

The simplicity of such a circuit (four MOSFETs per pixel, as shown in Fig. 1.11) did allow its integration on a very small surface and therefore a very small pixel size, but it could not avoid some limitations. The charge "seen" by each pixel is integrated for hundreds of microseconds on a capacitor to minimize the serial electronics noise. A multiplexed readout is then performed, at a prefixed time, by addressing the pixels one after the other.

This approach works under the following conditions:

(a) Each pixel must draw very low current to avoid saturating the readout amplifier and to contribute significantly to the parallel electronics noise.
(b) The data rate should be limited to 1 kHz.
(c) The device must be continuously "interrogated," i.e. it cannot react to an external trigger.

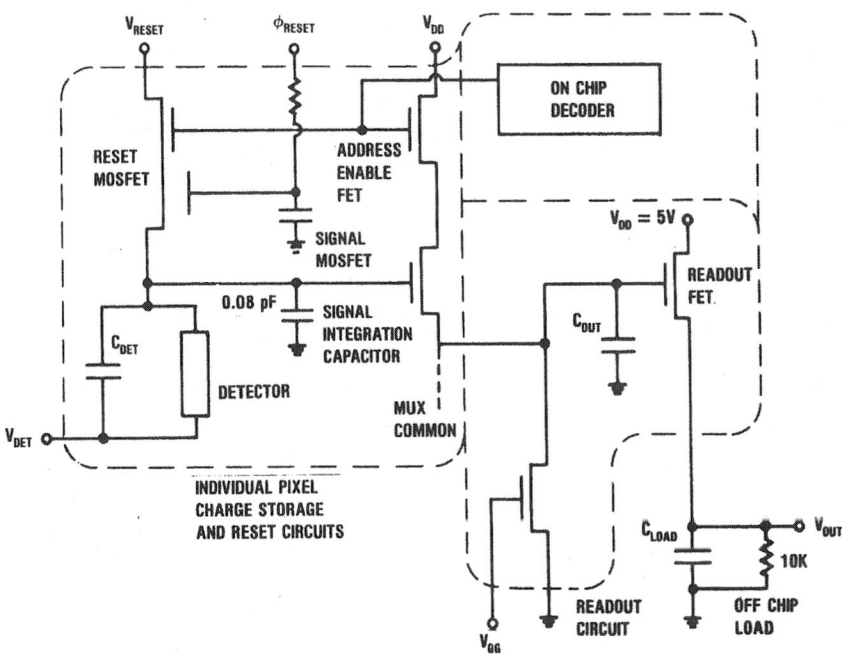

Fig. 1.11. Scheme of the circuit used by Gaalema [24] to detect X-rays with the first hybrid pixel detector

The above limitations, tolerable perhaps for imaging sensors (as, usually, images do not change too rapidly with time), were not acceptable for particle physics applications. The virtue of [24] has been to show the possibility and to indicate the potential of the hybrid pixel detector.

Designs of pixel readout circuits with much shorter integration times and the ability to react to external trigger were developed in the early 1990s, on both sides of the Atlantic, by groups developing vertex detectors for the Superconducting Super Collider (SSC) [25] in the United states and the Large Hadron Collider (LHC) [26] in Europe.

After the stop of the SSC program in 1994 the development of pixel detectors did continue, almost exclusively for a few years, at CERN, the laboratory hosting the LHC. To fully exploit the discovery potential of this collider, designed to reach unprecedented particle collision energies and beam intensities, detector technologies had to be pushed far beyond the current knowledge. An R&D program was therefore launched by CERN in 1990 and, under the control of a scientific committee, some 50 projects were financed, monitored, and completed over a 5-year period. One of those projects [5] had the merit to push pixel detector developments up to their first use in experiments.

To simplify the task, application in fixed target experiment was considered first. In this case, due to the large Lorentz boost which the particles receive, most of the tracks of interest go inside a forward cone of ≈300-mrad aperture. The area a pixel detector system should cover is therefore small (few to hundreds of square centimeters) and both the mechanics support and the services (cables, cooling system) can be massive, provided they are outside the forward cone where tracks are measured. Moreover, the easy and frequent access to the experimental setup possible in fixed target experiments is a definite advantage for a detector still in its infancy.

Among the fixed target experiments, heavy ion collisions have been considered first as they give birth to a very large number of particles in the final state. The three-dimensional information provided by pixel detectors is, in this case, a very powerful tool to disentangle the event topology and correctly reconstruct all tracks.

1.2.1.1 The OMEGA Pixel Detectors

The first pixel matrix, OmegaD,[5] had 1,024 pixels of size $75 \times 500\,\mu m^2$ [27] (16 columns and 64 rows). The readout chip, connected to the mating sensor through solder bumps as illustrated in Fig. 1.12, was realized in CMOS technology. Each channel had a continuously sensitive preamplifier followed by an asynchronous comparator and a digital delay line through which the

---

[5]The heavy ion experiments using the CERN Omega spectrometer were the first to use pixel detectors

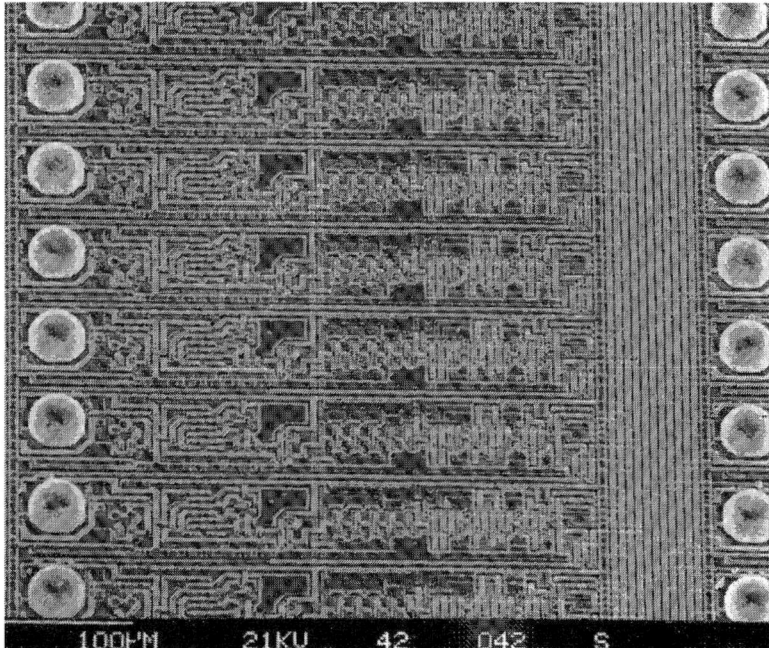

Fig. 1.12. Scanning electron microscope picture of part of the OmegaD readout chip. Solder bumps of 38-μm diameter are deposited at the input (on the *left* of the picture) of each pixel circuit; vertical busses for data extraction are also visible at the output (on the *right* of the picture) of the 75 × 500 μm^2 readout channels. The beginning of a new column of pixel circuits is barely visible on the far right of the picture

discriminated signals travel waiting for an external trigger. When the trigger is received all pixels with a coincident delay line signal are read out (see Sect. 3.4.2).

The performance of this chip was quite encouraging and, for the first time, particle tracks could be seen in the CERN experiment WA94 [28] as shown in Fig. 1.13.

The power dissipation was 30 μW per pixel (i.e. ≈1 mW/mm^2); the electronics noise was just below 100 e$^-$ rms while the threshold variation between channels was around 500 e rms. This last number dominates the noise of the chip even if it is not an intrinsic noise contribution.

This observation already points to a problem: one individual pixel circuit can have excellent performance, but it is difficult to get by design a correspondingly good response uniformity over the whole chip area.

Both technological limitations (e.g. electronics parameter variations over the chip area) and design choices (e.g. sensitivity to voltage drops along busses) contribute to nonuniformities. These sources of fluctuations cannot

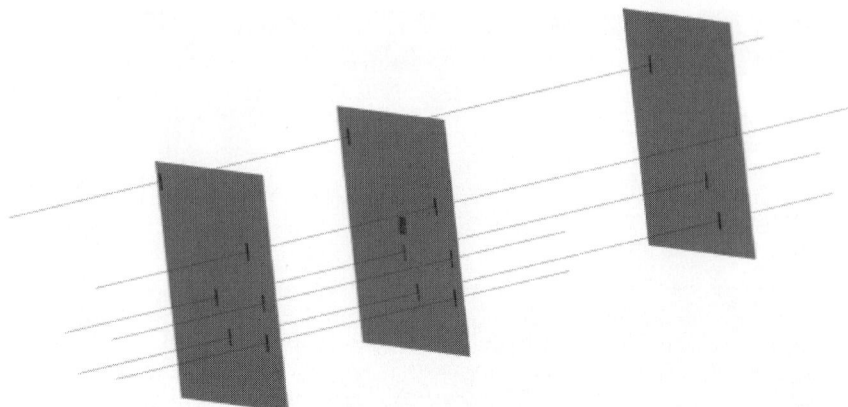

Fig. 1.13. Online event display of tracks coming from a sulfur–sulfur interaction in the CERN experiment WA94. Hits in three pixel detectors placed at distances of 2 and 3 cm and the corresponding reconstructed tracks are shown. They demonstrate the high efficiency and the low noise of these detectors

be completely eliminated and modern front-end chips dedicate part of the pixel cell area to trimming circuits. This became possible with the adoption of deep submicron (DSM) technologies featuring 0.25-μm structure sizes, while it was not practical in the 3 μm technology used for the OmegaD chip.

The OmegaD design evolved over several years as illustrated in [29], keeping the same architecture based on one delay line per pixel, but using smaller and smaller feature size, as permitted by the continuously evolving CMOS fabrication technologies. The number of transistors per pixel has correspondingly been increased from ≈80 to ≈500, which allowed better control circuits and proper tuning capability on critical parameters and therefore better response uniformity. The wafer size did also increase, which has some influence on the hybridization techniques described in Chap. 4. OmegaD was fabricated on 4-in. wafers with 3-μm feature size, and its last descendant, the ALICE1 chip [30], was fabricated on 8-in. wafers with 0.25-μm feature size.

Also the radiation hardness has greatly profited from the CMOS technological trend. Smaller feature size also means thinner silicon oxide layers, which, in turn, implies smaller parameter shifts caused by radiation in MOS transistors [31, 32]. The adoption of DSM technologies resulted in pixel circuits able to still operate after more than 300 kGy [29], almost a factor of 1,000 more than similar pixel circuits developed earlier with 3-μm feature size [27].

Several generations of pixel chips have been necessary to evolve from OmegaD to ALICE1. The intermediate chips, intensively used in heavy ion experiments, have allowed one to gain considerable experience on system aspects.

Moving from a "proof of principle," where 3 single-chip detectors have to operate for some hours close to a beam, to a system made of 84 multichip detectors to be used in experiments lasting months requires one to solve new problems. In general, there is the need to keep the detector specifications stable during production through the control of the critical fabrication processes and during the experiment through the control of the critical environmental and operational parameters. This was done for the experiment WA97 [30], where a telescope of seven pixel planes, each having 72,576 pixels and covering $\approx 29\,\text{cm}^2$, was built and operated. The production control did turn out to be especially complex as many fabrication processes (e.g. high-density interconnectivity, radiation hard electronics) were at the edge of their technical feasibility. Only few industries were interested in the small market represented by particle physics applications. The production yield for good multichip detectors[6] was about 35% [8].

Over the last three decades, experimental particle physics progressively moved from fixed target to colliding beam accelerators as these bring several advantages, the most relevant being the significantly higher energy that can be reached in the collision between elementary particles. Particles produced in a colliding beam accelerator are spread out over a very large solid angle, which has then to be covered by large area detectors.

Space resolution, granularity, and radiation resistance make pixel detectors ideal as vertex detectors and they must therefore be the first device encountered by the particles emerging from the interactions. Since the information carried by particles is deteriorated when passing through matter, minimal material has to be used to support and operate pixel detectors around a colliding beam accelerator.

1.2.1.2 The DELPHI Pixel Detector

The first use of pixel detectors in a colliding beam environment was promoted by Delpierre and collaborators [33] for the DELPHI experiment at the large electron positron (LEP) accelerator at CERN. The LEP energy increase above the threshold of W^+W^- production required to also increase the DELPHI angular acceptance to cover, with precise tracking, a polar angular range between 25 and 12°. Two crowns of pixel detectors, shown in Fig. 1.14, have been added to the microstrip vertex detector already operational in the experiment in order to extend its acceptance to the forward and backward regions [34].

The total area covered by the 1.2 million pixels is 0.15 m^2, and the power density required by the electronics is about 0.03 W/cm^2. This allows cooling of each detector crown by means of a single cooling pipe running on the inner support ring of the crown itself. This choice, together with the careful design

---

[6]These detectors, also called *ladders*, are made out of one sensor and six chips bump-bonded side by side and have 6,144 pixels

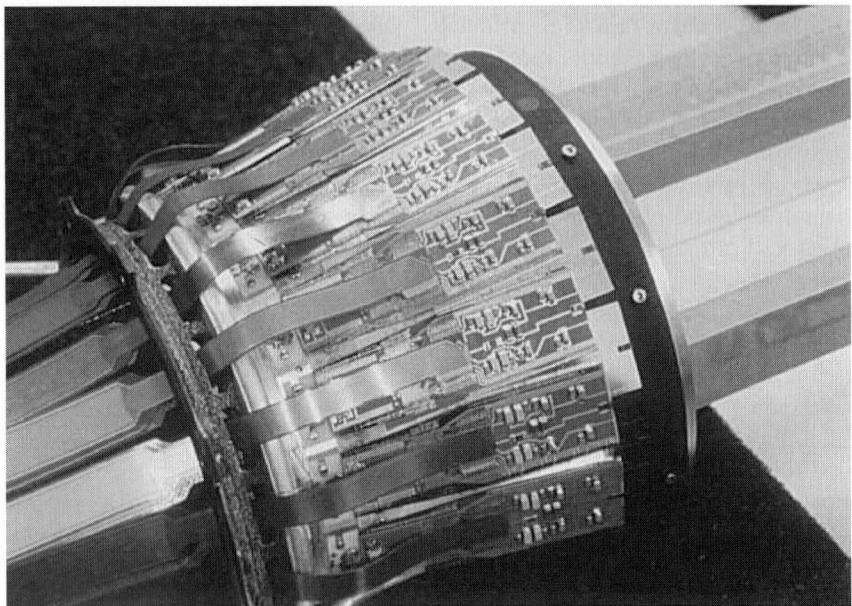

Fig. 1.14. The pixel forward detector of the DELPHI experiment. The kapton circuits necessary to connect the pixel modules to the readout electronics are clearly visible

and the proper selection of materials for the support structure, minimizes the dead material encountered by the particles at the cost of a temperature gradient of $\approx 10°$ over the sensor area.

The pixel modules have been designed to minimize material too. Their layout, shown in Fig. 1.15, allows us to easily fit the crown shape: ten 24×24 and six 16×24 pixel arrays cover the sensor area, resulting in a detector layout which is narrower toward smaller radii. Readout busses were, for the first time, implemented on the edges of the sensor substrate, but a higher than expected voltage drop on some lines finally did require the addition of an external kapton circuit. This was the first attempt for a more compact layout and then evolved in the multichip module deposited approach [35] to pixel module construction, which will be described in some more detail in Chap. 6.

The relatively large size of the pixels ($330 \times 330\,\mu m^2$) allows one to use an industrial standard for the bump-bonding process [36]. Detectors are operated at a typical threshold of 10 ke$^-$, which is comfortably high compared to the few hundred electrons noise, but sufficient to have an efficiency exceeding 99%. "Hot" pixels (i.e. those responding >1% of the triggers at the nominal threshold) total to 0.3% and were masked off using the appropriate feature in the front-end electronics. The use of pixel detectors has allowed the DELPHI experiment to almost double the track reconstruction efficiency below 25°.

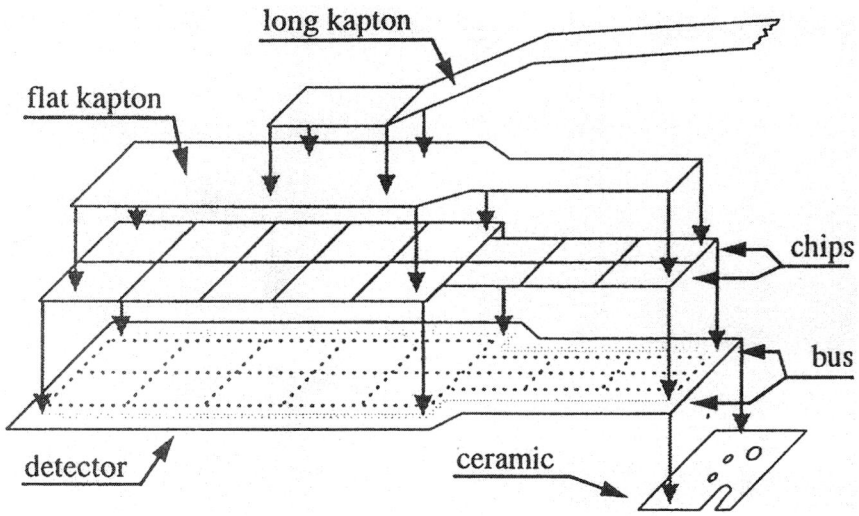

Fig. 1.15. Exploded view of a DELPHI pixel module

The overall production yield for good modules was 36% [11], very close to the value already found for the Omega ladders [8], but with a clear trend toward higher yields at the end of the production period, when the critical fabrication steps were better controlled.

The DELPHI pixel project was realized under stringent schedule constraints and some shortcuts, necessary to be ready on time, did not allow its complete optimization. This project remains, nevertheless, a very important step in the development of large area pixel detectors as some of the key issues (light and precise mechanical supports, cooling system, quality control of each production activity, etc.) have been addressed and solved for the first time.

The first generation of experiments using hybrid pixel detectors has allowed the evaluation of the potential and the limits of this new technique. As the results were largely positive, a large community of researchers started to develop pixel detectors tailored to the needs of high luminosity colliders. The description of this second generation pixel detectors will be given in Chap. 5.

1.2.2 Other Applications

Even if the pixel detectors were born for the needs of particle physics, they are potentially very useful in other domains where fast imaging with penetrating radiation is necessary.

Some applications in medicine, biology, and astrophysics will be illustrated in Chap. 5. Their success will greatly depend on the cost per square centimeter of detector. This is presently quite high ($\approx$€500), but can be sensibly reduced if large-scale applications could be envisaged and if some of the special requirements needed for particle physics applications (e.g. tens of kilogray radiation hardness) could be dropped. The high-density connectivity is a critical and expensive production step that is typical of the hybrid pixel design and not very much used in other applications. Some new pixel developments, presented in some detail in Chap. 6, try to avoid this step joining more intimately sensors and electronics. This also means compromises in performances, but may simplify the production steps and finally open the door to a still wider range of applications.

2 The Sensor

Abstract. The *sensor* is the part of the detector system where the interaction with the radiation takes place, and which "delivers" the signal to the readout electronics. The signal is produced by ionization in the sensor material, which is in almost every case silicon. This chapter summarizes some basic features of semiconductors, with emphasis on silicon. The properties and possible designs of silicon pixel sensors are discussed in detail and a short overview of the processes necessary for silicon sensor production is given. The chapter closes with the discussion of sensor materials other than silicon.

2.1 Introduction

Semiconductor sensors have been used in spectroscopic applications since the early 1960s. They use the same detection principle as gas-filled ionization chambers but the medium is the semiconductor crystal. The main advantage is that the energy for producing an electron–hole pair is only 3.6 eV in silicon while about 20 eV is required for gas ionization and therefore a much better energy resolution could be reached.

In the late 1970s semiconductor sensors became common as tracking devices in particle physics. In these applications silicon is the preferred material because it is the most intensively studied semiconductor and its electrical properties are well known. Section 2.2 briefly summarizes some aspects on the device physics relevant for the application of silicon as a sensor material. An overview of important sensor properties is given in Sect. 2.3 while the sensor concepts themselves are discussed in Sect. 2.5. Another reason for the choice of silicon is the fact that it is cheaply available in large quantities and its processing is well developed thanks to the progress in electronics industry. The processing steps necessary for sensor production are briefly discussed in Sect. 2.6.

As a crystalline material silicon is sensitive to radiation damage. However, this radiation-induced degradation of the sensor properties is well investigated and predictable as presented in Sect. 2.4. The radiation hardness requirements of particle physics experiments have all been met so far. Other semiconductor materials either have not reached a sufficient improvement in radiation hardness (e.g. GaAs) or are still not available in the necessary (large) wafer sizes (e.g. diamond) and quality (e.g. CdTe or SiC). However, semiconductor compounds containing elements with a higher atomic number than silicon are appreciated in nonparticle physics applications like X-ray imaging due to their better quantum efficiency. As this book is mainly focused on particle physics application of pixel detectors the emphasis will be put on silicon while other materials will be discussed only briefly in Sect. 2.7.

2.2 Device Physics and Fundamental Sensor Properties

This section will recall some aspects of semiconductor physics important for a semiquantitative understanding of the sensor behavior. It is not at all intended to cover the field of semiconductor physics which has been treated in many textbooks [37–44] to which the reader is referred. Furthermore, the general idea of silicon radiation sensors, especially of strip detectors, has already been discussed in a number of publications as for example [4, 45] and is commonly known.

2.2.1 Carrier Concentration

In this section the charge carrier concentration in intrinsic and doped semiconductors is discussed. A piece of semiconductor is called *intrinsic* if the concentration of impurities is negligible compared to the thermally generated free electrons and holes. The number of free electrons n can be calculated by multiplying the density of states in the conduction band $N(E)$ with the probability of their occupation given by the Fermi–Dirac function $F(E)$ [42]:

$$n = \int_{E_\mathrm{C} \equiv 0}^{\infty} N(E) F(E) \, \mathrm{d}E , \qquad (2.1)$$

where E_C is the energy of the lower bound of the conduction band arbitrarily set to zero. The density of states (see Fig. 2.1b) in the conduction band can be calculated as [42]

$$N(E) \, \mathrm{d}E = 4\pi \left(\frac{2 m_\mathrm{n}}{h^2} \right)^{2/3} E^{1/2} \, \mathrm{d}E , \qquad (2.2)$$

where m_n is the effective mass of electrons. The Fermi level is usually located close to the center of the band gap, which is (at room temperature) more than $10\,kT$ below the lower edge of the conduction band. Therefore the Fermi function (see Fig. 2.1c) can be approximated by an exponential function in the conduction band:

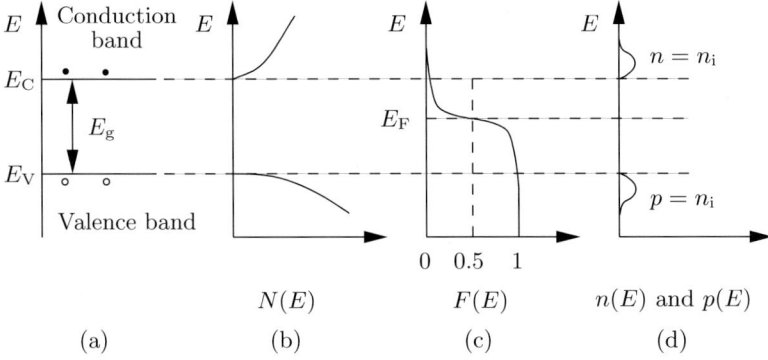

Fig. 2.1. Intrinsic semiconductor: (**a**) schematic band diagram, (**b**) density of states, (**c**) Fermi function, and (**d**) carrier concentration

$$F(E) = \frac{1}{1 + e^{(E-E_\mathrm{F})/kT}} \approx e^{-(E-E_\mathrm{F})/kT}. \tag{2.3}$$

Substituting (2.2) and (2.3) into (2.1) and calculating the integral yields the concentration of free charge carriers (see Fig. 2.1d):

$$n = \underbrace{2\left(\frac{2\pi m_\mathrm{n} kT}{h^2}\right)^{3/2}}_{N_\mathrm{C}} e^{-(E_\mathrm{C}-E_\mathrm{F})/kT}, \tag{2.4a}$$

$$p = \underbrace{2\left(\frac{2\pi m_\mathrm{p} kT}{h^2}\right)^{3/2}}_{N_\mathrm{V}} e^{-(E_\mathrm{F}-E_\mathrm{V})/kT}, \tag{2.4b}$$

where n (p) are the concentration of free electrons (holes), m_n (m_p) the effective mass of electrons (holes), E_C (E_V) the energy of the conduction (valence) band, and E_F the Fermi energy. The quantity N_C (N_V) is called *effective density of states* in the conduction (valence) band with a value of 2.8×10^{19} cm^{-3} (1.04×10^{19} cm^{-3}) [40]. The product of electron and hole concentrations

$$np = n_\mathrm{i}^2 = N_\mathrm{C} N_\mathrm{V}\, e^{-E_\mathrm{g}/kT} \tag{2.5}$$

is independent of the Fermi level and therefore also independent of the doping concentration. The energy gap E_g is defined as the difference between the lower edge of the conduction band, and the upper edge of the valence band, $E_\mathrm{C} - E_\mathrm{V}$. As the number of free electrons in an intrinsic semiconductor equals the number of holes (2.5) can be used to calculate the intrinsic carrier concentration to be

$$n_\mathrm{i} = 1.45 \times 10^{10}\ \mathrm{cm}^{-3}$$

in silicon at 300 K [40]. The Fermi level itself is defined by the requirement that the semiconductor in total is electrically neutral. The *intrinsic Fermi level* E_i can therefore be calculated by equating (2.4a) with (2.4b), leading to

$$E_i = \frac{E_C - E_V}{2} + \frac{3kT}{4} \ln\left(\frac{m_p}{m_n}\right). \tag{2.6}$$

As the second term of the sum is in the order of 0.01 eV at room temperature, the intrinsic Fermi level is located very close to the middle of the band gap.

The material used for semiconductor devices is in most cases not intrinsic but doped with a very small fraction of other materials to alter its conductivity. The elements used to dope silicon are either from the third group of the periodic table (e.g. boron) or from the fifth group (e.g. phosphorus, arsenic, antimony), so that they have either one valence electron less or more than silicon. The latter ones release their "extra" electron easily into the conduction band and are therefore called *donors*. This can be described by introducing a shallow energy level in the forbidden energy region of the band gap for each doping atom introduced. Donor levels are very close to the conduction band. For phosphorus the difference is 0.045 eV and for arsenic 0.054 eV. This means that at room temperature practically all donor states are ionized and the concentration of electrons, n, equals the concentration of donor atoms, N_D, if no other doping is present. Equating (2.4a) with N_D and using the fact that the electron concentration is intrinsic when the Fermi level equals the intrinsic level, one can calculate the shift of the Fermi level toward the conduction band:

$$E_F = E_i + kT \ln\left(\frac{N_D}{n_i}\right). \tag{2.7}$$

Introducing this elevated Fermi level into (2.4b) one immediately notices that the increase of the electron concentration is accompanied by a decrease of the hole concentration as already expressed in (2.5). Therefore silicon doped with donors is called *n-material*. The electrons are *majority* carriers and the holes are *minority* carriers

The same considerations can be made for silicon doped with boron. Boron has one electron less than silicon in the outer shell, and easily traps an electron from the valence band to establish a covalent bond to the neighboring silicon atom. This produces a free hole and boron is called an *acceptor*. Acceptor levels are very close to the valence band; for boron the difference is 0.045 eV. At room temperature practically all boron atoms are ionized and the acceptor concentration N_A and hole concentrations p are almost equal. In acceptor-doped material the Fermi level is shifted toward the valence band:

$$E_F = E_i - kT \ln\left(\frac{N_A}{n_i}\right). \tag{2.8}$$

The concentration of holes as majority carriers is much higher than the concentration of the electrons which are the minority carriers. Acceptor-doped silicon is therefore called *p-material*.

In reality, semiconductors contain both donors and acceptors, but, in most cases, one of the dopants dominates. In such a case the effective doping concentration is used, which is defined as the absolute value of the difference between the donor and the acceptor concentration.

2.2.2 Charge Generation and Recombination in Silicon

The basis for the detection of particles or radiation is their interaction with the sensor material. A part of the energy absorbed is used for the generation of electron–hole pairs that can be detected as electrical signals. Free charge carriers are also thermally generated, leading to the so-called leakage current. This generation is described by the Shockley–Read–Hall model for indirect semiconductors like silicon.

2.2.2.1 Charged Particles

Most particle physics experiments must measure trajectories of charged particles. This section discusses the amount of charge generated by those particles when crossing a layer of silicon.

Bethe–Bloch Formula

Charged particles deposit a part of their energy through many scattering processes with electrons of the absorbing material along the particle track. This process is dominant for particles heavier than electrons and described by the *Bethe–Bloch formula* [4, 46, 47]:

$$-\left\langle \frac{\mathrm{d}E}{\mathrm{d}x} \right\rangle = Kz^2 \frac{Z}{A} \frac{1}{\beta^2} \left(\frac{1}{2} \ln \frac{2m_e c^2 \beta^2 \gamma^2 T_{\max}}{I^2} - \beta^2 + \cdots \right), \qquad (2.9)$$

with

$\frac{\mathrm{d}E}{\mathrm{d}x}$	energy loss of the particle usually given in $\frac{\mathrm{eV}}{\mathrm{g/cm}^2}$
K	$4\pi N_\mathrm{Av} r_e^2 m_e c^2 = 0.307075$ MeV cm^2
z	charge of the traversing particle in units of the electron charge
Z	atomic number of absorption medium (14 for silicon)
A	atomic mass of absorption medium (28 for silicon)
$m_e c^2$	rest energy of the electron (0.511 MeV)
β	velocity of the traversing particle in units of the speed of light
γ	Lorentz factor $1/\sqrt{1-\beta^2}$
I	mean excitation energy (137 eV for silicon)

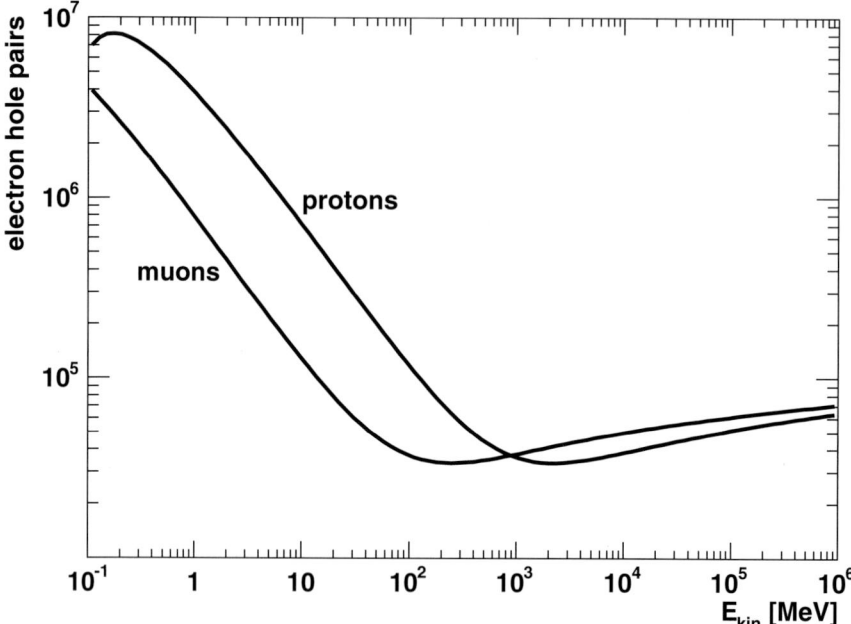

Fig. 2.2. Number of electron–hole pairs generated in a 300-μm-thick silicon layer

The dots at the end indicate the presence of additional correction terms, like for example the density correction for high particle energies and the shell correction for lower energies as is discussed in [46]. The kinetic maximum energy which can be transferred to an electron by a particle of mass M is

$$T_{\max} = \frac{2m_e c^2 \beta^2 \gamma^2}{1 + 2\gamma m_e/M + (m_e/M)^2} \ . \tag{2.10}$$

For particles much heavier than the electron the quadratic term in the denominator can be neglected. Usually the linear term is also disregarded. However, for high energies the Lorentz factor γ can reach the same order of magnitude as the mass ratio m_e/M, leading to a significant error. The number of electron–hole pairs generated in silicon as a function of the particle energy is plotted in Fig. 2.2. Among the particle properties only the charge explicitly appears in (2.9). The particle mass enters only into (2.10) but is implicitly also contained in (2.9) in the factors β and γ when scaling the energy loss with the particle's energy. Note that the dependence of (2.9) on the properties of the absorbing medium is fairly weak, as $Z/A \approx 1/2$ for most materials and the other material-dependent quantities appear in the logarithmic term.

For low energies the $1/\beta^2$ term in (2.9) is dominant and the stopping power decreases with increasing energy. At a particle's velocity β of about 0.96 ($\beta\gamma \approx 3$) a minimum is reached. At higher energies the logarithmic term leads to a slow rise again, which is eventually cancelled by the density correction (not included in Fig. 2.2). A particle with an energy loss in the minimum of the Bethe–Bloch formula is called a *minimum ionizing particle* (m.i.p.). The value of the minimum depends on the square of the particle charge but very weakly on the particle mass. Due to the flatness of the curve this expression is often also used for all particles with $\beta > 0.96$ (or $\beta\gamma > 3$).

In the very low energy range when the particle's velocity is comparable to the velocity of the orbital electrons of the absorption medium, the energy loss reaches a maximum and drops rapidly toward lower energies. This behavior is not described by the Bethe–Bloch formula 2.9 and its use is therefore limited to the range of particle velocities $\beta > 0.1$.

The number of electron–hole pairs generated by a traversing particle can be calculated by dividing the deposited energy by the mean energy needed for ionization, w, which for silicon is about 3.6 eV. The difference between w and the band gap of 1.12 eV generates phonons, which in the end will dissipate as thermal energy.

Landau Distribution

It is important to mention that the ionization is subject to statistical fluctuations and the value returned by (2.9) is only the average value of the so-called *Landau distribution* plotted in Fig. 2.3. So if a particle is not stopped in the sensor, the response varies around the peak of the distribution with a significant probability of high signals. Due to this tail the average value is higher than the most probable value of the distribution. The fluctuation around the maximum of this distribution becomes higher for thinner sensors. This has to be taken into account when designing the dynamic range of the readout circuit for these devices. The main reason for the Landau fluctuation is the rare but measurable occurrence of the so-called δ-electrons or knock-on electrons which obtain enough energy by the interaction to become ionizing particles themselves. As the direction of the knock-on electron is typically perpendicular to the direction of the incoming particle, δ-electrons lead to irregular charge clouds and degrade the spatial resolution. A detailed treatment of the energy deposited in silicon by high-energy particles is given, e.g., in [48, 49].

Shape of Ionization Path

High energetic particles penetrate the detector with nearly constant velocity, causing a uniform ionization along the path. A minimum ionizing particle with unit charge generates a signal of about 23,000 electrons (most probable value) in 300 µm of silicon.

If the total charge can be collected in about 5 ns, it will produce a current of the order of 1 µA which by far exceeds the leakage current of a single cell in a pixel detector.

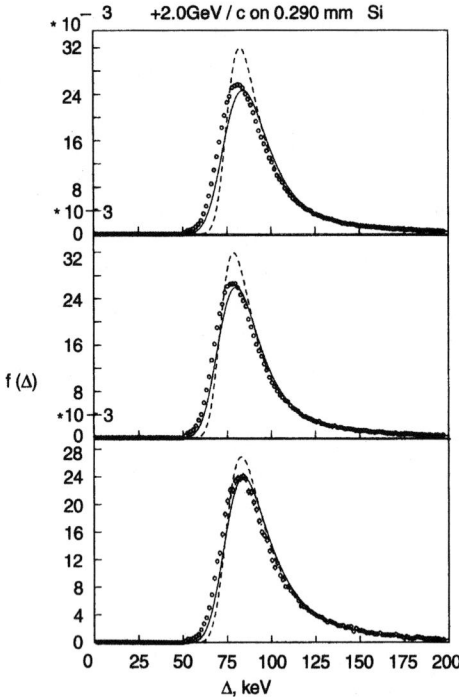

Fig. 2.3. Measured energy loss distributions (points) for 2 GeV positrons (*top*), positive pions (*middle*), and protons (*bottom*) traversing a 290-μm-thick silicon detector. The Landau function (*dashed lines*) and a more refined model (*solid lines*) are shown for comparison [48]

β-Particles produce a uniform charge cloud as long as their velocity remains relativistic. For this reason β-sources are often used in the laboratory to test silicon detectors.

For low-energy particles (or in the case of thick absorbers) the velocity of the particle is significantly reduced by the energy loss. This leads to an increased ionization as visible in Fig. 2.2 and again to a further reduction of the velocity. If the particle is stopped in the medium, most of its energy is deposited near the end of the trajectory. This local energy deposition (*Bragg peak*) is especially interesting in medical applications when protons (or other charged hadrons) are used to deliver energy to a well-specified volume covered by healthy tissue that should be affected as little as possible.

Another example of short-range particles are α-particles. Due to their high charge and their low speed the penetration depth in silicon is only a few micrometers, with the concentration of electron–hole pairs peaking close to its stopping point.

2.2 Device Physics and Fundamental Sensor Properties

Multiple Scattering

When the particle transverses the detector it is deflected by many small-angle scatters. The deflection is mainly caused by the Coulomb interaction of the charged particle with the nuclei. For hadrons the strong interaction also gives a contribution. The scattering angle of the projectile after many interactions when leaving the detector follows roughly a Gaussian distribution [50] with an rms of

$$\theta_{\text{plane}}^{\text{rms}} = \frac{13.6 \text{ MeV}}{\beta pc} z \sqrt{\frac{x}{X_0}} \left[1 + 0.038 \ln\left(\frac{x}{X_0}\right)\right], \qquad (2.11)$$

where the angle θ is expressed in rad, the particle momentum p in MeV, and the velocity β in units of the velocity of light c. The charge number of the projectile is z and x/X_0 is the thickness of the absorption medium in units of the radiation length. The radiation length X_0 of silicon is 9.36 cm which corresponds to a superficial density of 21.82 g/cm^2 [51].

If one needs to calculate the scattering angle of a sample composed of several materials and layers, one should add all layers in units of the radiation length and apply (2.11) only once.

A pixel detector built for LHC has a thickness of about 2% of a radiation length per layer, changing the trajectory of a 1 GeV particle by an angle $\theta_{\text{plane}}^{\text{rms}}$ of $\approx 0.1°$.

2.2.2.2 Electromagnetic Radiation

Silicon detects electromagnetic radiation from the infrared up to the X-ray range. This qualifies silicon for a wide range of imaging applications starting from the commercial photo cameras up to biomedical applications (see Chap. 5). However, this quite unique feature is not used in particle physics where the main task is to detect high energetic charged particles.

Photons interact with the sensor mainly via three processes, namely photoelectric effect, Compton effect, and pair production. The main difference to the interaction with charged particles is that the photon is either absorbed completely (photoelectric absorption, pair production) or scattered by a (relatively) large angle (Compton). A monochromatic photon beam penetrating through material is therefore not changed in energy but attenuated in intensity according to

$$I(x) = I_0 \, e^{-x/\mu}, \qquad (2.12)$$

with I_0 being the beam intensity before and $I(x)$ after traversing a medium of thickness x. The attenuation length μ is a material property of the absorbing medium and depends on the photon energy. It is shown for some semiconductor materials in Fig. 5.20. The probability for photon interaction in a 300-μm-thick silicon layer is plotted in Fig. 2.4 as a function of the photon

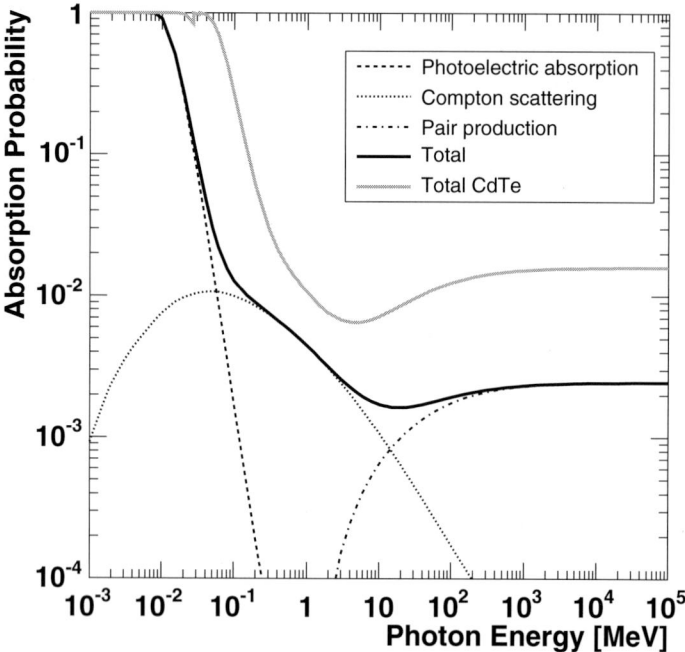

Fig. 2.4. Probability of photon absorption for 300 μm silicon as function of the photon energy. Contributions from different processes are indicated. The total absorption probability for 300 μm CdTe is also given for comparison (Data from [52])

energy. To indicate the benefit from high-Z materials the total absorption probability for cadmium telluride is also plotted.

The photoelectric effect is the absorption of a photon by an atomic electron which is hence moved into the conduction band. It is the dominant process at low photon energies, in silicon below about 100 keV as shown in Fig. 2.4. Its cross section is very strongly dependent on the nuclear charge Z of the absorbing material:

$$\sigma_{\text{Photo}} \propto Z^n, \quad (2.13)$$

with n varying between 4 and 5 depending on the photon energy [53]. For this reason high-Z materials like cadmium telluride (see Sect. 2.7) are preferred for X-ray detection.

At higher energies the photoelectric cross section drops down several orders of magnitude and scattering processes become more important. The cross section of Compton scattering is only linearly dependent on Z. At energies exceeding twice the electron mass pair production also contributes to the cross section (with $\sigma \propto Z^2$) and becomes the only important process at energies exceeding 10 MeV.

2.2 Device Physics and Fundamental Sensor Properties

Silicon is used for photon detection up to energies of about 100 keV (see Sect. 5.3). The use of other (higher Z) semiconductors can shift this value considerably (see Sect. 2.7).

Fano Factor

The average number of electron–hole pairs, N, generated for a constant amount of absorbed energy, e.g. caused by a monoenergetic beam of X-rays (or low range particles like α's), can be calculated by dividing the absorbed energy E by the average energy necessary to produce one electron–hole pair, w:

$$N = \frac{E}{w}. \tag{2.14}$$

The energy w required to create an electron–hole pair is about 3.6 eV in silicon. This value, which is more than three times larger than the band gap of 1.12 eV, is a material property and does hardly depend on the type and energy of the radiation if the latter is much larger than the band gap. The difference generates phonons, which in the end will dissipate as thermal energy.

The fraction of deposited energy that is used for electron–hole separation and phonon generation is subject to fluctuations which cause N to vary by

$$\langle \Delta N^2 \rangle = FN = F\frac{E}{w}, \tag{2.15}$$

with F being the *Fano factor* [54]. If the fraction of the absorbed energy used for the electron–hole pair generation would be fixed, their number N would also be fixed and the value of F would be zero. The Fano factor F, however, is, according to theoretical works, in the order of 0.1 for most semiconductors [55, 56] and determines the best possible energy resolution of semiconductor sensors in spectroscopic applications. The experimental determination of this quantity is difficult. For silicon the measured values range from 0.07 [57] over 0.124 [58, 59] to 0.16 [60, 61]. F is also a function of the temperature and, below 1 keV, also of the energy of the incoming photon.

2.2.2.3 Generation and Recombination

In thermal equilibrium the concentration of charge carriers is constant in time and obeys the mass action law (2.5) due to the balance of generation and recombination of charge carriers. The thermal generation rate G_{th} of charge carriers does not depend on the concentration:

$$G_{\text{th}} = \frac{n_i}{\tau_g}, \tag{2.16}$$

with τ_g being the generation lifetime. The recombination rate is proportional to the product of the charge carrier concentrations, np. In the case of low-level injection, i.e. when the majority carrier concentration is practically unchanged, the recombination is in fact limited by the concentration of the minority carriers, leading to

$$R = \frac{p}{\tau_{r,n}} \quad \text{for n-material},$$
$$R = \frac{n}{\tau_{r,p}} \quad \text{for p-material}, \tag{2.17}$$

with $\tau_{r,n/p}$ being the recombination lifetime in n- and p-type semiconductor, respectively. The numerical values of τ_g and τ_r can differ a lot.

In the presence of excess carriers the product of the electron and hole concentration exceeds the value given in (2.5). These excess carriers might be introduced by injection or radiation. After injection or radiation has stopped, a thermal equilibrium is reached again by enhanced recombination proportional to the concentration of the minority excess carriers. This leads to an exponential decay with the characteristic time τ_r.

In the case of removal of carriers the product of the carrier concentrations will fall below n_i^2 and generation, which will be unaffected by this change, dominates as the recombination rate will be very low due to the lack of carriers. Generation increases the carrier concentration and therefore also the recombination rate. This also leads to an exponential return to the equilibrium condition but with time constant τ_g. If the generated carriers are continuously removed, like in a reversely biased diode, the carrier concentration product will stay permanently below n_i^2 and the equilibrium state is never reached. The result is a steady generation current of the drained carriers.

The charge carrier generation and recombination process is very different for direct semiconductors like for example gallium arsenide and indirect semiconductors like for example silicon. In indirect semiconductors a direct band-to-band recombination is highly suppressed as holes at the top of the valence band and electrons at the bottom of the conduction band carry different crystal momentum. A direct transmission conserving both energy and momentum is not possible without lattice interaction. The dominant Schockley–Read–Hall process involves localized energy states in the forbidden energy region acting as stepping stones. They are caused by crystal imperfections or unwanted impurities. The carrier lifetimes can be calculated using the concentration, energy levels, and other characteristics of these generation–recombination centers. They are inversely proportional to the concentration of generation–recombination centers and the thermal velocity. The latter quantity is proportional to the square root of the temperature, which influences the temperature dependence of the leakage current (see Sect. 2.2.4). For a complete treatment of the Schockley–Read–Hall mechanism the reader is referred to the original literature [62,63] or one of the textbooks mentioned earlier [4,37,40,42,44].

2.2.3 Transport of Charge Carriers

The separation and movement of the charge carriers (electrons and holes) to the electrical contacts induces signal pulses which are detected by the readout electronics. In this section we will briefly discuss the charge carrier transport mechanism in semiconductors. Free charge carriers can be considered as free particles with a mean kinetic energy of $\frac{3}{2}kT$. They move randomly with a mean thermal velocity of about 10^5 m/s. From time to time they are scattered on the lattice or on impurity atoms or other scattering centers. Their mean free path is of the order of 0.1 μm, which translates to a mean time τ_c between two collisions of about 10^{-12} s. The traveled distance averaged over many charge carriers in equilibrium condition is zero. Superimposed to the statistical movement there can be diffusion due to a concentration gradient or a drift due to an external electrical field.

2.2.3.1 Diffusion

The random movement of the charge carriers implies that in the case of a gradient in the carrier concentration it is more probable that a carrier from a high-concentration region arrives at a low-concentration region than vice versa. This effect, which spreads the distribution, is called diffusion. The diffusion current per unit area is described by

$$\boldsymbol{J}_{\text{n,diff}} = -D_n \boldsymbol{\nabla} n = -\frac{kT}{e} \mu_n \boldsymbol{\nabla} n \quad \text{for electrons,}$$

$$\boldsymbol{J}_{\text{p,diff}} = D_p \boldsymbol{\nabla} p = \frac{kT}{e} \mu_p \boldsymbol{\nabla} p \quad \text{for holes,} \quad (2.18)$$

where $\boldsymbol{\nabla} n$ and $\boldsymbol{\nabla} p$ are the gradients of the electron and hole concentration. The diffusion constants $D_{n,p}$ are related to the mobilities via the Einstein equation $D_{n,p}/\mu_{n,p} = kT/e$ included in (2.18).

2.2.3.2 Drift

In the presence of an external electric field the charge carrier is accelerated between two random collisions. The direction is given by the field leading to an average drift with a velocity given by

$$v_n = -\frac{e\tau_c}{m_n} \boldsymbol{E} = -\mu_n \boldsymbol{E} \quad \text{for electrons,}$$

$$v_p = \frac{e\tau_c}{m_p} \boldsymbol{E} = \mu_p \boldsymbol{E} \quad \text{for holes,} \quad (2.19)$$

with v being the average drift velocity, e the electron charge, m the effective mass of electrons (holes), and E the electric field. The mobility μ is a function of τ_c and therefore dependent on temperature, doping concentration, and the concentration of other lattice imperfections.

The drift velocity is only proportional to the electric field when the latter is small and the velocity gained due to the electric field is small compared to the thermal velocity of the charge carriers. At higher fields the charge carriers acquire a higher acceleration. As the mean free path is not altered by the field, the number of random collision per unit time becomes higher. This counterbalances a further acceleration and finally leads to a saturation of the drift velocity at saturation values $v_{s,n}$ and $v_{s,p}$. This effect of velocity saturation (also referred to as "mobility degradation") can be described by parameterizing the mobility as a function of the electric field [64]:

$$\mu = \frac{v_s/E_c}{\left[1 + (E/E_c)^\beta\right]^{1/\beta}}, \qquad (2.20)$$

with the fitted parameter sets

$$\left.\begin{array}{l} v_s = 1.53 \times 10^9 \times T^{-0.87} \text{ cm/s} \\ E_c = 1.01 \times T^{1.55} \text{ V/cm} \\ \beta = 2.57 \times 10^{-2} \times T^{0.66} \end{array}\right\} \text{ for electrons}$$

and

$$\left.\begin{array}{l} v_s = 1.62 \times 10^8 \times T^{-0.52} \text{ cm/s} \\ E_c = 1.24 \times T^{1.68} \text{ V/cm} \\ \beta = 0.46 \times 10^{-2} \times T^{0.17} \end{array}\right\} \text{ for holes}$$

and T being the absolute temperature and E the absolute value of the electrical field. This model describes the mobility with an accuracy of about 5%.

The mobilities of electrons and holes resulting from (2.20) are plotted in Fig. 2.5. In a typical (not irradiated) silicon detector the applied voltage is in the order of 150 V and the *average* field is about 5,000 V/cm, which is about the value where the mobility starts to drop. Indeed, the electric field inside a silicon detector is not constant, as will be shown in Sect. 2.2.4.

The low field mobility for intrinsic silicon at 300 K is given as [65]

$$\mu_n = 1,415 \pm 46 \text{ cm}^2/(\text{Vs}),$$

$$\mu_p = 480 \pm 17 \text{ cm}^2/(\text{Vs}).$$

Electrons are three times more mobile than holes, which makes the holes more prone to trapping especially in the irradiated material.

The drift current per unit areas J_drift can be calculated by multiplying the drift velocity by the carrier concentration n (p) and the electron charge e:

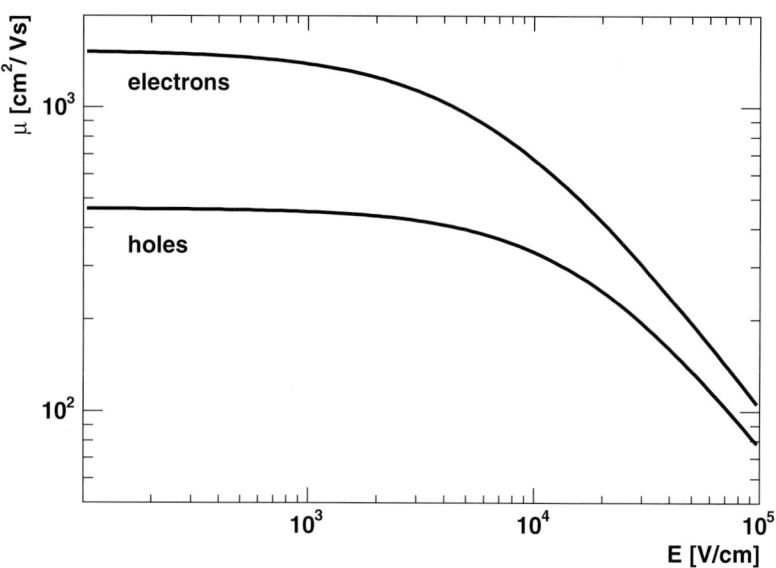

Fig. 2.5. Mobility as function of the electric field calculated with (2.20)

$$\boldsymbol{J}_{\text{n,drift}} = -en\mu_{\text{n}}\boldsymbol{E} \quad \text{for electrons,}$$
$$\boldsymbol{J}_{\text{p,drift}} = ep\mu_{\text{p}}\boldsymbol{E} \quad \text{for holes.} \tag{2.21}$$

The proportionality factor between current density and electric field is the conductivity σ or the specific resistance ρ of the material:

$$\sigma \equiv \frac{1}{\rho} = en\mu_{\text{n}} \quad \text{for electrons,}$$
$$\sigma \equiv \frac{1}{\rho} = ep\mu_{\text{p}} \quad \text{for holes.} \tag{2.22}$$

The 2-kΩ cm n-doped material typically used for silicon detectors has a phosphorus concentration of about 2×10^{12} cm^{-3}.

2.2.3.3 Effect of Magnetic Field

Semiconductor detectors can be operated inside a magnetic field. The magnetic field influences the drift of the signal charge and therefore the properties of the sensor. An electron or a hole moving in an electric field experiences, in the presence of a magnetic field $\boldsymbol{B}$, the Lorentz force $\boldsymbol{F} = \pm e\left(\boldsymbol{E} + \boldsymbol{v} \times \boldsymbol{B}\right)$ with the sign of the charge carrier under consideration. The drift direction will deviate from the direction of the electric field by the Lorentz angle θ_{L} according to

$$\tan\theta_{L,n} = \mu_{Hall,n} B_\perp xs \quad \text{for electrons,}$$
$$\tan\theta_{L,p} = \mu_{Hall,p} B_\perp \quad \text{for holes,} \tag{2.23}$$

with $B_\perp$ being the magnetic field perpendicular to the drift velocity. The Hall mobility μ_{Hall} differs from the drift mobility by the very weakly temperature dependent Lorentz factor r:

$$\mu_{Hall} = r\mu,$$

with values for r of 1.15 for electrons and 0.72 for holes at 0°C. For doping concentrations below 10^{14} cm^{-3} the value of r is independent of the doping concentration.

According to (2.20) the mobility decreases strongly with increasing electric field; the same applies to the Lorentz angle as indicated by (2.23). In a device with a nonconstant field distribution the carriers therefore drift along a curved trajectory. The displacement of the signal charge during their drift in the magnetic field can be significant and has to be considered when designing a detector system, in particular when fixing the readout pitch of the sensor in the magnetic field.

Especially in sensors operated in a radiation environment the sensor bias has to be increased during the lifetime of the sensor. Therefore the impact of bias and radiation on the Lorentz angle has been studied [66–68]. As the electric field in the sensor usually exceeds 10^3 V/cm, when the mobility of electrons start to decrease drastically (see Fig. 2.5), the Lorentz angle for a given magnetic field is determined by the bias voltage and the choice of charge carriers.

2.2.4 The pn-Junction

The reversely biased pn-junction is the basic building block of silicon sensors. It builds up an electric field that collects the signal charge and suppresses the leakage current, an important noise source. A pixel sensor is a reversely biased pn-diode with a highly segmented cathode or anode.

2.2.4.1 Thermal Equilibrium

If there is a transition between n-doped and p-doped material, some of the majority charge carriers of one side diffuse into the differently doped side due to the concentration difference. They will recombine with the majority carriers producing a region close to the junction which is depleted from free charge carriers. As in this *depletion zone* the acceptor and donor ions are left without their reversely charged free carrier, this region is electrically charged and is therefore also called *space charge region*. The space charge region causes an electrical field counteracting the diffusion that can be characterized by the so-called *built-in voltage* V_{bi} as illustrated in Fig. 2.6. In thermal equilibrium

2.2 Device Physics and Fundamental Sensor Properties

diffusion and drift current cancel each other. For an abrupt pn-junction with constant doping concentrations on both sides the built-in voltage is

$$V_{bi} = \frac{kT}{e} \ln\left(\frac{n_{0,n}\, p_{0,p}}{n_i^2}\right) \approx \frac{kT}{e} \ln\left(\frac{N_D N_A}{n_i^2}\right), \qquad (2.24)$$

with $n_{0,n}$ being the electron concentration in the n-doped side and $p_{0,p}$ the hole concentration in the p-doped side. In most cases a complete ionization of donors and acceptors can be assumed and electron and hole concentrations can be replaced by the concentrations of donors, N_D, and acceptors, N_A. The value of the built-in voltage is just the difference of the Fermi potentials between the n- and p-doped material as indicated in Fig. 2.6.

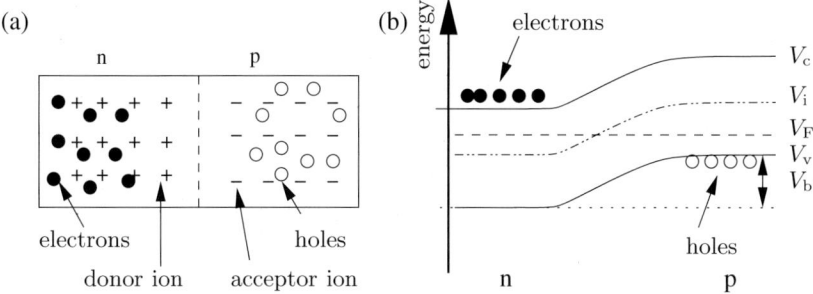

Fig. 2.6. Schematic illustration of a pn-junction (**a**) and its band diagram to illustrate the formation of the built-in voltage (**b**)

2.2.4.2 Reverse Bias

An external voltage applied in the same direction as the built-in voltage will remove further majority carriers from either side and will extend the space charge region. The junction is now *reversely biased* and the reverse voltage applied will be defined as positive in the following.[1] The width of the depletion zone, the electric field, and the potential as function of the applied voltage can be calculated by solving the one-dimensional Poisson equation. With the same assumptions as used for (2.24) one obtains

$$W = x_n + x_p = \sqrt{\frac{2\varepsilon_0 \varepsilon_{Si}}{e}\left(\frac{1}{N_A} + \frac{1}{N_D}\right)(V + V_{bi})}, \qquad (2.25)$$

---

[1] In this chapter we follow the convention that a reverse bias is positive and a forward bias is negative. This leads to additional minus signs in the coming equations compared to those in the standard textbooks, but is more convenient as sensors are operated reversely biased.

where W is the total width of the depletion zone, x_n and x_p its part on the n- and p-side respectively, and V the externally applied voltage. In silicon sensors the junction is usually realized by a shallow and highly doped ($N_\text{A} > 10^{18}$ cm^{-3}) p$^+$-implant in a low-doped ($N_\text{D} \approx 10^{12}$ cm^{-3}) bulk material. Therefore the term $1/N_\text{A}$ in (2.25) can be neglected, meaning that the space charge region is reaching much deeper into the lower doped side of the junction. In addition, the built-in voltage of the order of 0.5 V is small compared to typical operation voltages exceeding in most cases 50 V and can therefore also be neglected. Both assumptions lead to

$$W \approx x_\text{n} \approx \sqrt{\frac{2\varepsilon_0\varepsilon_\text{Si}}{eN_\text{D}}V}, \qquad (2.26)$$

which is widely used for calculating the depletion depth.

The space charge, potential, and electric field in the direction perpendicular to the junction are plotted in Fig. 2.7. The electric field reaches its maximum value of

$$E_\text{max} = 2V/W \approx \sqrt{\frac{2eN_\text{D}}{\varepsilon_0\varepsilon_\text{Si}}V} \qquad (2.27)$$

directly at the junction. From there it decreases linearly to the end of the space charge region where it vanishes. The linearity is due to the constant doping concentration while the slope is proportional to the doping concentration. The full depletion voltage V_depl is the voltage needed to extend the depletion region over the complete thickness d of the device. It is one of the most important sensor parameters as it often defines the minimum operating voltage which according to (2.26) is proportional to the square of the wafer

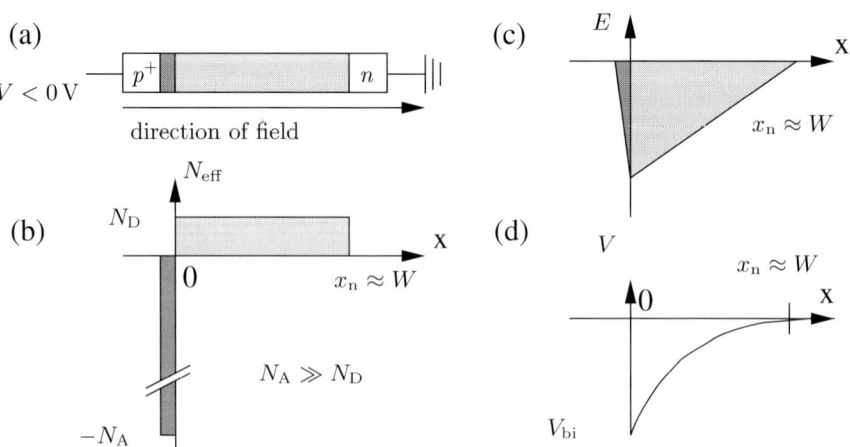

Fig. 2.7. (a) Depletion region, (b) space charge, (c) electric field, and (d) for an abrupt p$^+$n-junction with $x_\text{n} \approx W$

2.2 Device Physics and Fundamental Sensor Properties

thickness. If the applied bias exceeds the full depletion voltage, the device is said to be *overdepleted* and a constant field of $(V - V_{\text{depl}})/d$ is added at each point in the device.

If a free charge enters the space charge region of a reversely biased pn-junction, it will be removed by the electric field. The current caused by radiation is the signal that should be detected by a charge-sensitive amplifier. In absence of radiation there is always a steady *leakage* or *dark current* which has various components. One is due to the diffusion of free carriers from the undepleted volume into the sensitive space charge region. More important is the thermal generation at generation–recombination centers at the surface of the device and in the depleted volume. The latter usually dominates the leakage current of an unstructured pn-junction and is proportional to the depleted volume:

$$J_{\text{vol}} \approx -e\frac{n_i}{\tau_g}W \approx -e\frac{n_i}{\tau_g}\sqrt{\frac{2\varepsilon_0\varepsilon_{\text{Si}}}{eN_D}V}, \qquad (2.28)$$

with J_{vol} being the volume generation current per unit area, τ_g the carrier generation lifetime, and n_i the intrinsic carrier concentration.

The temperature dependence of the volume generation current is hidden in the intrinsic charge carrier concentration – see (2.4a), (2.4b), and (2.5) – and the generation lifetime, which leads to[2]

$$J_{\text{vol}} \propto T^2 e^{-E_g(T)/2kT}. \qquad (2.29)$$

A useful rule of thumb for rough estimations is to assume that the volume current doubles every 8 K.

If the reverse bias is increased to very high values, electrical breakdown occurs at the junction which is, according to Fig. 2.7, the region of the maximum field. For an abrupt and planar junction the breakdown voltage V_{bd} can be estimated to [40, 42–44]

$$V_{\text{bd}}\,[\text{V}] \approx 5.3 \times 10^{13} \times N_D^{-3/4}, \qquad (2.30)$$

with N_D being the doping concentration of the substrate in units of cm^{-3}. However, this approximation is valid only for substrates with an initial doping concentration above roughly $2 \times 10^{15}\,\text{cm}^{-3}$. In detector grade material that usually has an initial doping concentration of the order of 10^{12}–$10^{13}\,\text{cm}^{-3}$ breakdown occurs much earlier than predicted by (2.30) due to crystal imperfections close to the junction and higher fields at the implant corners.

---

[2] If experimental data of irradiated sensors are fitted to (2.29), a too high value for the band gap E_g is obtained [69]. This can be understood if the additional dependence of the generation lifetime on the energy of trap levels and temperature is considered

2.2.4.3 Forward Bias

If a pn-junction is under forward bias, i.e. the n-side is negatively biased with respect to the p-side, the current voltage characteristic will be determined by the minority carriers. Using the mass action law (2.5), one can replace the term $p_{0,p}/n_i^2$ or $n_{0,n}/n_i^2$ in the expression for the built-in voltage (2.24) by the reverse concentrations of the minority carriers $1/n_{0,p}$ or $1/p_{0,n}$ and rearrange it to

$$n_{0,p} = n_{0,n}\, e^{-eV_{bi}/kT} \quad \text{for electrons,}$$
$$p_{0,n} = p_{0,p}\, e^{-eV_{bi}/kT} \quad \text{for holes.} \tag{2.31}$$

This means that the carrier concentrations at the edge of the depletion zone are linked via the electrostatic potential difference V_{bi}. Although (2.31) was derived for the thermal equilibrium condition, one expects this condition also to hold if the electrostatic potential is changed by an applied voltage V:

$$n_p = n_n\, e^{-e(V_{bi}+V)/kT} = n_{0,p}\, e^{-eV/kT} \quad \text{for electrons,}$$
$$p_n = p_p\, e^{-e(V_{bi}+V)/kT} = p_{0,n}\, e^{-eV/kT} \quad \text{for holes,} \tag{2.32}$$

where n_n, n_p, p_p, and p_n are the electron and hole concentrations at the edge of the depletion zone in the nonequilibrium condition and the forward bias defined as negative. Equation (2.32) contains the condition of low injection, meaning that the majority carrier concentrations remain constant up to the boundary of the depletion region: $n_n \approx n_{0,n}$ and $p_p \approx p_{0,p}$. The diffusion current of the minority carriers will be proportional to their deviation value in thermal equilibrium, which exists far away from the junction and therefore an exponential increase of the forward current results to

$$J = J_s \left(e^{-eV/kT} - 1\right), \tag{2.33}$$

with J_s being the saturation diffusion current for reverse bias [4,37,40,42,44].

2.2.5 Surface Barrier

A metal brought on top of a not too highly doped semiconductor crystal forms a surface barrier with rectifying properties similar to those of a pn-junction. In fact the first semiconductor devices used in the beginning of the twentieth century were such rectifiers. Germanium surface barrier counters are widely used in X-ray spectrometers.

Metals differ from semiconductors by their conduction band being partially filled with electrons or the valence band overlapping the conduction band. The Fermi level representing the occupancy probability of one half is therefore located inside the conduction band. The concentration of free charge carriers is high enough to prevent the built-up of a static electric field inside the metal. Interaction with the environment can be described by surface charges. A characteristic quantity of a metal is its work function $e\phi_m$, representing the energy necessary for removing an electron from the Fermi level to the vacuum. In a semiconductor the corresponding work function $e\phi_s$ is the difference between the Fermi level E_F and the vacuum. The energy necessary to release an electron from the conduction band into the vacuum is called electron affinity χ. It is the energy difference between the vacuum and the lower edge of the conduction band.

Figure 2.8 illustrates the process of barrier formation. The semiconductor is of n-type and uniformly doped. The work function of the metal is assumed to be larger than the one of the semiconductor (which is the case, e.g., for Ag, Au, and Cu but not for Al). In Fig. 2.8a both materials are separated and the vacuum level is used as a reference. If both materials are brought into contact the electrons in the semiconductor's conduction band move into the metal due to their higher energy and leave behind the positively charged donor atoms. The electrons form a very thin layer at the metal surface and create an electric field in the semiconductor which counteracts the electron drift. The state of equilibrium is reached when the Fermi levels in both materials are the same. As the vacuum level has to be continuous the bands in the semiconductor are thus bent upward by the built-in voltage

$$V_{bi} = e(\phi_m - \phi_s).$$

The potential barrier as viewed from the metal toward the semiconductor is given by

$$\phi_{Bn} = \phi_m - \chi = V_{bi} + \frac{1}{e}(E_c - E_F).$$

It stays unchanged if an external voltage is applied.

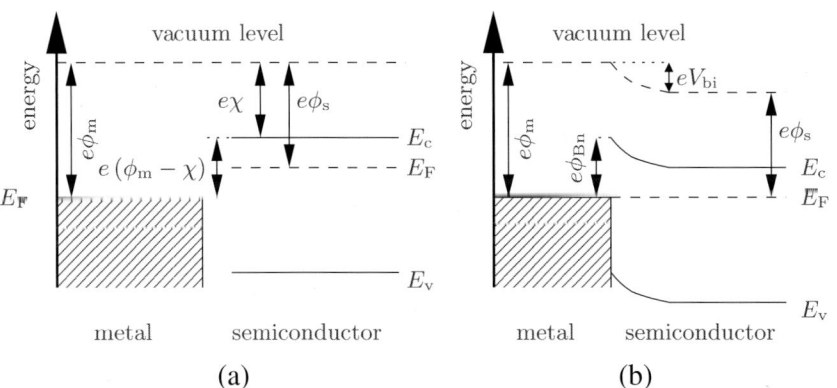

Fig. 2.8. Band diagrams of a contact between a metal and an n-type semiconductor (with $\phi_m > \phi_s$). (**a**) Both materials are isolated from each other. (**b**) The materials are in contact and thermal equilibrium has established

If a voltage of the same polarity as the built-in voltage is applied to the surface barrier, the depletion region will extend further into the semiconductor. The depletion depth can be calculated in the same way as was done for the pn-junction, leading to (2.26) for biases much above the built-in voltage.

2.2.6 Metal Oxide Semiconductor Structure

The sandwich of metal, silicon oxide, and semiconductor, called MOS structure (Fig. 2.9), is the basic building block of the MOS field effect transistors (MOSFETs) which are widely used in microelectronics. As they have been described in detail in [41] or other textbooks [40, 42, 44] it is justified to give only a simplified description here. In strip sensors MOS structures are often used as coupling capacitors while in charge coupled devices they are used for charge storage and transfer. In pixel detectors, which are directly coupled to the readout electronics, understanding MOS structures is important to deal with the surface regions between the pixels and with the consequences of radiation-induced surface damage. Furthermore, MOS capacitors are powerful tools for surface characterization which often is part of the testing procedure.

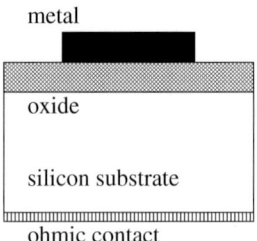

Fig. 2.9. Schematic cross section of a MOS structure

The operation of a MOS structure is illustrated in Fig. 2.10. In case a bias is applied between the silicon substrate and the metal, which is often also called gate contact, no direct current will flow thanks to the insulating oxide layer. Thermal equilibrium can therefore be assumed separately in both regions: silicon and metal. This means that the Fermi level remains constant in each of the regions. Due to the potential change at the surface of the silicon the bands will be moved relative to the Fermi level, causing a change in the distribution of free charge carriers. This is the *surface field effect*. Due to a different work functions in metal and silicon and the presence of fixed (mostly positive) charges in the dielectric mainly close to the interface, a deflection of the bands already exists at zero bias. The *flat band voltage* V_{FB} caused by both effects has to be compensated by an applied voltage to again obtain unbent, flat bands and a constant charge carrier distribution up to the silicon's surface:

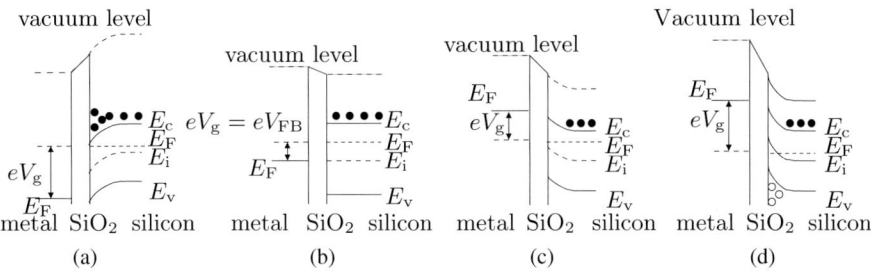

Fig. 2.10. Band diagram of a MOS structure in accumulation (**a**), flat band condition (**b**), depletion (**c**), and inversion (**d**). The *filled circles* indicate electrons, and the *open circles* holes

$$V_{\text{FB}} = \underbrace{\phi_{\text{m}} - \phi_{\text{s}}}_{=:\phi_{\text{ms}}} - \frac{ed}{\varepsilon_0 \varepsilon_{\text{ox}}} N_{\text{ox}}, \qquad (2.34)$$

with ϕ_{m} and ϕ_{s} being the work functions of the metal and semiconductor, d the thickness of the oxide, and N_{ox} the number of oxide charges per unit area. The difference between the work functions ϕ_{ms} for aluminum and silicon is in the order of 0.8–1.1 V for p-type and 0.2–0.4 V for n-type silicon. The oxide charge N_{ox} used in (2.34) is meant as the sum of all different kinds of immobile charge which can appear in the MOS system. A more specific terminology is explained, e.g., in [70]. The oxide charge depends very much on the oxidation process and crystal orientation. For $\langle 1,0,0\rangle$ silicon values as low as several 10^{10} cm^{-2} can be achieved while the charge per area in $\langle 1,1,1\rangle$ material is about an order of magnitude higher. After a dose of several kilograys due to ionizing radiation a saturation value of some 10^{12} cm^{-2} is reached.

In the following a MOS structure on n-type silicon is considered with the silicon held at zero potential and a voltage applied to the gate contact. If the applied voltage is more positive than the flat band voltage V_{FB}, the bands are bent down and electrons are attracted forming an *accumulation layer* as shown in Fig. 2.10a. As the flat band voltage is usually negative this is also the case if zero voltage is applied. In some cases the conducting accumulation layer is unwanted and has to be suppressed.

If the applied bias exactly counterbalances the flat band voltage, all bands are flat (Fig. 2.10b) and the electron concentration is constant in the whole silicon substrate.

If an external voltage more negative than the flat band voltage is applied, the bands are bent up and the electrons of the n-type silicon are pushed into the bulk (Fig. 2.10c). At the surface of the silicon a depletion zone is formed giving this state the term *depletion*.

If the bias becomes more and more negative, the bands will continuously be bent up while the Fermi level in the silicon remains unchanged. At a given point the intrinsic level will cross the Fermi level (Fig. 2.10d). If the

Fermi level is lower than the intrinsic level, the material is p-conductive and *inversion* occurs. The holes filling the inversion layer are either supplied from a p-region close by, as in a MOS transistor, or by thermal generation. Depending on the generation lifetime τ_g it might need some time until the accumulation layer is filled. This nonequilibrium state is called *deep depletion*. It can artificially be maintained by draining the holes of the inversion, e.g., into a close negatively biased p-contact. The depletion depth in the state of deep depletion is [44]

$$W_{\text{deep depletion}} = \frac{\varepsilon_{\text{Si}} d_{\text{ox}}}{\varepsilon_{\text{ox}}} \left[\sqrt{1 + \frac{2(V - V_{\text{FB}})}{eN_d \varepsilon_0 \varepsilon_{\text{Si}}} \left(\frac{\varepsilon_0 \varepsilon_{\text{ox}}}{d_{\text{ox}}} \right)^2} - 1 \right]. \quad (2.35)$$

If an inversion layer is formed and the voltage is decreased further, the hole concentration at the surface will reach and finally exceed the electron concentration in the bulk and *strong inversion* is reached. At this point the depletion layer will no longer grow having reached its maximum stable value:

$$d_{\max} = \sqrt{\frac{4\varepsilon_0 \varepsilon_{\text{si}} \Psi_B}{eN_D}}, \quad (2.36)$$

with $\Psi_B = \frac{E_F - E_i}{e} = \frac{kT}{e} \ln \frac{N_D}{n_i}$ being the difference between the Fermi and the intrinsic level far away from the surface. The voltage at which strong inversion occurs is called, following MOS electronics nomenclature, threshold voltage V_{th}:

$$V_{\text{th}} = V_{\text{FB}} - 2\Psi_B - \frac{d_{\text{ox}}}{\varepsilon_0 \varepsilon_{\text{si}}} \sqrt{4eN_D \varepsilon_0 \varepsilon_{\text{si}} \Psi_B} . \quad (2.37)$$

The same considerations are also valid for p-type MOS structures but with different signs according to the different polarity of the free charge carriers and different position of the Fermi level.

2.2.7 Punch Through

This section discusses the electrical potential of floating implants like the one shown in Fig. 2.11. If a "small" backside voltage is applied, the depletion zone will extend around the grounded implant on the left as shown in Fig. 2.11a. The floating p$^+$-implant will only be surrounded by its intrinsic depletion zone. Its potential will only differ from the undepleted n-type bulk by the built-in voltage, that can be calculated using (2.24). The junction is in thermal equilibrium and no net current is flowing into or out of the implant. If one further increases the backside voltage, the floating implant is following the backside potential up to the point when the depletion zone of the grounded p$^+$-implant reaches the floating one (Fig. 2.11b). Now the potential difference between the two p$^+$-implants is large enough that holes

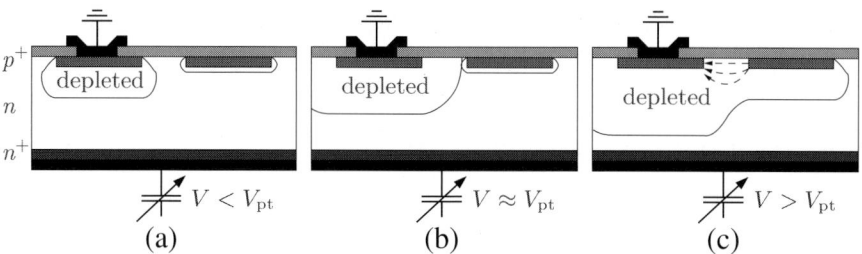

Fig. 2.11. Illustration of punch through. Only the left implant is connected to ground. (**a**) The applied voltage is below the punch-through voltage and the unconnected implant on the right is on backside potential (plus V_{bi} of the junction). (**b**) The depletion zone has just reached the floating implant. (**c**) The potential of the right implant is fixed by the punch-through hole current (*arrows*) toward the grounded implant

from the floating implant can overcome the potential barrier of the junction and the positive space charge and flow to the grounded implant. From this point on, the potential of the floating implant is no longer determined by the backside voltage but rather by the grounded p$^+$-implant. If the backside bias is increased further, the potential of the right p$^+$-implant stays (almost) constant and its depletion zone starts to grow (Fig. 2.11c). The holes from the volume generation current of this space charge region are collected by the unconnected implant and flow from there into the grounded electrode as indicated by the arrows in Fig. 2.11c.

The potential of the unconnected implant is called the *punch-through voltage* V_{pt} and is mainly determined by the distance of the implants and the substrate doping. A useful and, in most cases, sufficient approximation is to use the voltage needed to reach the floating implant with the depletion zone from the closest connected implant, the situation illustrated in Fig. 2.11b.

However, there is also a weak dependence of the punch-through voltage on the volume current, the backside voltage, and the state of the silicon–oxide interface (oxide charge, surface potential). For an exact calculation one has to determine the potential along the "channel" and solve the diffusion equation as done, e.g., in [4, p. 213].

Punch through can also be used for biasing of AC-coupled strip sensors. The punch through structure needs less space than a poly or implanted resistor and does not require an additional technology step. However, in strongly irradiated devices this biasing method leads to an additional noise contribution [71, 72]. In pixel sensors which are directly coupled to the electronics punch through structures can be used to get a high resistive access to the pixel implants for wafer pixel tests (see Sect. 2.5).

Many sensor designs use floating guard rings which bias themselves via punch-through. Field plates can in addition be used to manipulate the surface potential and to increase the punch through voltage and consequently the voltage drop between adjacent rings (see Sect. 2.5).

2.3 Pixel Sensors and Their Properties

In the previous sections the general properties of silicon have been discussed. This section discusses the use of silicon as detection material for ionizing radiation.

2.3.1 Different Types of Silicon Sensors

There are many types of silicon detectors which differ from each other mainly in the shape of the electrodes. The simplest geometry possible is sketched in Fig. 2.12. A large area p^+-implantation is placed in an n-bulk. When applying a positive bias to the backside, which has an ohmic contact provided by an n^+-implantation and an aluminum layer, the depletion zone starts to grow from the junction into the bulk. Signal charge liberated by an ionizing particle will be collected by the field and can be detected by both electrodes of the diode. If possible, one connects the preamplifier to the electrode on ground potential as shown in Fig. 2.12. Such sensors are not able to measure the position of the particle precisely. If spatial information is to be obtained, one of the electrodes or both have to be segmented. If the segmentation is coarse and the shape of the elements is more or less equal in both directions on the sensor's surface, the device is called *pad* detector. The distinction between pads (large) and pixels (small) is quite arbitrary. As already discussed in Chap. 1, this

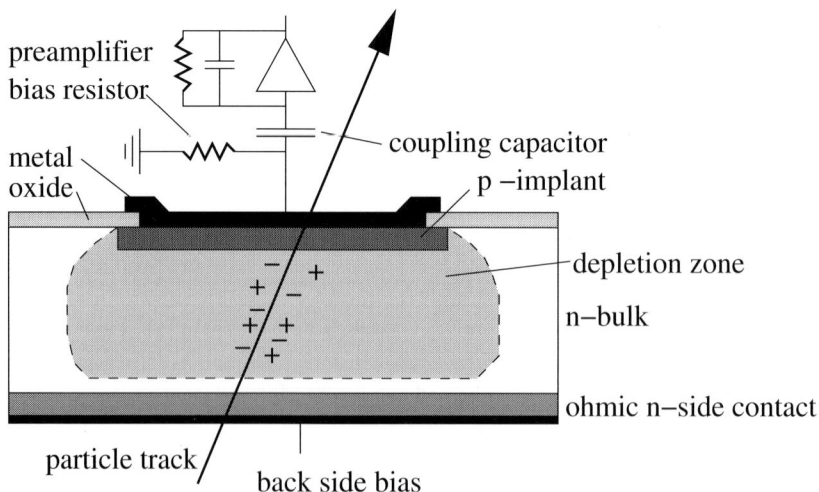

Fig. 2.12. Schematic cross section of a simple silicon pad sensor

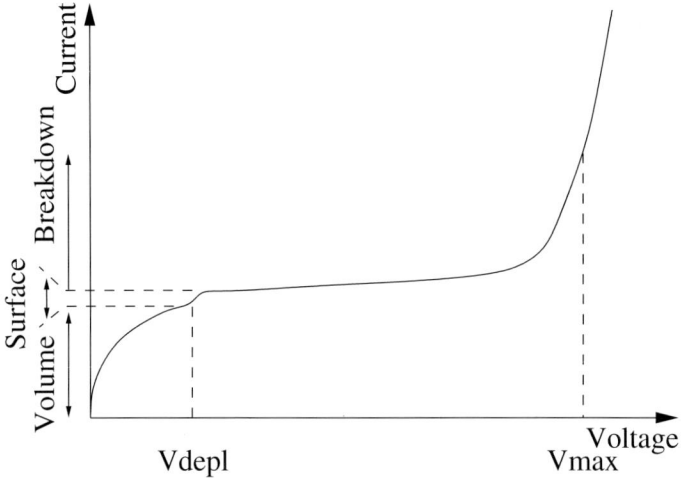

Fig. 2.13. Typical shape of a silicon sensor's *IV*-curve with indication of the origin of different current contributions

book uses the term *pad* if the size of the cells is sufficiently large and their number sufficiently small that all of them can be connected by wire bonds either directly onto the cell or onto a bond pad routed to the device edge. One of many examples where pad detectors were used for tracking is the UA2 experiment at the CERN-SppS collider [73]. Very large pad detectors with no segmentation on the device level were used, for example, as sampling elements in the H1 Plug calorimeter at the HERA storage ring at DESY [74].

If the cell size is below the millimeter range, the connections to the preamplifier with wire bonds becomes impractical and the bump-bond technique (see Chap. 4) must be used. In this case one speaks of a *hybrid pixel* detector.

When the electrodes are largely asymmetric (e.g. several cm long and a few tens of μm wide) the device is called *strip* detector. Wire bonding is the natural connection technique in this case. Most of the silicon detectors used today in particle physics are strip detectors. Moreover, very often in strip detectors the coupling capacitor and the bias resistor (indicated in Fig. 2.12) of each channel are integrated onto the sensor; i.e., strip detectors are often AC-coupled. In pixel sensors with cell dimensions of several hundred micrometers in both directions this integration is not easily possible and because of the small leakage currents per channel also not necessary. Pixels have a direct connection to the readout electronics or, in other words, they are DC-coupled. In the following only pixel detectors will be described.

2.3.2 Leakage Current and Maximum Operation Voltage

The leakage or dark current is flowing in the absence of external effects like particles or light if a reverse bias is applied. In addition to volume generation current, surface generation and avalanche breakdown contribute to the leakage current. A typical current-to-voltage characteristic of a reversely biased silicon sensor is sketched in Fig. 2.13. For voltages below full depletion the current increases according to (2.28) proportional to the square root of the bias. When the space charge region reaches a structured backside, an additional surface contribution arises. After full depletion the IV-curve displays a plateau region in which the current increase is very small before electrical breakdown occurs at very high voltages. As breakdown is at the beginning restricted to certain regions of high electric field, the increase of the current starts smoothly. If the voltage is increased further, hard breakdown occurs that eventually destroys the device. This sudden current increase by several orders of magnitude is not visible in Fig. 2.13 due to the linear y-axis.

It is clear that the sensor has always to be operated at voltages well below this hard breakdown but it is not generally possible to give a strict rule which defines the maximum operation voltage possible. High current is problematic in two respects. A general high-current level which emerges especially after irradiation-induced bulk damage may cause thermal runaway as explained in Sect. 2.4.1.8. Furthermore, local electrical breakdown, which is sometimes also referred to as microdischarge, leads to a very strong current increase in a restricted number of channels. This normally drives the preamplifier of these channels into saturation. These noisy channels have to be masked in order not to jam the whole readout system. The regions of this localized breakdown grow with further increase of the sensor bias until a hard breakdown occurs. The maximum possible operation voltage (which is sometimes also referred to as breakdown voltage) is given by both considerations, the maximum power dissipation and the maximum acceptable number of noisy pixels. Whenever possible the operation voltage should be restricted to the plateau region of the curve shown in Fig. 2.13.

IV-curves are a very powerful tool for sensor testing. Almost all possible problems in the sensor production process lead to a deviation of the curve from the expected shape. The current–voltage characteristics of an arbitrary set of large $(8.8\,\text{cm}^2)$ pixel sensors taken as part of an acceptance testing procedure is shown in Fig. 2.14. Only four of the ten sensors shown are "good" sensors. They display a pronounced plateau region of the current and a breakdown voltage above 200 V. Four sensors display a strong current increase around a bias voltage of 100 V, which was measured to be roughly their full depletion voltage. This indicates that there are defects on the "ohmic" side, the n-side, of the sensors, like scratches or aluminum spikes. Both cause charge injection when the space charge region reaches these defects located at the side apart from the junction. Two sensors break down already at about 50 V. They obviously have a problem located on the p-side, the side of the junction.

If one wants to measure IV-curves of pixel sensors, one immediately has to face the problem how to contact the structured (i.e. pixelated) side. After

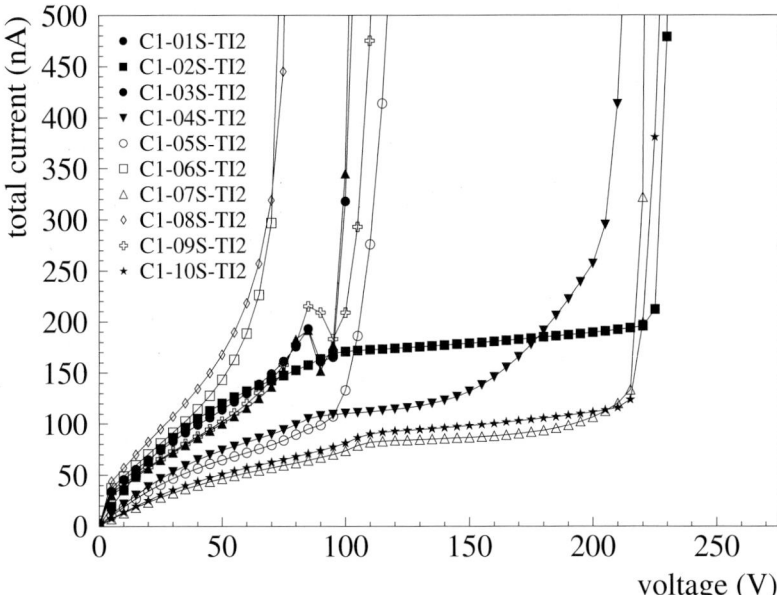

Fig. 2.14. *IV*-measurements of an unselected batch of pixel sensors of about 8.8-cm² size (Data from [75])

bump bonding and flip chipping the sensor to a readout chip, the pixels are grounded via the chip and the measurement can easily be performed. However, one might want to perform such a measurement as part of the testing procedure prior to bump bonding when the sensor is still on the undiced wafer. In this case a special feature for the pixel biasing has to be implemented on the pixel sensor as discussed in Sect. 2.5.

2.3.3 Full Depletion Voltage and Substrate Doping

As already mentioned in Sect. 2.2.4 the full depletion voltage V_{depl} is the voltage needed to extend the depletion zone over the whole thickness of the sensor. In normal operation only the charge produced in the depleted volume can be detected; therefore the maximal signal charge is detectable when full depletion is reached. For an unstructured diode this parameter depends only on the sensor thickness d and the substrate doping N_D and can easily be found out using (2.26):

$$V_{\text{depl,diode}} = \frac{eN_D d^2}{2\varepsilon_0 \varepsilon_{\text{Si}}}. \tag{2.38}$$

Although the depletion voltage for structured devices can differ from (2.38), as discussed in Sect. 2.5, (2.38) gives a good estimate of this important sensor property.

In principle, the full depletion voltage could be determined from an *IV*-measurement as volume current does no longer increase when full depletion is reached. This would be quite convenient as such a measurement is, in any case, performed in order to measure the breakdown voltage. However, transition from the regime where the current is proportional to the square root of the voltage and the regime where the current is constant is not precisely measurable. For this reason the full depletion voltage is usually determined via a *CV*-measurements as the backside capacitance displays a stable plateau. The capacitance per unit area of a pn-diode is calculated as two conductive plates separated by the depletion zone of width $W(V)$:

$$C(V) = \frac{\varepsilon_0 \varepsilon_{Si}}{W(V)} \approx \begin{cases} \sqrt{\frac{\varepsilon_0 \varepsilon_{Si} e N_D}{2V}} & \text{for } V < V_{\text{depl}}, \\ \frac{\varepsilon_0 \varepsilon_{Si}}{d} & \text{for } V > V_{\text{depl}}. \end{cases} \quad (2.39)$$

In the upper right part of (2.39) the expression for the depletion depth (2.26) was inserted, while for the fully depleted state the sensor thickness d was used. The full depletion voltage can be determined by plotting $1/C^2$ versus V. Both branches can be fitted by a straight line. Their intersect is usually referred to as the full depletion voltage as shown in Fig. 2.15. The frequency at which the capacitance measurement is performed has to be chosen carefully. Especially in irradiated detectors the concentration of free charge carriers in the "undepleted" volume is very low and they are less mobile. At frequencies of several 10 kHz the capacitance is then hardly bias dependent as the contribution of the free charge carriers in the bulk to the total capacitance is small. Therefore *CV*-characteristics of irradiated detectors should be measured at very low frequencies (e.g. 100 Hz).

To apply this technique to pixel detectors one needs direct access to all pixel cells. If the sensor contains a biasing structure to measure the *IV*-curves as described in Sects. 2.5.2.4 and 2.5.3.2, this can also be used for the *CV*-measurements. As the biasing structure causes extra serial capacitances and resistances the *CV*-curves measured are not as easy to interpret as the description above suggests, but the full depletion voltage can still be determined. In absence of such a biasing structure the full depletion voltage is often measured with a planar diode processed on the same wafer as the sensor under study. However this measurement is only meaningful if the material parameters are identical; e.g., the sensor is not irradiated. Furthermore, it has to be noted that there can be large variations of the substrate doping within the wafer.

After attaching the sensor to the readout chip a *CV*-measurement is, in most cases, not possible any longer as the high-frequency signals of the capacitance meter are disturbed by the electronics.

Most methods to determine the depletion voltage after attaching the readout electronics use the detector response to radiation. One way to study the propagation of the depletion zone is to analyze the response to high energetic

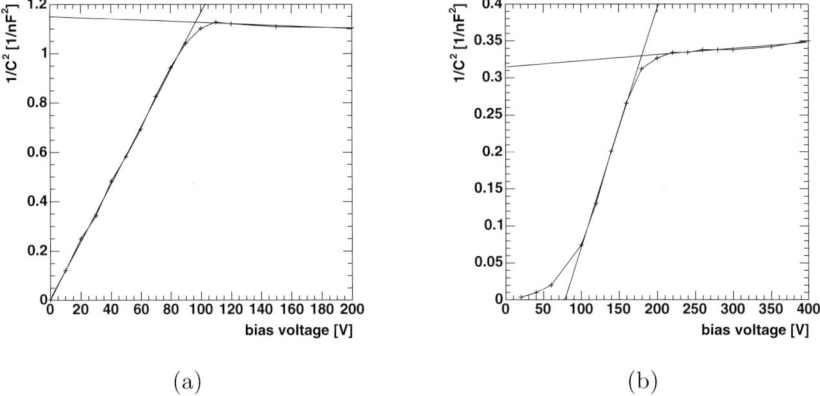

Fig. 2.15. *CV*-characteristics of two double-sided strip detectors. One is unirradiated (**a**) and one irradiated to 1.2×10^{14} n_{eq}/cm^2 (**b**). The full depletion voltage was determined from the intersection points of the two straight lines to 95 and 180 V respectively [72]

particles penetrating the sensor under a shallow (or grazing) angle [76] as illustrated in Fig. 2.16. The number of pixels hit by the particle is a direct measurement of the depth of the depletion region. This means that it is not necessary to fully deplete the device as in the other methods mentioned. This is particularly interesting if, in highly irradiated samples, breakdown occurs before full depletion is reached. To obtain a good resolution on the depletion depth it is important to have the analog information of the pulse height, especially for the first and last pixel of a cluster.

Figure 2.17 shows the results of such a measurement of an n^+-in-n- sensor (see Sect. 2.5.3) irradiated to a fluence of $\Phi = 10^{15}$ n_{eq}/cm^2. The origin of the horizontal axis in the upper part of Fig. 2.16 corresponds to the point

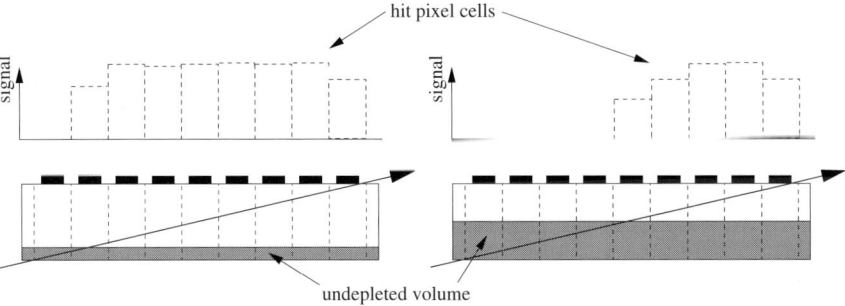

Fig. 2.16. The grazing angle method for the study of sensor depletion. An ionizing beam crosses several pixels at a shallow angle. The cluster size is used to determine the depletion depth. The situation for two different depletion depths is shown

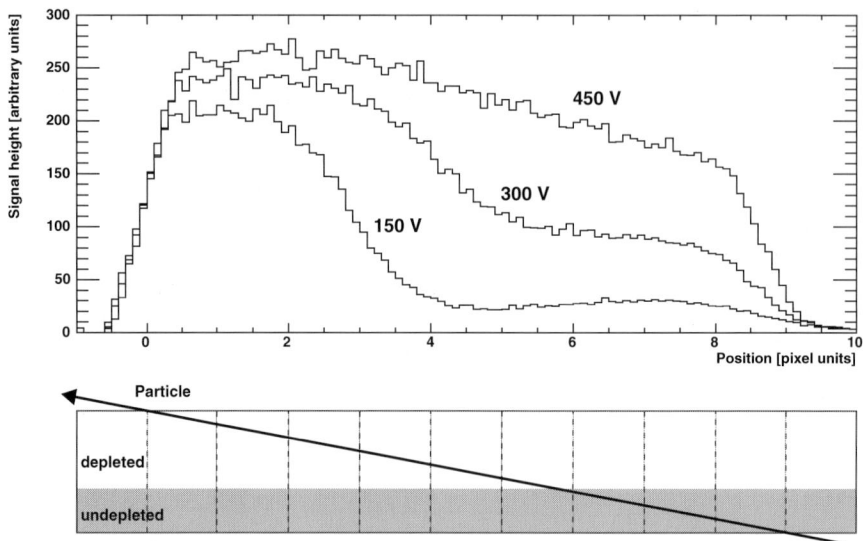

Fig. 2.17. Measurement of the depletion depth in a test beam setup using tracks hitting a highly irradiated sensor ($\Phi = 10^{15}\,\mathrm{n_{eq}/cm^2}$) using the grazing angle method. The evolution of the sensitive volume with increasing bias is visible (data from [76, 78])

of intersect between the particle track and the sensor surface. The x-axis is also a measure of the y-position in the sensor as indicated in the sketch below. The fine binning was possible by using a beam telescope projecting the particle position with subpixel resolution. The curve taken with a bias of 150 V shows a clear drop of the signal around a position of 3. Knowing the pitch of the sensor (in this case 125 μm) and the angle of the track (15°) it is possible to calculate the depletion depth to roughly 100 μm. Figure 2.17 also indicates that the depletion zone in an irradiated detector is not well defined. Especially the curve taken at a bias voltage of 300 V shows that signal is also collected in the "undepleted" region. This is due to the double peak structure of the electric field discussed in Sect. 2.4.1.6. The fact that the curve taken at 450 V still shows a slope is due to charge trapping which is also discussed in Sect. 2.4.1.7.

Although this method gives very nice results, the large effort necessary to operate a test beam rules out this technique for testing large quantities of devices.

A much simpler method, which can also be used in laboratory, is to plot the response to a β-particle, e.g. from a ^{90}Sr source, versus the sensor bias voltage. When the full depletion voltage is reached the signal height saturates. An external trigger (e.g. a scintillator placed below the silicon detector) should be provided unless the readout chip has a self-triggering capability.

Even without analog information a full depletion measurement is possible, if a high energetic γ-source with a point-like energy deposition is used, e.g. the 60-keV X-ray line of ^{241}Am. In most cases the γ passes the sensor without interaction and therefore the absorption probability is about the same every where in the sensor. The rate of signals delivered by the electronics will be proportional to the sensitive volume [79]. This method allows one to measure the full depletion voltage without analog information using only the hit rate. However, a self-triggering capability of the chip is needed.

It has to be mentioned that these three methods only work if the junction depleting the sensor is located at the side of the pixels as is the case in p$^+$-in n sensors and in n$^+$-in-n sensors which have undergone the radiation-induced type inversion.

2.3.4 Pixel Capacitance

The pixel capacitance is important for two reasons. The total capacitance of each pixel determines the noise of the preamplifier [80,81], which is discussed in detail in Sect. 3.3.5. The capacitance between a pixel and its next neighbor cell determines the cross talk between pixels as discussed in Sect. 3.2.1.4. The total capacitance is made of:

- Capacitance to the backside
- Sum of the capacitances to the neighbor pixels
- Capacitance to the ground plane of the (closely spaced) readout chip
- Other (small) contributions, like the capacitance of the bumps

The contributions mentioned last are small and difficult to quantify. Therefore they will not be discussed. The first three points are subject of the coming sections.

2.3.4.1 Capacitance to Backplane and Readout Chip

The capacitance to the backside is determined by the pixel area A, the sensor thickness d, and the dielectric constant of the sensor material, ε, and can be estimated using the well-known formula for a parallel plate capacitor:

$$C = \varepsilon_0 \varepsilon_{\text{Si}} \frac{A}{d} . \tag{2.40}$$

This expression neglects any effects due to fringe field, however. For a typical pixel cell of $20,000\,\mu\text{m}^2$ on 300-μm-thick silicon this gives a contribution of about 7 fF.

A capacitive coupling of the same order of magnitude is obtained between the sensor and the readout chip if an air gap (bump height) of 20 μm is assumed.

2.3.4.2 Interpixel Capacitance

The capacitance to the neighbor pixels, the so-called *interpixel capacitance*, gives by far the largest contribution to the total capacitance. As the sum of the capacitance to all neighbors is approximately proportional to the perimeter, square-shaped pixels are of advantage. In addition, the gap size between the pixel implants determines the interpixel capacitance. To a smaller degree the properties of partly conducting surface regions also play a role.

An analytical analysis using a very much simplified geometry limited to square-shaped pixels and considering the silicon as a perfect dielectric [82] predicts a total capacitance of about 60 fF for a $125 \times 125\,\mu m^2$ pixel with 20-μm gap.

Device simulations which perform a numerical finite element analysis of the differential equations describing the electrical properties of the semiconductor are more reliable. This allows in principle the extraction of the capacitance. However, the computing resources necessary to run a full three-dimensional device simulation, which is needed to describe the situation in a pixel device, are quite large. In the case of nonsquare pixels with a high aspect ratio the interpixel capacitance is dominated by the neighbors sharing the long edge. In such a case a two-dimensional simulation is a good approximation. The absolute values obtained with such studies also have a large error but the relation between different pixel configurations are portrayed correctly [83, 84]. The values reported for the capacitance between neighboring $50 \times 400\,\mu m^2$ pixels are between 30 and 100 fF depending on the implant dimensions.

The measurement of such small capacitance requires some attention. Especially the placement of several probe needles onto very small area on the sensor generates large stray capacitances whose fluctuations caused by very small movements of the probe holders are much higher than the value to be measured. Usually the measurement is performed with specially designed test structures that allow an easy access to all pixels adjacent to the one under test via wire bonds. The reported values for the capacitance to all neighbors are between 30 and 80 fF for $50 \times 400\,\mu m^2$ p-stop isolated n-in-n pixels [84].

Another method is to bump bond a special router chip to the sensor instead of the readout electronics. This allows one to measure many pixels in parallel. The value measured is much higher and therefore less sensitive to systematic effects. The values reported are about 26 fF for $150 \times 150\,\mu m^2$ p-stop isolated n-in-n pixel with a very large gap [85, 86]. A large variety of n-in-n pixel devices with a pitch of $125 \times 125\,\mu m^2$ have been investigated. The measured values cover a wide range between 20 fF for p-stop isolated devices with a large gap (at very high bias voltage) and 80 fF for p-spray isolated devices with a gap of 20 μm.

So for the pixel dimensions typical of the present particle physics experiments one has to expect a total pixel capacitance of $\lesssim 100\,fF$ for square pixels of about $125 \times 125\,\mu m^2$ (close to the choice of the CMS experiment) and

$\lesssim 200\,\text{fF}$ for long and narrow pixels of $50 \times 400\,\mu\text{m}^2$ (choice of the ATLAS experiment).

2.3.4.3 Cross Talk

A signal charge deposited on one pixel can, via capacitive coupling also induce a signal on its neighbors. This cross talk is discussed in great detail in Sect. 3.2.1.4. The worst case scenario is that the total pixel capacitance is completely dominated by the interpixel capacitance to only two next neighbors. This is the case in long and narrow pixel cells, which are arranged in a "normal" pattern as shown in Fig. 2.23a. However, even in this worst case scenario the dominant next neighbor capacitance is (for the most unfavorable pixel layout) in the order of $\lesssim 100\,\text{fF}$. As the effective input capacitance of the preamplifier is normally in the order of $\gtrsim 2\text{--}5\,\text{pF}$, the total cross talk will always be $\lesssim 5\,\%$.

2.3.5 Charge Motion and Signal Formation

The electrical signal used for the particle detection is generated on the collecting electrodes by the drift of the signal charge in the electric field. Hence a signal is already detectable when the charge starts to move and not only when it arrives at the collecting electrode. The instantaneous current induced on an electrode by the infinitesimal movement of a charge e with the drift velocity $\boldsymbol{v}$ is given by [87]

$$i = e\boldsymbol{E}_\text{w}\boldsymbol{v} , \tag{2.41}$$

where $\boldsymbol{E}_\text{w}$ is the so-called *weighting field*. It is different from the electric field inside the sensor that causes the drift and is obtained by applying unit potential to the electrode under consideration and zero potential to all others. To calculate the charge Q induced on an electrode by a charge e drifting in the time interval $[t_1, t_2]$ from position $\boldsymbol{x}_1$ to $\boldsymbol{x}_2$, one has to integrate (2.41) over the time of charge collection:

$$Q = \int_{t_1}^{t_2} i(t)\,\text{d}t = e\,[\phi_\text{w}(\boldsymbol{x}_1) - \phi_\text{w}(\boldsymbol{x}_2)] , \tag{2.42}$$

where ϕ_w is *weighting potential* also obtained by raising the electrode under consideration to unit potential, setting all others to zero, and solving the Poisson equation.

In a pad detector (with the lateral dimensions of the collecting electrode exceeding much the wafer thickness) the potential (far from the device edges) varies only in one dimension. The weighting potential is a linear function of the depth (see Fig. 2.19a), while the electric potential causing the drift displays a parabolic shape due to the space charge (see Fig. 2.7). The linearity of the weighting potential has mainly two consequences:

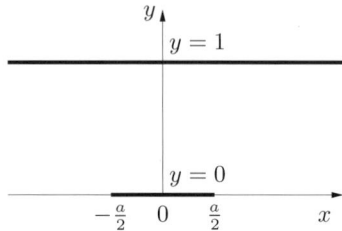

Fig. 2.18. Electrode geometry for the calculation of the weighting potential. The electrodes are marked with a *thick line*

- The charge induced is the same for any part in the drift path.[3]
- If an electron–hole pair is generated in the middle of the detector, the hole approaching the electrode induces the same signal as the electron departing from it.

However, in pixel detectors where the size of the collecting electrode is comparable or even smaller than the wafer thickness this picture changes drastically. Figure 2.18 shows the configuration of the electrodes in one pixel cell, with a sensor thickness arbitrarily set to 1. The collecting electrode set to unit potential is the one on the bottom. The solution of the Poisson equation for this configuration can be found using the Schwarz–Christoffel conformal transformation [88]. The weighting potential can be written, for instance, as

$$\phi_\mathrm{w} = \frac{1}{\pi} \arctan \left[\frac{\sin(\pi y) \sinh\left(\pi \frac{a}{2}\right)}{\cosh(\pi x) - \cos(\pi y) \cosh\left(\pi \frac{a}{2}\right)} \right], \qquad (2.43)$$

where the positive branch of the inverse tangent function is used. This potential is plotted for two values of the electrode width in Fig. 2.19b,c. The smaller the collecting electrode, the larger becomes the region in which the weighting potential approaches zero. This means that very little signal is induced on the collecting electrode while the charge drifts in this region. At the same time signal is induced on the neighbors of the pixel electrode under consideration. The closer the charge approaches the collecting electrode, the more narrow gets the region where the charge is induced while the signals sign reverses in the remote electrodes. This is expressed in the increasing gradient of the weighting potential. When all the charge has arrived at the collecting electrode, the integral of the induced signal has the value of charge collected, while it vanishes for all the neighbors. As the charge collection in silicon sensors is very fast (about 5 ns) the bipolar signal at the neighbor pixels is usually not visible.

---

[3] As the drift velocity changes due to the variation in the electric field, the current will not be constant in time

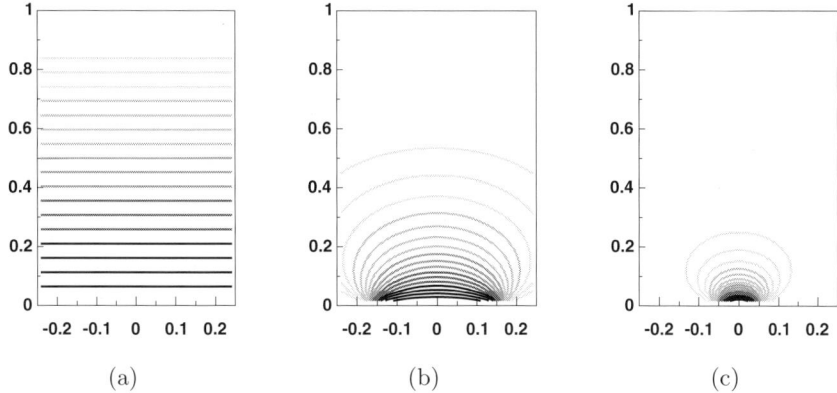

Fig. 2.19. Plot of the weighting potential. Two infinite parallel plates (**a**), and a collecting electrode of 1/3 (**b**) and 1/10 (**c**) times the wafer thickness

So the main differences of a highly segmented pixel detector compared to a "large area" pad detector are often called *small pixel effect* and can be summarized in two points:

- Most of the signal is induced in the last part of the charge drift path.
- Charge carriers drifting toward the backplane do not contribute significantly to the signal.

These effects are especially important in the case of irradiated sensors when a significant part of the signal charge is trapped (stops drifting). If the trapping probability is equal throughout the sensor, most of the trapped charge is trapped before inducing a significant fraction of its signal on the collecting electrode. This makes such devices more sensitive to charge trapping than, e.g., large area pad detectors.

2.3.6 Spatial Resolution

Spatial resolution is mainly determined by the pixel pitch. The choice of the readout mode (analog or single threshold binary), the reconstruction algorithm, and the amount of charge sharing between pixels also play a role.

2.3.6.1 Binary Readout

In the case of single threshold binary readout electronics the theoretical expectation for the spatial resolution is easy to determine. To perform this calculation for a pixel detector with pitch p centered around position 0, the following assumptions must be valid (see Fig. 2.20):

- The threshold is adjusted in such a way that only one pixel per particle track fires.

- Only particles hitting the detector between $-p/2$ and $p/2$ trigger a signal in pixel 0.
- The detector is hit by a uniform density of particles, $D(x) = 1$.

Then the average difference between the "real" impact position x_r and the measured impact position $x_\mathrm{m} = 0$ is given by

$$\sigma^2_\mathrm{position} = \frac{\int_{-p/2}^{p/2} (x_\mathrm{r} - x_\mathrm{m})^2 D(x_\mathrm{r}) \,\mathrm{d}x_\mathrm{r}}{\int_{-p/2}^{p/2} D(x_\mathrm{r}) \,\mathrm{d}x_\mathrm{r}} = \frac{\int_{-p/2}^{p/2} x_\mathrm{r}^2 \, 1 \mathrm{d}x_\mathrm{r}}{\int_{-p/2}^{p/2} 1 \,\mathrm{d}x_\mathrm{r}} = \frac{p^2}{12} \quad \text{and}$$

$$\sigma_\mathrm{position} = \frac{p}{\sqrt{12}} \,. \tag{2.44}$$

So with the simplest configuration a spatial resolution of about $0.28\,p$ can be reached.

Usually the threshold of the readout electronics is set as low as possible without getting a too high rate of noise hits. This means that two (or more) pixels can be triggered by the same particle if the signal charge is shared between the pixels. The group of pixels showing a signal from the same particle is called *cluster*. The occurrence of clusters with more than one pixel improves the resolution. This is usually the case when the track passes in between the two pixels as indicated with s in Fig. 2.20b. For the events triggering two pixels one expects a resolution of $s/\sqrt{12}$, and for the events triggering only one pixel, $(p-s)/\sqrt{12}$. The optimal average spatial resolution is obtained when

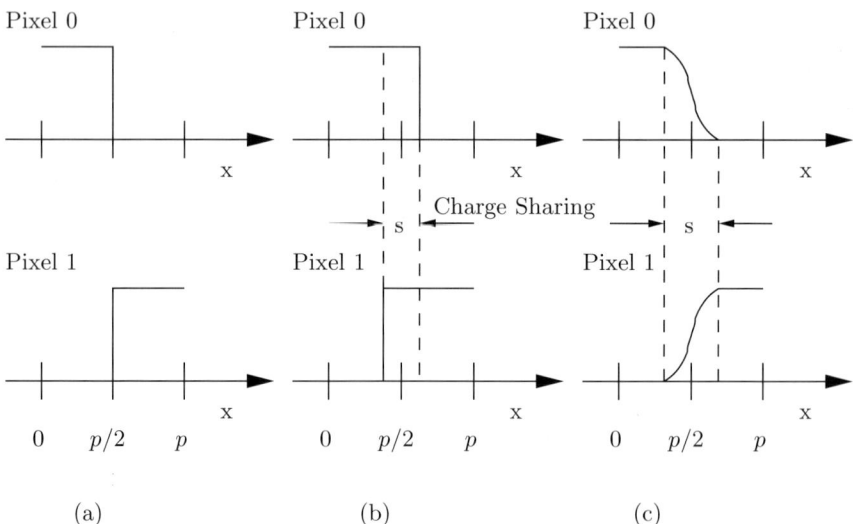

Fig. 2.20. Signals in two adjacent pixels as function of the impact position x. Detector with binary readout without charge sharing (**a**), with binary readout with charge sharing (**b**), and with analog readout (**c**).

$s = p/2$, or when the number of one-pixel clusters and two-pixel clusters equals. However, one is not able to distinguish between one track triggering two channels and the much less probable case of two tracks hitting the sensor within $2p$.

2.3.6.2 Analog Readout

An additional improvement of the spatial resolution can be reached with an analog readout that delivers a signal proportional to the collected charge. Such a situation is displayed in Fig. 2.20c. In the region where both pixel cells show a signal, quite elaborate interpolation algorithms, as e.g. the so-called η-function described in [89], can be used. A rule of thumb for the best possible resolution is to divide the pitch by the signal-to-noise ratio of the analog signal. However, in the region where only one pixel fires, one is still limited to a resolution of $(p-s)/\sqrt{12}$. For this reason one tries to expand the region of charge sharing, s, as far as possible by tilting the detector and using the Lorentz deflection. The fraction of double hits is limited by the smallest possible pixel threshold.

2.3.6.3 Measurement Example

The behavior discussed in the previous sections is regularly observed in test beam measurements where the position reconstructed from the pixel data is compared with the prediction of a high-resolution telescope of silicon strip detectors. Figure 2.21 shows the distributions of the reconstruction errors (the so-called residuals) obtained from the same data. The reconstruction algorithms represent the situations illustrated in Fig. 2.20. In this example a 285-μm-thick pixel detector with a pitch p of 125 μm was situated normal to a high-energy pion beam [90]. A magnetic field of 3 T caused a deflection of the collected electrons by a Lorentz angle of about 17° [67].

In the case of the residuals shown in Fig. 2.21a the reconstructed position was defined by the center of the pixel with the highest signal. By this definition the occurrence of double hit clusters is excluded. The shape of the distribution corresponds well with the expectation of a box folded with a Gaussian distribution caused by the charge diffusion. Its rms is about 38 μm, which is roughly the pitch divided by $\sqrt{12}$.

The second row of histograms in Fig. 2.21 corresponds to the situation illustrated in Fig. 2.20b. A pixel threshold was introduced and adjusted to about 3,600 ke$^-$, the value at which (in this case) the number of one-pixel and two-pixel clusters roughly equals. For one-pixel clusters the center of the hit pixel was defined as the reconstructed position, while for two-pixel clusters it was the border between the two. In both cases a box-like distribution with an rms of about 20 μm was obtained (see Fig. 2.21b, c), which represents the expected improvement of about a factor of 2.

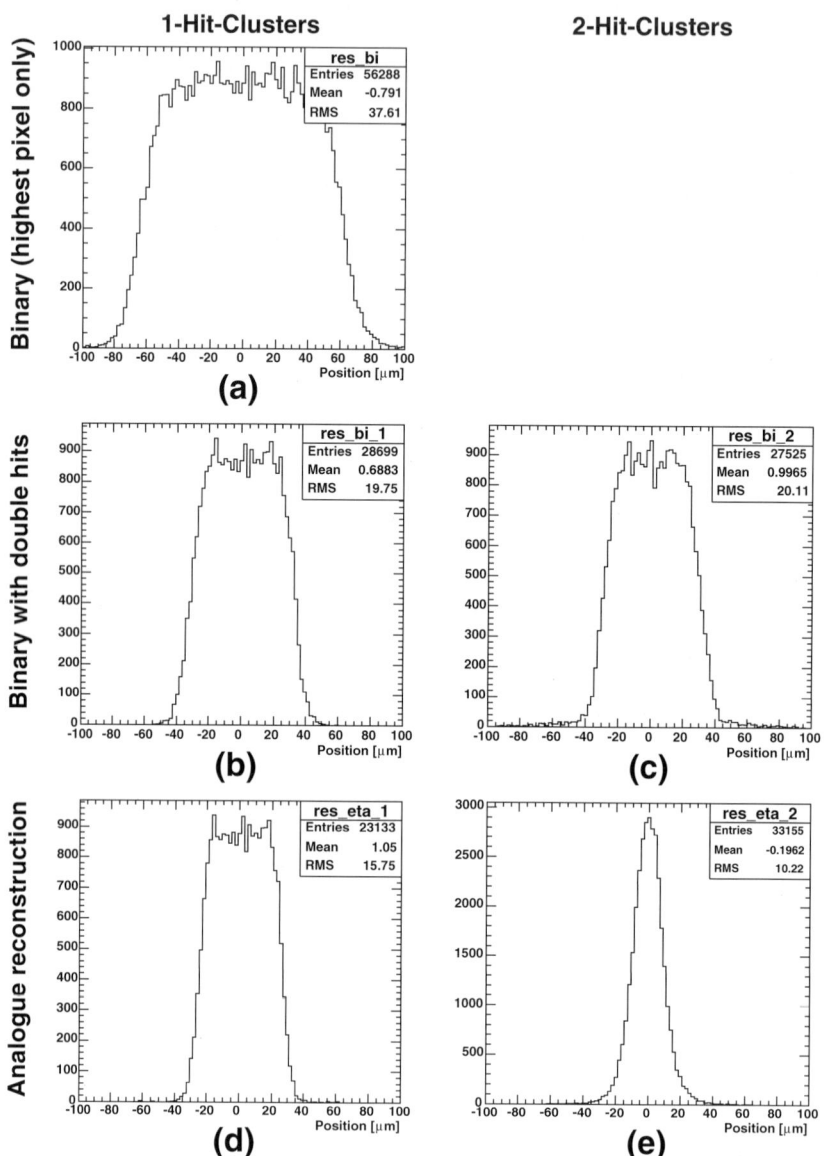

Fig. 2.21. Residual distribution of a pixel detector with a pitch of 125 mm for three different reconstruction algorithms. Detector with binary readout without charge sharing (**a**); with binary readout with charge sharing: single hit clusters (**b**) and double hit clusters (**c**) analog readout: single hit clusters (**d**) and double hit clusters using a η-algorithm for interpolation (**e**) (data form [90])

The residuals obtained with an analog reconstruction algorithm are shown for one-pixel clusters (Fig. 2.21d) and two-pixel clusters (Fig. 2.21e). The threshold was set to a low value of about $2,000\,e^-$. The obtained fraction of double hit clusters in the present example is about 60%. In the case of one-pixel clusters, again, the center of the hit pixel was defined as the reconstructed position. The rms of the box-shaped residual distribution for the single hits (see Fig. 2.21d) is about $16\,\mu m$, which is about the expected value if one assumes the width of region without charge sharing, $(p-s)$, to be about 40% of the pitch.

For the two-pixel clusters an analog interpolation with a η-function [89] was used. In this case the residual distribution has roughly a Gaussian shape (see Fig. 2.21e). The rms of this distribution[4] is about $10\,\mu m$. The average spatial resolution is the weighted mean of both histograms which is about $12.5\,\mu m$. Although in this example the spatial resolution was increased by a factor of 3 by moving from the most simple reconstruction algorithm to an analog interpolation, it has to be kept in mind that an analog readout requires a careful consideration of all factors limiting the performance (lowest possible threshold, noise, dynamic range of the ADCs, etc.).

2.3.6.4 Charge Sharing

Charge sharing is mostly affected by the position and angle of the track with respect to the sensor surface as shown in Fig. 2.22a–c. As all tracks come from a small region close to the primary interaction, the sensor modules can be tilted in a way to obtain the desired impact angle and average cluster size. This is done, e.g., in the barrel region of the ATLAS pixel detector in a so-called shingled geometry [12] and in the disk region of the CMS pixel detector, resulting in a turbine-like geometry [13].

In the presence of a magnetic field, which is the case in almost every particle physics experiment, the drift path of the signal charge is deflected by the Lorentz angle, which also may increase charge sharing as indicated in Fig. 2.22d.

In principle it is also possible to compensate for the Lorentz drift as shown in Fig. 2.22e, concentrating the charge mostly on one pixel. This might be necessary if the signal-to-noise ratio does not guarantee a good detection efficiency of double hits, but will limit the space resolution to about $0.28p$.

In Fig. 2.22 only one of the two dimensions of the pixel detector is shown. Of course charge sharing can also happen in the perpendicular direction. Clusters of size 4 are possible when a perpendicular track hits the corner between four pixels. While the total signal of 23,000 electrons deposited by a particle crossing a fully depleted detector is large, the charge seen in one channel can

---

[4] Often the σ of a Gauss function fitted to the residual distribution is quoted as spatial resolution, which would be $9.2\,\mu m$ in this example. This, however, neglects the non-Gaussian tails of the distribution mainly caused by δ-electrons.

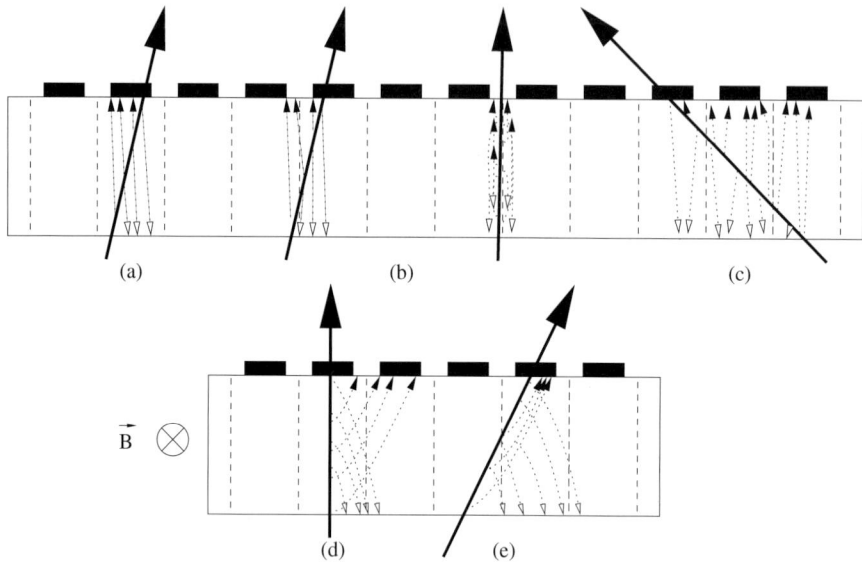

Fig. 2.22. The formation of clusters in a silicon detector. The different types of arrows denote the polarity of the charge carriers (the filled ones are electrons; the open ones holes) while the *big arrows* indicated the particle tracks. Clusters are determined by the position and angle of the track: (**a**) size 1, (**b**) size 2, (**c**) size 3. The presence of a magnetic field leads to a Lorentz shift: (**d**) the track is perpendicular to the sensor surface and (**e**) tilted by the Lorentz angle

become uncomfortably small if the signal is distributed over many pixels. Partially depleted operation, which might be necessary after irradiation, and trapping further decrease the collected signal charge. In order to guarantee track efficiency, at least one pixel in a cluster must collect enough charge to overcome its threshold. High operation voltage, for example, can provide a deeper space charge region. The Lorentz angle decreases with irradiation due to the higher voltage necessary to operate the detector [66, 67]. This reduces charge sharing of the type shown in Fig. 2.22d and therefore degrades the spatial resolution. At the same time a reduced charge sharing provides a higher signal per channel and increases the probability that a hit is registered.

2.3.6.5 Pixel Pattern

A possibility of increasing the spatial resolution in one of the directions without changing the pixel size is to arrange the pixels in a bricked pattern as shown in Fig. 2.23b. This is especially useful when the pixel shape is not square but rectangular with a high aspect ratio. If two pixels from different rows are triggered in the same cluster, the effective pitch along the horizontal axis is halved. At the pixel corners only three pixel cells join, reducing the probability for four-pixel clusters. Further advantage of such a pattern is that the capacitance to the next neighbor responsible for cross talk is halved in one direction, which is, in the case sketched, the more critical direction. However, the bricked pattern is not easily compatible with a mirrored bump pad geometry favored by most readout chips[5] (see Fig. 2.23). While in the "normal" pattern the bump pads can be placed inside the pixel area, in the bricked pattern always some pads lie on top of the neighbor pixel and have to be routed via a metal line. If a single metal process is used, this leads to a high cross-coupling capacitance. The bricked design is a viable option for the MCM-D approach as the pads have to be fed anyhow through several layers of dielectrics (see Sect. 6.2.1).

(a) (b)

Fig. 2.23. Two possible pixel configurations: (**a**) normal (**b**) bricked

2.3.7 Radiation Hardness

The term *radiation hardness* is not well defined and often used with completely different meanings. The degradation of the sensor properties is a steady process and the decision if a sensor is still functional depends on the requirements of the specific application. If a pixel detector is mainly used for pattern recognition and track finding, the detection efficiency and the noise background rate are the most important properties. Radiation damage in the sensors degrades the signal collection as discussed in Sect. 2.4. This can partly be counterbalanced by increasing the bias voltage. However, increasing the bias might reduce the charge sharing between neighboring pixels and by this also reduce the accuracy of position measurement. So it is difficult to decide when a detector is no longer useful. Some factors limiting sensor operation in a high-radiation environment are discussed in Sect. 2.5, together with the different sensor concepts.

In general there are two independent strategies to improve the radiation hardness of the sensors that can be followed in parallel. The first one is to study the material properties after irradiation as presented in Sect. 2.4 and to optimize the device design with respect to this postradiation parameters as shown in Sect. 2.5. The other approach, not covered by this book, is to

---

[5]Most readout chips have a pattern as shown in Fig. 2.23a. This geometry ensures that the analog part of the electronics (close to the bump pad) is as distant as possible from the digital part (far from the bump pad) of the neighbor cell. Cross talk is reduced by this arrangement, as explained in Chap. 3.

modify the material properties of silicon by specific impurities so as to reduce the impact of radiation to crucial material parameters. An example is the enrichment of silicon with oxygen discussed in Sect. 2.4.1.

2.4 Radiation-Induced Effects on Silicon

Due to their high granularity pixel detectors are used as tracking devices in the innermost regions of collider experiments. The high track density in combination with high repetition rates gives a fluence of about 2×10^{11} particles per square centimeter and hour of peak-luminosity in the innermost pixel layer of the LHC experiments and is responsible for considerable damage induced to the sensors and to the electronics.

Radiation-induced effects (radiation damage) are usually divided into bulk and surface defects. The former are caused by the displacement of crystal atoms while the latter include all effects in the covering dielectrics and the interface region. The most important surface effect is the increase of the oxide charge which saturates after some kilograys to values of about $10^{12}\,\text{cm}^{-2}$. At higher hadron fluence, bulk damage also becomes important. The main effects are:

- Increase of the leakage current
- Change of the space charge in the depleted region (*type inversion* or *space charge sign inversion*) and subsequent increase of the full depletion voltage
- Charge trapping

Other materials than silicon have been investigated for radiation hardness but most data are available for silicon. Silicon will be discussed in this section, and other materials will be mentioned briefly in Sect. 2.7.

2.4.1 Bulk Damage

As silicon and other semiconductors are crystals it is obvious that their properties will change with radiation. High energetic particles will not exclusively interact with the electron cloud producing the electrical signal but also with the nuclei, often displacing them out of the lattice position. This produces crystal imperfections which may be electrically active and hence change the electric properties of the material. In the inner regions of high rate experiments, e.g. at the TEVATRON or LHC, the concentration of crystal defects will, after some years of operation, exceed the initial substrate doping concentration. At this moment the material properties will be mainly determined by the defects. Since the late 1980s radiation-induced effects have been systematically studied [91–95]. Even though it is not yet possible to correlate all changes in the operation parameters of the sensors with specific microscopic defects, the changes have been studied and parametrized in detail. These parametrizations allow a prediction of all important sensor parameters for a given radiation scenario. We will restrict ourselves in the following to an overview of these macroscopic changes.

2.4.1.1 Origin of Bulk Damage

Bulk damage is caused by interaction of the incident particles with the nuclei of the lattice atoms. In contrast to ionization such interactions are not reversible in most cases. To remove a silicon atom from its lattice position a minimum recoil energy E_d of about 25 eV is required [96]. Electrons need an energy of at least 260 keV in order to provide such a recoil energy in a collision, while protons and neutrons, because of their higher mass, require only 190 eV. If the recoiling silicon atom gets enough energy through the collision, it can cause further defects. In case this energy exceeds about 2 keV the atoms loose most of their energy in a very localized area, creating a cluster of defects. According to simulations these clusters are believed to have an inner diameter of about 10 nm surrounded by an about 200-nm-wide area with a lower defect density [97]. As already indicated above, different particles interact differently with the silicon lattice. As charged particles scatter via electromagnetic interaction with atom nuclei that are partially screened by their electron clouds, they produce more point defects and less clusters than neutrons which only feel the nuclear and not the electromagnetic interaction. However, the differences between the types of interaction are leveled out by the secondary interactions of knocked out silicon atoms. To be able to compare the damage caused by the different types of particles with different energies, radiation damage is scaled with the *nonionizing energy loss* (NIEL). This quantity summarizes all energy deposited in the crystal which has not been used for the fully reversible process of ionization. Neutrons of 1 MeV are used as reference particles. The fluence Φ_{phys} of an arbitrary type of particle causes the same NIEL as the fluence Φ_{eq} of 1 MeV neutrons. The energy-dependent *hardness factor* κ of a certain type of particle which converts the "physical" fluence Φ_{phys} into the neutron equivalent fluence Φ_{eq} can be calculated according to [98]. The experimental determination of the hardness factors is done via the normalization of the leakage current [99]. For the most frequently used irradiation facilities measured values are available. For example the hardness factor κ of 24 GeV protons (which are provided by the CERN-PS) is 0.62 [99, 100] or for 193 MeV pions (provided by the PSI ring cyclotron) is 1.14 [99]. In the following text all fluences are given in units of neutron equivalent fluence, $n_{\text{eq}}/\text{cm}^2$.

The primary defects caused by irradiation, silicon vacancies, and interstitials are not stable; i.e., they are able to move through the crystal. This movement can lead to an annealing if defects meet during their migration through the crystal. But also secondary point defects with other defects already present in the crystal can be formed, which might be stable and display different electrical properties. Point defects in general cause energy levels in the band gap whose position can be measured by different spectroscopic methods. They can be charged and, depending on the position of their energy levels, have an impact on the space charge in the depletion zone. As the mobility of the defects is strongly temperature-dependent it is clear that radiation-induced changes of sensor properties show a complex annealing behavior due to the many possible secondary defects.

2.4.1.2 Leakage Current

The energy levels in the band gap caused by the crystal defects act as generation–recombination centers. They lead to a decrease of the generation lifetime τ_g and an increase of the volume generation current I_{vol} proportional to the fluence Φ:

$$\frac{1}{\tau_g} = \frac{1}{\tau_{g,\Phi=0}} + k_\tau \Phi \,, \tag{2.45}$$

with k_τ being the *lifetime-related damage rate*, or

$$\frac{I_{vol}}{V} = \frac{I_{vol,\Phi=0}}{V} + \underbrace{\alpha \Phi}_{\Delta I_{vol}/V} \,, \tag{2.46}$$

with V being the depleted volume and α the *current-related damage rate*. Because of the relation between generation lifetime and volume generation current expressed in (2.28), the two damage constants are also related:

$$\alpha = e n_i k_\tau \,.$$

The lifetime-related damage rate k_τ is more fundamental than the current-related damage rate α as it does not depend on the intrinsic charge carrier concentration and therefore not on the temperature at which the measurement was performed. However, one usually quotes α as this is the parameter directly determined from the current–voltage measurements.

The damage constant α is independent of the initial resistivity of the silicon, the concentration of other dopants like oxygen or carbon, and the production process of the sensor [101].

After irradiation the leakage current anneals with time as shown in Fig. 2.24. This strongly temperature-dependent annealing behavior was in the past [93] fitted to a sum of exponential functions with a relative amplitude and different "decay" times and could be interpreted as several defects which anneal with different time constants. This parametrization describes the measured data for annealing times at room temperature of less than one year.[6] For longer annealing times or high annealing temperatures no saturation of the annealing has been observed and therefore the time evolution

---

[6]In [93] two parameter sets are given. According to [9] the one for not-type-inverted sensors should be used also for type-inverted sensors

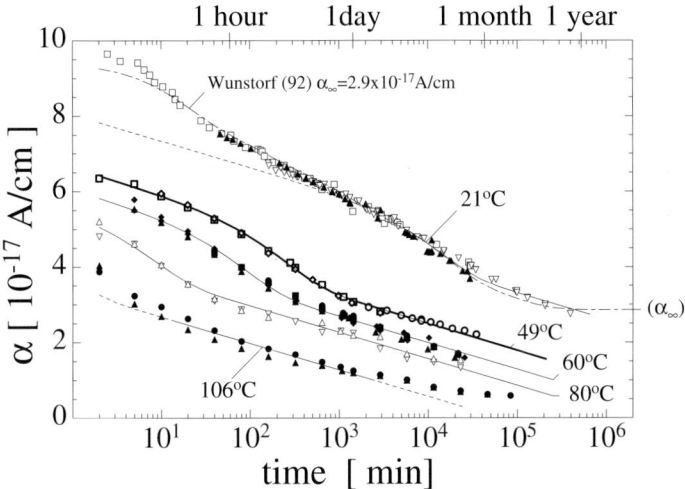

Fig. 2.24. Current-related damage rate α as function of the cumulated annealing time. The *dash-dotted line* is a fit according to [93]. The *solid lines* are fits according to (2.47) [9]

of the current-related damage rate has been described using an additional logarithmic term [9]:

$$\alpha(t) = \alpha_i \exp\left(-\frac{t}{\tau_i}\right) + \alpha_0 - \beta \ln\left(\frac{t}{t_0}\right), \quad (2.47)$$

with t_0 arbitrarily set to 1 min. The dependence on the annealing temperature T_a is hidden in[7]

$$\frac{1}{\tau_i} = k_{0,i} \exp\left(-\frac{E_i}{kT_a}\right) \quad (2.48)$$

and in α_0. The fit parameters were determined to be

$\alpha_i = (1.26 \pm 0.06) \times 10^{-17}$ A/cm
$k_{0,i} = 1.2^{+5.3}_{-1.0} \times 10^{13}$ s^{-1}
$E_1 - (1.11 \pm 0.05)$ eV
$\beta = (3.07 \pm 0.18) \times 10^{-18}$ A/cm
$\alpha_0 = -(8.9 \pm 1.3) \times 10^{-17}$ A/cm $+ (4.6 \pm 0.4) \times 10^{-14}$ AK/cm $\times 1/T_a$

It has to be mentioned that other formulations of (2.47) are possible, suggesting the annealing to be a first-order process with a temperature-independent α_0.

---

[7] It should be stressed that T_a denotes the temperature at which the sensors are annealed or stored and is not to be confused with the temperature T at which the current–voltage curve, to determine the damage constant α, is taken

2.4.1.3 Effective Doping

In the previous equations the doping concentration was equated with the concentration of donors N_D or acceptors N_A assuming that on each side one of those is so much dominant that all other contributions may be neglected, which is the case for new detectors. If several dopings and electrically active defects are present, these numbers have to be replaced by a quantity called *net doping* or *effective doping* N_eff, which is the difference of all donor-like states and all acceptor-like states and can be determined from the full depletion voltage using (2.26):

$$|N_\text{eff}| = \frac{2\varepsilon_0 \varepsilon_\text{si} V_\text{depl}}{ed^2} . \tag{2.49}$$

As the effective doping concentration is, according to the definition, positive for n-material and negative for p-material, all formulas contain only its absolute value $|N_\text{eff}|$.

In the studies performed to determine the fluence and time dependece of the full depletion voltage simple unstructured diodes were used and the full depletion voltage was deduced from CV-measurements as described in Sect. 2.3.3. This method assumes a constant space charge which is not given for highly irradiated sensors where the field shows a double peak as mentioned in Sect. 2.4.1.6. In this case the numbers derived form the measurements are effective or average numbers.

2.4.1.3.1 Fluence Dependence

As radiation-induced point defects can be charged, it is not surprising that the effective doping concentration changes with irradiation. The fluence dependence of effective doping and of the full depletion voltage are plotted in Fig. 2.25. Starting with n-doped material it decreases up to a certain fluence of the order of $\Phi = (2\text{--}5) \times 10^{12}\,\text{cm}^{-2}$ at which the space charge almost vanishes. With further irradiation the absolute effective doping concentration increases again dominated by acceptor-like defects with a negative space charge. The depletion behavior is now like in p-material. As a consequence of this *type inversion*, or more correctly *space charge sign inversion*, the pn-junction moves from the p^+-side of the sensor to the n^+-side and the space charge region grows from there. This behavior has been proven using short-range α-particles [93].

Due to the mobility of defects the net doping concentration changes after the end of the irradiation. The time evolution of the effective space charge at an environment temperature of 60° C is shown in Fig. 2.26. As the defects and their reactions are not yet understood in detail, a phenomenological parametrization has been performed. The most accepted description of this behavior is the so-called *Hamburg model* [9, 102, 103]:

$$N_\text{eff} = N_{\text{eff},\Phi=0} - \underbrace{[N_\text{C}(\Phi) + N_\text{a}(\Phi, T_\text{a}, t) + N_\text{Y}(\Phi, T_\text{a}, t)]}_{\Delta N_\text{eff}(\Phi, T_\text{a}, t)} . \tag{2.50}$$

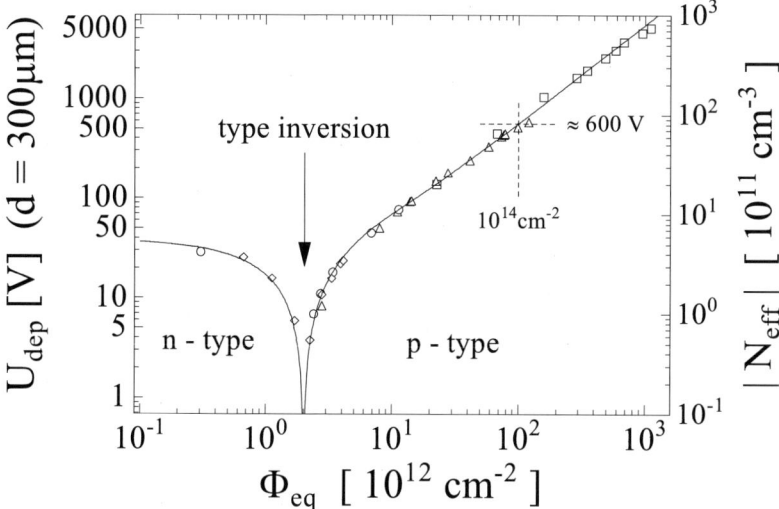

Fig. 2.25. Change of the full depletion voltage of a 300-μm-thick silicon sensor and its absolute effective doping versus the normalized fluence, immediately after the irradiation [93]

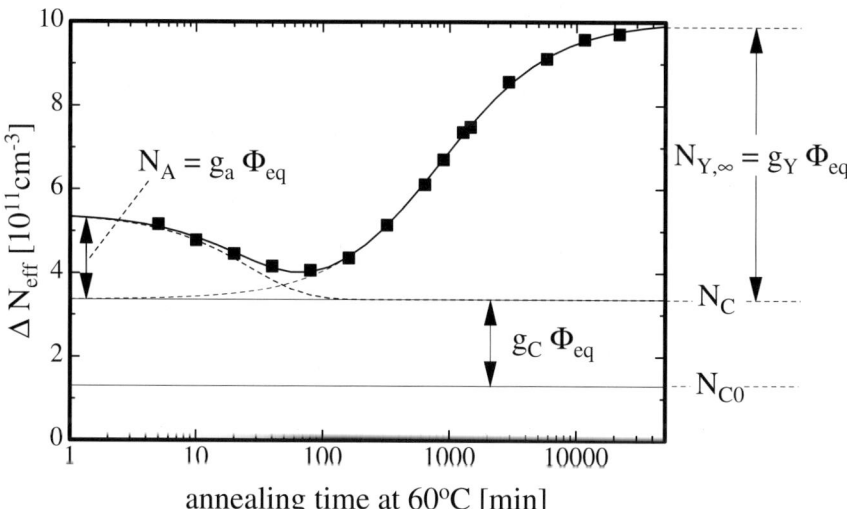

Fig. 2.26. Typical annealing behavior of the irradiation-induced changes of the effective doping concentration ΔN_{eff} at a temperature of 60°C after irradiation with a fluence of 1.4×10^{13} cm^{-2} [102]

Here $N_C(\Phi)$ describes the fluence dependence of the effective doping and contains only the fluence Φ as parameter. The other two terms describe the change of the effective doping after the irradiation and are therefore also dependent on the storage temperature T_a.

Stable Damage

The expression N_C in (2.50) denotes the *stable damage*, which depends neither on annealing time nor on temperature:

$$N_C(\Phi) = N_{C,0}\left(1 - e^{-c\Phi}\right) + g_c\Phi. \tag{2.51}$$

The first term in (2.51) characterizes the deactivation of the initial donor states. The initial concentration of removable donors, $N_{C,0}$, differs from the effective donor concentration $N_{\text{eff},\Phi=0}$ measured before irradiation. This is interpreted as a partial donor removal, while part of the initial donors stay electrically active even after very high fluences. References [9, 103, 104] report a fraction of removable donors for different samples of $N_{C,0} = (0.6-0.9) \times N_{\text{eff},\Phi=0}$. The removal constant c also shows a large fluctuation for different materials, and values in the range of c= $(1-3) \times 10^{-13}$ cm² have been quoted. However, the product of both values, called *initial donor removal rate*, is constant and has, for float zone material with standard oxygen content, a value of $N_{C,0} \times c = (7.5 \pm 0.6) \times 10^{-2}$ cm^{-1} [9].

The parametrization (2.51) given for the stable damage assumes that a prioriy no acceptor states are contained in the silicon materials, which is true for most materials used. Initial acceptors will also be removed exponentially with the fluence. A term describing this can be introduced according to [93]. The deactivation of doping atoms at low fluences has been directly observed with four probe resistance measurements. The removal constants were determined to be $c_d = 2.42 \times 10^{-13}$ cm² for phosphorus and $c_a = 1.98 \times 10^{-13}$ cm² for boron [105]. These values are in the same range as those quoted above which were fitted to measurements of the full depletion voltage.

The second term in (2.51), $g_c\Phi$, describes the creation of acceptor-like defects proportional to the fluence. The introduction rate is $g_c = (1.49 \pm 0.04) \times 10^{-2}$ cm^{-1} [9]. These defects are acceptor-like in a sense that they lead to a negative space charge and hence to an increase of the full depletion voltage. However, they do not lead to an increase of the conductivity of the material, because the levels caused by these defects are deep in the band gap. The resistivity of undepleted silicon after type inversion is about the intrinsic resistivity of silicon [106, 107].

Short-Term Annealing

The second term in (2.50), N_a, specifies the *short-term* or *beneficial annealing* parametrized by

$$N_\mathrm{a} = \Phi \sum_i g_{\mathrm{a},i}\, e^{-t/\tau_{\mathrm{a},i}(T_\mathrm{a})} \approx \Phi g_\mathrm{a}\, e^{-t/\tau_\mathrm{a}(T_\mathrm{a})} . \qquad (2.52)$$

Short annealing times (of the order of hours or less) described in [93] are not relevant for the operation of the sensors. Therefore all but the longest decay times can be neglected as done in the right part of (2.52). The amplitude g_a of the exponential function was experimentally determined to be $g_\mathrm{a} = (1.81 \pm 0.14) \times 10^{-2}\,\mathrm{cm}^{-1}$ [9]. The temperature-dependent decay time $\tau_\mathrm{a}(T_\mathrm{a})$ can be expressed by the Arrhenius relation

$$\frac{1}{\tau_\mathrm{a}(T_\mathrm{a})} = k_{\mathrm{a},0}\, e^{-E_\mathrm{a}/kT_\mathrm{a}} ,$$

with $k_{\mathrm{a},0} = 2.4^{+1.2}_{-0.8} \times 10^{13}\,\mathrm{s}^{-1}$ and $E_\mathrm{a} = (1.09 \pm 0.03)\,\mathrm{eV}$.

Reverse Annealing

The last contribution N_Y in (2.50) is the so-called *reverse annealing* term which describes the increase of the full depletion voltage after some weeks at room temperature. Here several parametrizations are possible depending on the underlying model. Although it is commonly agreed that the reverse annealing is a first-order process, the experimental data are best fit by [9]

$$N_\mathrm{Y} = \underbrace{g_\mathrm{Y} \Phi}_{N_{\mathrm{Y},\infty}} \left(1 - \frac{1}{1 + t/\tau_\mathrm{Y}}\right) , \qquad (2.53)$$

with $g_\mathrm{Y} = (5.16 \pm 0.09) \times 10^{-2}\,\mathrm{cm}^{-1}$ and a fluence-independent time constant

$$\frac{1}{\tau_\mathrm{Y}} = k_{\mathrm{Y},0}\, e^{-E_\mathrm{Y}/kT_\mathrm{a}}$$

containing the parameters $k_{\mathrm{Y},0} = 1.5^{+3.4}_{-1.1} \times 10^{15}\,\mathrm{s}^{-1}$ and $E_\mathrm{Y} = (1.33 \pm 0.03)\,\mathrm{eV}$.

2.4.1.4 Oxygen-Enriched Material

Extensive work has been done to evaluate whether the post-radiation properties can be affected by specific impurities other than boron, phosphorus, or arsenic. It was found out that the enrichment of the silicon substrate with oxygen, which is believed to capture vacancies in stable and electrically neutral point defects, leads to a superior postradiation performance [100, 109]. This observation mainly concerns the stable damage N_C and the reverse annealing N_Y and is restricted to the irradiation with charged hadrons.

The stable damage is mainly affected via the acceptor introduction rate g_c, which is reduced by a factor of 4 with respect to standard float zone material to a value of $g_\mathrm{c} = 5.3 \times 10^{-3}\,\mathrm{cm}^{-1}$ as shown in Fig. 2.27a. This means that the increase of the full depletion voltage after irradiation-induced type inversion is much smaller in oxygenated than in standard float zone material. The reverse annealing is also altered by the oxygenation. Instead of the linear fluence dependence of $N_{Y,\infty}$ one gets a saturation function for this component, as shown in Fig. 2.27b, reducing the effect of reverse annealing by a factor of 2 for a fluence of $\Phi = 6 \times 10^{14}\,\mathrm{cm}^{-2}$. Furthermore, the time constant for the reverse annealing is about doubled.

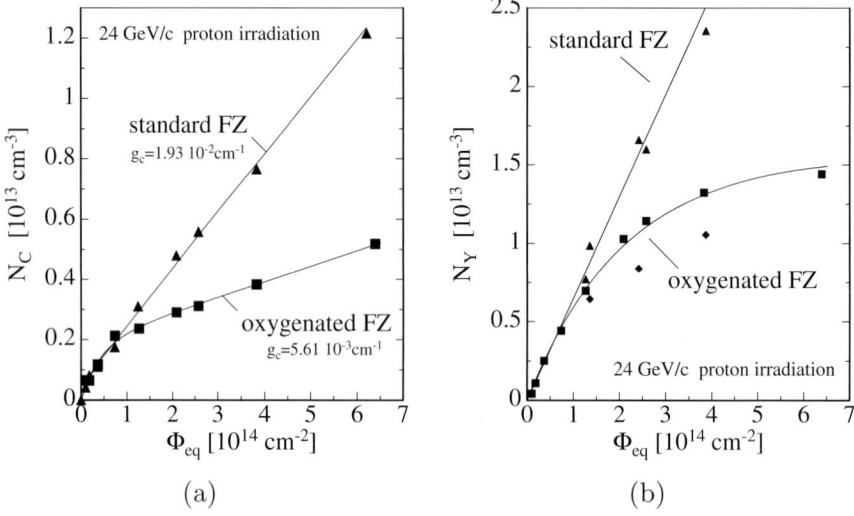

Fig. 2.27. Comparison of oxygenated (DOFZ) and standard (FZ) material. Damage parameters N_C describe the stable damage (**a**) and the parameters N_Y describe the amplitude of reverse annealing (**b**) [109]

The reasons for the different behavior of oxygen-enriched silicon are not yet fully understood. The fact that no change of the damage parameters after neutron irradiation is observed might be a hint that point defects play a major role, while clusters which are mainly caused by neutrons are not affected. As the main fluence in the innermost part of the experiments at hadron colliders where pixel detectors are commonly located is due to charged pions, the use of oxygenated material is recommended there.

2.4.1.5 Noninverted Surface Layer

Regions close to the sensor's surface do not undergo the process of type inversion and remain n-type. This was found out independently from capacitance–voltage measurements of irradiated MOS structures [93] and from current–voltage measurements between two p^+n-diodes, e.g. two adjacent strips in a strip detector, which still showed a typical behavior of two back-to-back diodes [110].

The reason of this effect is still unclear. One possible explanation involves the mobility of the primary defects, vacancies, and interstitials, which might diffuse out of the crystal's surface before forming stable secondary defects. This would lead to a vanishing defect concentration at the surface and a reduced one in the neighbouring region [111]. However, there is some indication that the donors still present in this noninverted regions are not the initial phosphorus impurities but levels deeper in the band gap [112].

2.4.1.6 Electric Field in Heavily Irradiated Silicon

After irradiation-induced space charge sign inversion, the electric field inside the sensor differs from the ideal picture sketched in Fig. 2.7. One hint is the signal charge collection plotted as a function of the drift distance to the collecting electrode shown in Fig. 2.17. The n^+-side readout is highly segmented, meaning that practically only the electrons contribute to the signal. If one naively assumes the inverted bulk to be uniformly "p^+-doped," one would expect a clear separation between the depleted and undepleted volume. However there is clearly a signal from the "undepleted" region at a bias of 150 and 300 V. Furthermore, these signals become higher with increasing bias. This indicates the presence of an electric field in this region.

Studies using a short laser pulse penetrating only a few micrometers into the silicon and analyzing the time evolution of the pulse [113] resulted in the so-called *double junction* model, e.g., described in [114, 115]. This model is still not sufficient to describe the pulse shape in Fig. 2.17 and a further refinement is necessary as shown in [76].

2.4.1.7 Charge Trapping

Radiation-induced defects are responsible not only for generation–recombination centers increasing the leakage current and charged defects with dramatic influence on the full depletion voltage but also for trapping centers. Traps are mostly unoccupied in the depletion region due to the lack of free charge carriers and can hold or *trap* parts of the signal charge for a time longer than the charge collection time and so reduce the signal height. A parameter to describe trapping is the trapping time τ_t, which is inversely proportional to the concentration of traps and therefore inversely proportional to the fluence Φ [93]:

$$\frac{1}{\tau_t(\Phi)} = \frac{1}{\tau_{t,\Phi=0}} + \gamma\Phi. \tag{2.54}$$

The coefficient γ was measured to be $0.41 \times 10^{-6}\,\text{cm}^2/\text{s}$ for electrons and $0.60 \times 10^{-6}\,\text{cm}^2/\text{s}$ for holes after neutron irradiation. After irradiation with charged hadrons (protons and pions) this coefficient was found to be significantly larger: $0.56 \times 10^{-6}\,\mathrm{cm^2/s}$ for electrons and $0.77 \times 10^{-6}\,\mathrm{cm^2/s}$ for holes [116].

For most of the tracking devices used in particle physics this effect is much less of a problem than the other radiation-induced effects already mentioned. After a fluence of $10^{14}\,\mathrm{n_{eq}/cm^2}$ about 90 % of the signal charge in a 300-μm-thick detector can be collected. However, this number decreases to about 50% for $10^{15}\,\mathrm{n_{eq}/cm^2}$ and trapping will eventually limit the use of silicon detectors for fluences much beyond this number.

In spectroscopic applications trapping causes serious degradation as it affects the pulse height. This is especially the case for devices with long drift distances, like charge coupled devices or silicon drift detectors. Devices exposed to radiation, like satellite-based X-ray telescopes, have to be frequently recalibrated to take into account the effects of radiation-induced trapping centers.

2.4.1.8 Influence of Bulk Damage on Sensor Operation

The bulk damage has consequences on the operation of irradiated sensors. The increase of current and operation voltage leads to an increased power dissipation and heats the sensor. Higher temperature implies higher leakage current and therefore larger dissipated power. This is a positive feedback system that may quickly diverge (thermal runaway) unless prevented by proper cooling. The volume generation current as well as the full depletion voltage can be predicted for a given fluence through the equations presented above.

If the sensors will be irradiated above the level of type inversion, the increase of the net doping concentration proportional to the fluence leads to an increase of the full depletion voltage which can, in some applications, exceed thousand volts after some years of operation. As it is unpractical to increase the operation voltage into this range, one might choose to work with partially depleted sensors. However, for a given maximum operation voltage the depth of the depletion zone and therefore the electrical signal will decrease. The detector system has therefore to be designed in such a way that it still can work with the reduced signals, or the maximal operation voltage is still high enough to provide a sufficient signal.

The reverse annealing also turns out to be critical if the sensors are planned to be used for several years. The full depletion voltage will increase significantly after several weeks at room temperature. Therefore in the LHC experiments cooling will be continued during the shut-down periods. Some days or weeks without cooling for maintenance will be unavoidable, but they must be short enough to keep the sensors still in the phase of beneficial annealing. Oxygenated silicon will give some additional safety margin as the reverse annealing is reduced in such a material.

A damage projection of the full depletion voltage for the innermost barrel layer of the ATLAS pixel detector (b-layer) is shown in Fig. 2.28. The

2.4 Radiation-Induced Effects on Silicon 79

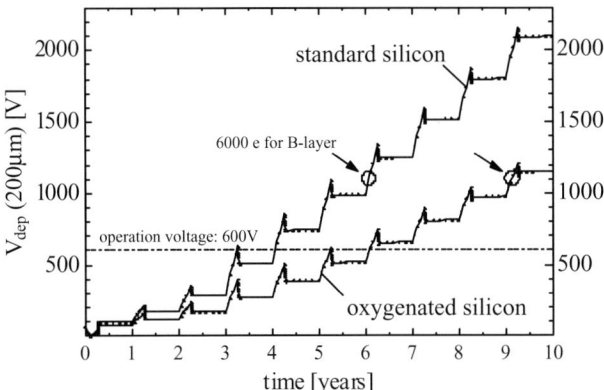

Fig. 2.28. Damage projection of the full depletion voltage for the innermost barrel layer of the ATLAS pixel detector (b-layer). Every year consists of a data-taking period, a maintenance period at room temperature (2 weeks), and a period with cooling but no LHC operation [109]

sensor thickness was planned to be 200 μm and therefore the full depletion voltage for this thickness was calculated. As the sensor can be operated partially depleted the point where the signal at a bias voltage of 600 V falls below 6,000 electrons is marked. This is the point where the sensor is no longer considered as properly working. The two lines represent the scenarios for oxygenated and standard float zone silicon. Each year is divided into three periods. The 100 days at the beginning of each year are the data-taking periods with the LHC running. During this time the full depletion voltage increases. The only exception is the beginning of the first year when it decreases up to the point of the space charge sign inversion. As the detector is cooled to about $-10°C$ during operation, all annealing effects are strongly suppressed. The data-taking period is followed by maintenance period of about 2 weeks when the cooling system cannot be operated. During this time annealing takes place. In the first years there is only a decrease of the full depletion voltage due to the beneficial annealing while in the later years reverse annealing also starts becoming visible. During the rest of the year when the cooling system is switched on but LHC is not running, there is practically no change of the full depletion voltage.

In this projection the advantage of oxygenated silicon is clearly visible. While a detector built with standard float zone silicon has to be replaced after 6 years, it can be operated for 9 years if oxygenated silicon is used. This is mainly due to its lower introduction rate of stable damage g_c. The reduced reverse annealing provides an additional safety margin if the duration of the maintenance without cooling has to be exceeded.

2.4.2 Surface Effects

In silicon the surface region is also sensitive to radiation. The term *surface damage* summarizes all defects in the overlaid dielectrics, e.g. the silicon oxide, and the interface between the silicon and the dielectric. As the crystal structure of silicon oxide is highly irregular, displacements of single atoms due to irradiation do not lead to macroscopic changes. Ionization in the oxide, however, is not fully reversible and may cause steady changes of the interface properties.

After a sudden creation of electron–hole pairs in the oxide layer, caused, e.g., by ionizing radiation, most electron–hole pairs recombine immediately. As electrons have high mobility in the oxide ($\mu_{n,\text{oxide}} \approx 20\,\text{cm}^2/(\text{Vs})$) they will be collected by any positively biased electrode close by. The holes have a very low mobility in the oxide ($\mu_{p,\text{oxide}} \approx 2 \times 10^{-5}\,\text{cm}^2/(\text{Vs})$) because of a large number of shallow hole traps. They move very slowly in the direction of the electric field in the oxide, hopping from one shallow trap into the next. If the holes arrive in the transition region between silicon and oxide, where many deep hole traps exist, they may be kept there permanently [117]. These traps are probably interstitial oxygen atoms that can capture holes [118].

Another source of positive oxide charges is trivalent silicon atoms with one dangling bond. The outer electron not involved in a chemical bond can easily be removed by scattering with electrons also resulting in a positively charged state [118].

These charges which are now fixed represent an additional contribution ΔN_{ox} to the total oxide charge N_{ox}, which, according to (2.34), leads to an increase of the flat band voltage:

$$\Delta V_{\text{FB}} = \frac{ed}{\varepsilon_0 \varepsilon_{\text{si}}} \Delta N_{\text{ox}} \,. \tag{2.55}$$

As the holes follow the field in the oxide it is not surprising that the radiation-induced shift of the flat band voltage of a MOS structure is dependent on the voltage applied to the surface as shown in Fig. 2.29. The effect is lowest if no bias is present and the direct recombination of electron–hole pairs is most probable. Highest flat band voltage shift is reached if the substrate is biased negatively with respect to the gate or oxide's surface and the holes can follow the field to the interface with its high trap concentration. As the number of traps is limited a saturation of the oxide charge after a dose of some kilograys is observed. For a high-quality thermally grown oxide a typical value of the order of $N_{\text{ox,sat}} \approx 3 \times 10^{12}\,\text{cm}^{-2}$ is reached [41, 117, 120]. The positive oxide charge has an influence on the electric field in the silicon bulk close to the surface and induces a compensating electron accumulation layer in n-type silicon and a depletion layer in p-type material.

A further effect of radiation is the generation of interface states leading to a surface generation current when the space charge region reaches the surface. This contribution to the dark current is proportional to the not implanted

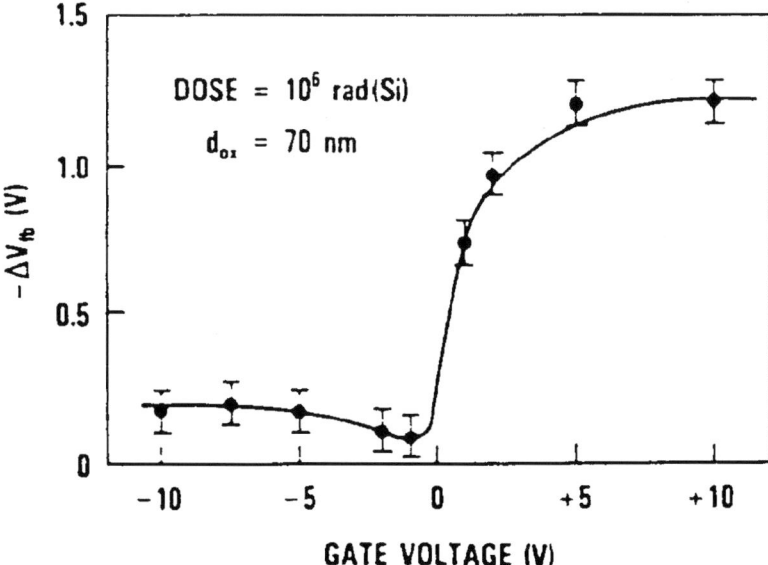

Fig. 2.29. Radiation-induced flat band voltage shift in dependence of the gate voltage during irradiation [119]

surface area of the sensor. Also in this case the trivalent silicon is believed to give the most important effect. Usually dangling bonds of silicon and oxygen are saturated with hydrogen by a suitable annealing procedure. The hydroxide ion (OH^-) can be separated from the silicon by radiation and is mobile in the oxide. It will be collected by a positively charged electrode [118].

Surface damage can be annealed at temperatures above 150°C. The process is driven by electrons from the silicon tunneling into the oxide. The interface traps can completely be annealed but part of the oxide charge is stable [93].

Surface damage has no direct consequences for the operation of the particle sensors. However, the design has to be adjusted in a way that the changes in the electric field due to the oxide charges do not influence the sensor performance as discussed in Sect. 2.5.

2.5 Sensor Concepts

In this section different sensor concepts are presented and their features are discussed. But first of all an overview of the possible combinations of substrate and electrode types is given.

2.5.1 Overview of Sensor Types

In a strip detector only one spatial coordinate per layer is measured. To obtain both coordinates from the same detector, both sides have to be segmentend. Pixel detectors in contrast measure both spatial coordinates on the same side of the sensor. Segmentation is therefore necessary only on one side of the sensor. In principle there is the choice of n-type and p-type substrate material and the choice of n^+ and p^+-type-doped collection electrodes which gives the four possible combinations listed in Table 2.1. In the end of the 1970s only n-type material was available in detector grade quality. Therefore almost all developments of silicon sensors were made using this type of material.

Table 2.1. All possible combinations of substrate and electrodes type

Readout electrode	Substrate	
	p-type	n-type
p^+	Double-sided process (expensive). No advantage over p^+-in-n	Typical single-sided processed sensor for most applications. See Sect. 2.5.2
n^+	Single-sided process. Not yet seriously developed. May be a cheaper replacement of n^+-in-n in future	Double-sided process necessary. Present "standard-device" if radiation hardness is required. See Sect. 2.5.3

2.5.1.1 p^+-Doped Electrodes in n^+-Doped Substrate

The first silicon sensors built in the late 1970s were realized by placing p^+-implantations in a high-resistive n-substrate. This so-called p^+-in-n^+ approach is shown in Fig. 2.30a–c. The ohmic side of the sensor is completely unstructured and no photolithographic process is necessary there. The main advantage of this approach is the simplicity and consequently the low costs. The p^+-pixels are electrically isolated from each other by design. Before irradiation, insulation is provided by the pn-junction. After the inversion of the bulk to p-like material the interpixel insulation is due to the electron accumulation layer caused by the increased oxide charges. As there is no interpixel insulation structure necessary that might cause high electric fields and subsequently electrical breakdown, such sensors usually display a higher breakdown voltage than n^+-in-n sensors. The high-voltage capability is determined by the edge termination via guard rings (see section below). Values exceeding 1000 V have been reached.

In an unirradiated device the depletion zone is growing from the junctions on the structured p^+-side (see Fig. 2.30a). The sensor can be operated

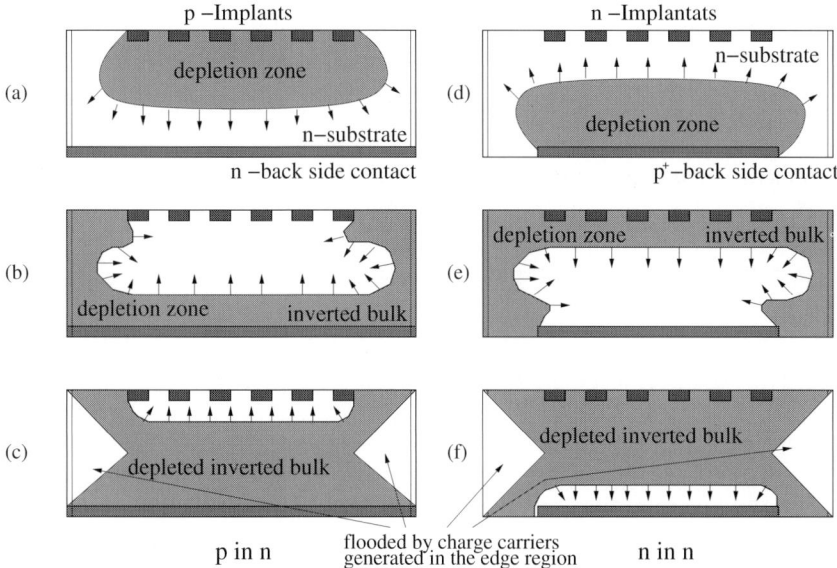

Fig. 2.30. Comparison of p$^+$-in-n (**a**)–(**c**) and n$^+$-in-n (**d**)–(**f**) sensors. Partial depletion (**a**) and (**d**) before and (**b**) and (**e**) after type inversion. In (**c**) and (**f**) the generation of electron–hole pairs in the heavily damaged edge region is also taken into account (After [72])

partially depleted with the high-field zone at the side of signal collection, the pixel implants. This is important if the backside is mechanically damaged and an extension of the depletion zone into the damaged region would lead to an unacceptable current increase. So even sensors with small backside scratches can be used. After an irradiation fluence of a few $10^{12}\,\mathrm{n_{eq}/cm^2}$ the substrate material converts from n-type to p-type. The unstructured back side is connected via the conductive cutting edge to the electron accumulation layer at the edge of the p$^+$-side. As the electrons act more or less like an n$^+$-implant, the depletion zone grows from both sides. An undepleted region in the center in electrical contact with the pixel region remains.[8] This situation is shown in Fig. 2.30b, neglecting the strong generation of charge carriers in the heavily damaged region of the cutting edge. When the applied voltage exceeds one quarter of the full depletion voltage, a potential minimum at the edge region of the sensor is formed into which the holes generated at the cutting edge are drained as sketched in Fig. 2.30c. The injected current is limited by the low conductivity of the region which is flooded by holes and therefore not depleted any more. As long as the depletion region has not yet reached the surface all pixels on the p$^+$-side are shorted via the inverted undepleted bulk material.

---

[8] As discussed in Sect. 2.4.1 this picture is much simplified. However it is sufficient to motivate the choices of the different sensor concepts

As this material behaves like a (bad) insulator, it is transparent to fast signals. This means that p$^+$- in-n sensors, irradiated over the point of inversion, still deliver spatial information when operated slightly underdepleted [72].

These p$^+$-in-n sensors are suitable for all applications where the radiation and temperature scenario predict that the full depletion voltage can always be applied. For a setup with an operation and storage temperature of about $-10°$ C the upper limit on the expected fluence in such a scenario is of the order of $(2\text{--}3) \times 10^{14}$ n$_{eq}$/cm^2. If the particle fluence expected will be much higher, this sensor concept becomes unsuitable and the n$^+$-in-n approach has to be followed.

2.5.1.2 n$^+$-Doped Electrodes in n-Doped Substrate

The pixel cells of such a sensor consist of n$^+$-implants in an n-substrate while the junction is located on the backside. The large area p$^+$-implantation forming the junction must not extend to the cutting edge of the sensor. The mechanical damage caused to the crystal by sawing the wafer and extracting the sensor makes the edge conductive. This would short the junction. A structured backside and, consequently, a double-sided processing is therefore mandatory. Additional precautions to suppress the electron accumulation layer shortening the n$^+$-pixels explained in Sect. 2.5.3 have to be taken. For the above reasons n$^+$-in-n sensors are at least a factor of 2 more expensive than p$^+$-in-n sensors and their use has to be carefully justified.

An unirradiated sensor starts to deplete from the p$^+$-side as shown in Fig. 2.30d. Before the space charge zone has reached the surface all pixels are connected via the undamaged and thus conductive bulk material. This is not transparent to a fast signal and the charge would be distributed over all channels. Before type inversion such a sensor cannot be operated partially depleted, which is an additional demand to the quality of the fabrication process. When full depletion is reached the electric field collecting the signal charge is lowest close to the pixels, and an overdepletion is desirable. However, after type inversion the depletion zone is growing from the segmented n$^+$-side which allows a strongly underdepleted operation of the sensors (see Fig. 2.30e). This is especially important for the pixel sensors in experiments at hadron colliders.

Another important advantage of the n$^+$-in-n approach is that the double-sided processing allows a guard ring design keeping all sensor edges at ground potential as explained in Sect. 2.5.3. This avoids a lot of problems arising when considering the whole system. High voltages could in fact lead to destructive sparking between sensor edge and readout chip or neighboring sensor modules.

2.5.1.3 n^+-Doped Electrodes in p-Doped Substrate

It is also possible to build n^+-in-p sensors. They collect electrons (higher mobility, less prone to trapping), would not undergo a type inversion, and only need a single-sided manufacturing process. The last feature is very important, because it makes this kind of sensors cheap and potentionally suitable for areas larger than 1–2 m². An n-side isolation is still necessary as the positively charged surface might induce an inversion channel between the pixels. So far the "n^+ in n" approach has been preferred in very high radiative environment because of the guard ring concept (see Sect. 2.5.3). However, future LHC upgrades might require a pixel detector also at radii larger than 10 cm. With increasing area the cost argument becomes more and more important.

2.5.1.4 p^+-Doped Electrodes in p-Doped Substrate

Placing p^+-doped electrodes in a p-doped substrate is just a theoretical possibility. This approach combines, in fact, the disadvantages of the n^+-in-n (expensive double-sided manufacturing process) and the p^+-in-n (hole collection) approach.

2.5.2 p^+ in n: Low-Cost Solution for Applications in a Low- or Medium-Radiation Environment

In this section different features of single-sided p^+n-sensors are presented. But before discussing design details of features in the silicon a short discussion of the layers on top of the silicon is inserted, as they determine the behavior of the region in between the pixels.

2.5.2.1 Electrical Potential at the Surface

The presence of an electron accumulation layer at the surface region has an important impact on the sensor behavior. It is very sensitive to the potential which is established on top of the passivation.

von Neumann Boundary Condition

The *von Neumann* boundary condition assumes that there is no electric field perpendicular to the surface at the top of the device or, in other words, that the electrical potential on top of the passivation is the same as on top of the silicon. This boundary condition, which is default in most device simulation packages, leads to high electric fields in the interpixel region, especially at the edges of the implants. It is a reasonable description for a new sensor handled always in a clean and dry environment immediately after application of the bias voltage. This state, however, is not stable. After some time a highly resistive film develops on top of the passivation, eventually charging up to the potential of the next metal line [121].

Gate Boundary Condition

The state when the potential of neighboring metal lines is established on the whole surface is, from the point of view of the potential distribution, equivalent to a metal gate covering the whole surface. This state is therefore frequently described as *gate* boundary condition. The region between the pixel implants can then be considered as a MOS structure influenced by the thickness of the oxide. This leads to lower electric fields as not all field lines from the backside have to end in the implant. In the presence of several dielectric layers on top of the silicon as indicated in Fig. 2.31, it is not always straightforward to decide which layer thickness is important. The first oxide layer on top of the silicon can always be considered as "gate" oxide due to its high quality. The layers on top of this oxide layer might partly be "conductive" and can usually be considered as "gate electrode."

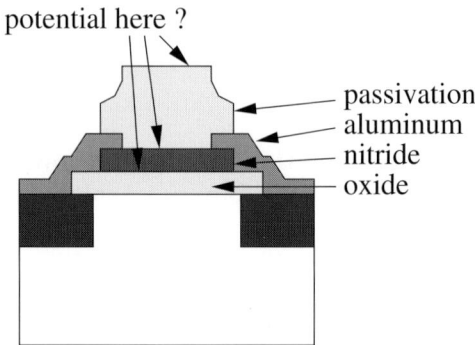

Fig. 2.31. Sketch of the various layers on top of the silicon (not to scale). The electrical potential to which the different layers adjust influences the electrical properties of the sensor. This potential depends on the properties of the materials themselves, on the environment humidity, and on the surface contamination (e.g. dust) and changes with time. As the factors mentioned above are difficult to control a prediction of the time scale for the onset of the potential is not always possible

Time Dependence

After the application of the bias, the boundary condition changes from a von Neumann-like situation to a gate-like. Therefore, IV-curves taken very slowly (i.e. in hours) show often less current than those taken fast (i.e. in seconds). The time needed for this transition can vary from milliseconds to hours. If metallization is overlapping the lateral pn-junction, the gate boundary condition is established instantaneously at the most critical spot and such sensors usually show nice and stable IV-curves. For other sensors this transition is very much dependent on the characteristics of the passivation layers and the environment. Especially in nitride-passivated devices operated in dry nitrogen it can take several hours. So sensors with nice characteristics measured at normal laboratory conditions (air humidity about 40–60%) might display an early breakdown at dry nitrogen atmosphere which dissapears after being biased for a few hours. This is especially problematic as most tests are usually performed under "normal" laboratory conditions while the detectors are often operated in dry nitrogen.

Oxygen-passivated sensors have also to be tested with great care. As the low temperature oxides used for passivation can contain humidity the gate boundary condition is usually established very fast, also in dry nitrogen atmosphere. However, the humidity evaporates very slowly within weeks or months. The potential change of the sensor behavior should also be checked in this long time scale.

2.5.2.2 Pixel Layout

A pixel cell usually consists on a rectangular-shaped p^+-implant in the n-substrate as shown in Fig. 2.32. While the pitch of the pixel cells is given by physics requirements for the spatial resolution, the width of the implant is a "layout decision." In this section the influence of the implant or gap width on the sensor performance is discussed.

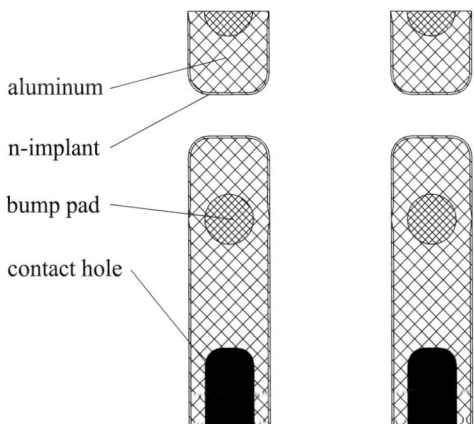

Fig. 2.32. Simple layout of a pixel cell

Capacitance

The pixel capacitance to the neighbors is dependent on the gap width between the implants. The capacitance can be calculated using device simulation packages. In Fig. 2.33 the dependence of the interpixel capacitance is

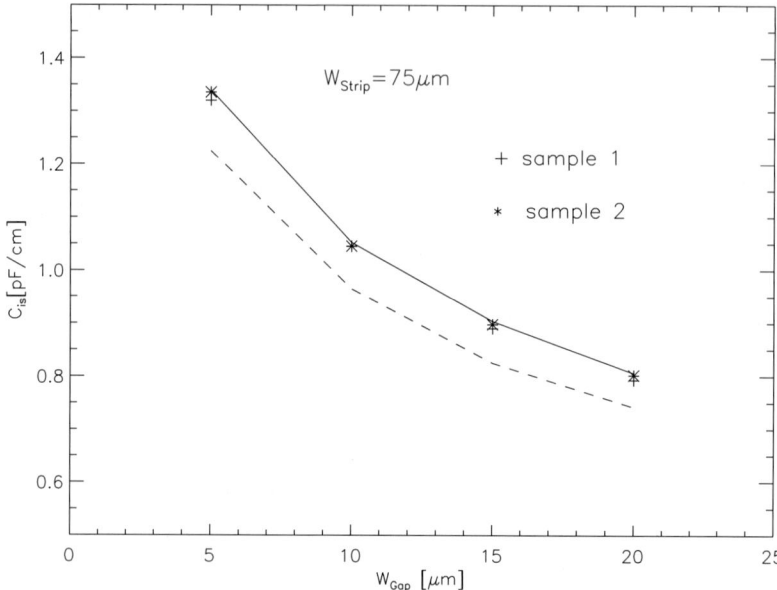

Fig. 2.33. Comparison of the interpixel capacitance for different gap widths, keeping the implant width constant (varying pitch). The markers indicate measurements on two samples of strip sensors. The lines show the results of a two-dimensional device simulation. The *dashed line* includes only the field lines in the silicon itself, while the *solid line* represents a simulation where the oxide layer covering the silicon and the surrounding air have also been taken into account [112]

plotted versus the gap between the implants while the implant width was kept constant [112]. The markers in the plot show values measured with two samples of about 1-cm-long macropixels (or mini strips). The value therefore has been normalized to a pixel length of 1 cm. Due to limits of computing power only a two-dimensional software package was used for this calculation, meaning that the other contributions to the total capacitance were neglected. The dashed line shows the result of a simulation where only the silicon itself was taken into account by the finite element analysis, leading to a value which is about 10% below the measured value. This is the difference one expects from the ratio of the dielectric constants in air (≈ 1) and silicon (≈ 11.8). If one also considers the air on top of the silicon, one obtains the solid line matching the measurements perfectly. The main message from this plot is that the interpixel capacitance, which dominates the total capacitance, is roughly inversely proportional to the gap width and therefore pixels with large gaps between their implants are preferred. One attempt to move this approach to an extreme was to reduce the implants of a pixel with an area of $50 \times 500\,\mu m^2$ to only five dots of $20 \times 25\,\mu m^2$ [122]. However, the noise was not measured to be significantly smaller than in other sensors with "moderately" large gaps, indicating that the curve shown in Fig. 2.33 flattens out at large values of gap width.

There are other considerations that limit the maximum possible gap width, namely the electric field, the full depletion voltage, and the charge collection.

Depletion

The two options sketched in Fig. 2.34 show qualitatively the difference of depletion behavior between a wide and a narrow gap option. If a positive

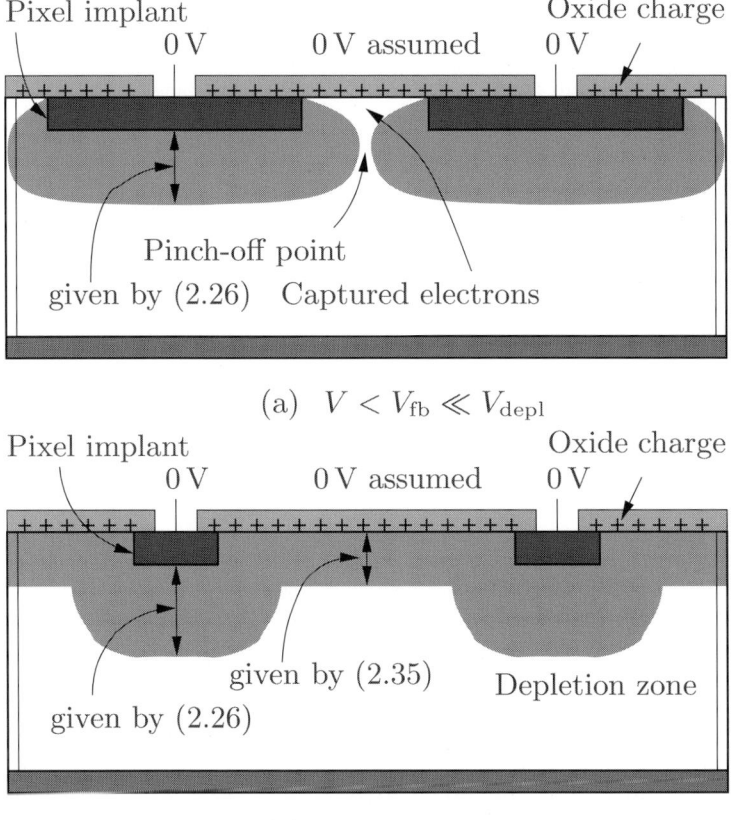

Fig. 2.34. Comparison between a small and a wide gap topology. (a) If the gap between the implants is small, the depletion zones growing from adjacent pixel merge before the flat band voltage V_{FB} is reached. An electron layer is pinched off from the bulk and captured close to the surface. (b) With a wide gap the surface is depleted when the bias voltage exceeds the flat band voltage (Drawn according to [4])

voltage is applied at the backside while holding the pixel implants at ground potential, the depletion region is growing into the depth of the sensor as well as laterally according to (2.26). The surface of the insulation oxide layer is assumed to be at ground potential as discussed in the previous paragraph. With increasing bias the voltage drop across the oxide also increases so that the electron accumulation layer caused by the oxide charge at the interface between silicon and oxide decreases and eventually disappears when the flat band voltage is reached. If the depletion regions of two adjacent pixel implants merge before the flat band voltage is reached, the channel between the surface and the backside contact is pinched off and the remaining electrons are caught at the surface as shown in Fig. 2.34a. Their potential is determined by the potential of the saddle point and therefore by the geometry and it is hardly dependent on the backside voltage even when the full depletion voltage is exceeded. The situation where the pinch-off occurs before the flat band voltage is reached will be referred to as *small gap* situation in the following. It has to be mentioned that this small gap situation can also occur at rather large distances between the implants if the flat band voltage is very high due to radiation damage of the interface and a thick oxide.

If at flat band voltage the depletion zones of the pixels are still unconnected, the depletion zone also grows from the silicon–oxide interface downward as sketched in Fig. 2.34b. As holes are drained by the pixel implants, an inversion layer cannot form and deep depletion below the interpixel region establishes with a depletion depth W according to (2.35). The influence of the MOS-like interpixel region becomes more important for smaller implant widths and larger gaps. The full depletion of a simple diode (zero gap) is determined by the pn-junction and can easily be calculated using (2.38). However, for a very narrow implant the depletion is dominated by the MOS-like interpixel region, leading to a higher value. The upper bound for a vanishing implant width can be estimated using (2.35):

$$V_{\text{depl,MOS}} = \frac{eN_D d^2}{2\varepsilon_0 \varepsilon_{\text{Si}}} + \frac{eN_D d\, d_{\text{ox}}}{\varepsilon_0 \varepsilon_{\text{ox}}} + V_{\text{FB}} , \qquad (2.56)$$

where d_{ox} is the thickness of the oxide. Equation (2.56) contains the suprising fact that in the case of very narrow implants the full depletion voltage is dependent on the thickness of the covering oxide which also enters into the flat band voltage V_{FB}.

The dependence of the full depletion voltage on the gap width and oxide thickness has been studied [123] for strip detectors using a two-dimensional device simulation package. The radiation-induced changes of the oxide charge N_{ox} were also taken into account. The result can be summarized as follows:

- For thin oxides (roughly 200–300 nm) the change of the full depletion voltage is moderate even for quite extreme geometries.

- In the case of an oxide thickness in the order of 1 μm the gap dependence of the full depletion voltage is still low in the preirradiated state but becomes a major issue when the oxide charge rises.

In this context it appears to be of advantage not to grow thick oxides in the style of CMOS field oxides in the region between the pixels but limit the thickness to several hundred nanometers.

Electric Fields

The same study [123] also investigated the influence of the gap width on the distribution of the electric field in the sensors. The electric field peaks at the edge of the pixel implants. The height of the peak is dependent on the gap width. It has been shown that metal covering the lateral pn-junction increases the breakdown voltage [124,125] by establishing the gate boundary condition instantaneously. For unirradiated devices (with a low interface charge) even extreme geometries will not be prone to breakdown, but after exposure to several kilograys, when the oxide charge is saturated, the electric field can reach the critical value of roughly 30 V/μm for geometries with (very) large gaps. This is the most critical situation concerning breakdown in the life cycle of such a sensor. With further irradiation the mobility of charge carriers will be reduced, decreasing also the tendency for breakdown. When the high field region of the junction (due to irradiation induced type inversion of the substrate) moves to the unstructured backside, breakdown becomes even less critical. The breakdown voltage of p^+- in-n sensors is then mostly determined by the guard rings and not by the pixel geometry.

Signal Collection

The size of the collecting electrodes, which are the p^+-implants, influences the signal formation as discussed in Sect. 2.3.5. For an unstructured pad detector with large electrodes the drift of both holes and electrons contributes to the signal. If the collecting electrode is segmented, only the drift of charge carriers close to it contributes to the signal as a consequence of the different shape of the weighting field. These are the holes in the case of p^+-in-n sensors moving to the p^+-implants. The smaller the pixel width (and the larger the gap), the smaller the area becomes in which most of the signal is created. So if very large gaps are used signal charge generated in the middle of the gap has to drift a longer distance and induces the electrical pulse later. Delayed signals have been observed for the extreme geometry mentioned above [122]. This also makes the sensor more sensitive to charge trapping, as holes trapped after 90% of their drift path have induced almost no signal. Holes are in addition more prone to trapping than electrons due to their lower mobility. Both facts have to be taken into account when designing pixel detectors which should sustain large radiation doses.

Summary

Summarizing this section one can say that the requirement of low capacitance favors large gaps between the implants while all other considerations point to small gaps. However, these considerations do not lead to a definite statement for the largest possible gap. Only if the sensor should be exposed to fluences above several times $10^{14}\,n_{eq}/cm^2$, when trapping becomes an important issue, the use of the smallest gap compatible with the noise requirements is recommended.

2.5.2.3 Guard Rings

The edge termination is a critical issue for the long-term stability of silicon detectors. The behavior of the edge depends very much on the potential of the sensor surface. As discussed before the surface of the device after some time always adjusts to the potential of a metal line close by. However, outside the sensitive region of the device the situation is a bit more complex as not the whole surface is at the same potential. The cutting edge is conductive due to the mechanical damage caused by the cutting procedure and will be at the backside potential, which is, according to Fig. 2.35, the bias voltage. The metal lines in the sensitive region are grounded. The surface potential between the sensitive region and the cutting edge will be somewhere in between. The real value is not predictable as it depends on the surface conductivity and its local variations and time dependence and therefore on environmental conditions like humidity, dust contamination, and so on. Figure 2.35 illustrates the two extreme cases which are both problematic. For simplicity the sensitive region is drawn as a simple unsegmented diode, as the segmentation has no impact on the edge behavior. In the case shown in Fig. 2.35a the ground potential has extended from the central region almost to the edge. This leads to a lateral extension of the depletion and when the space charge reaches the cutting edge the strong crystal damage which is present there acts as a very effective generation center and causes a dramatic increase of the leakage current.

The other extreme is when the surface potential adjusts to the scribe line potential as shown in Fig. 2.35b. As it is positive with respect to the potential of the grounded electrode an electron accumulation layer will be formed, reaching the lateral end of the junction. As the electron accumulation is conductive it will adjust to the backside potential and the full bias voltage drops over a very short distance, leading to high electric fields and a low breakdown voltage.

The purpose of *multiguard rings* is to establish a smooth voltage drop toward the cutting edge and to assure that the outermost ring is (in the ideal case) on the backside potential. No space charge region can then establish outside the outermost ring as sketched in Fig. 2.36. The rings bias themselves via the punch-through mechanism discussed in Sect. 2.2.7. The potential

2.5 Sensor Concepts 93

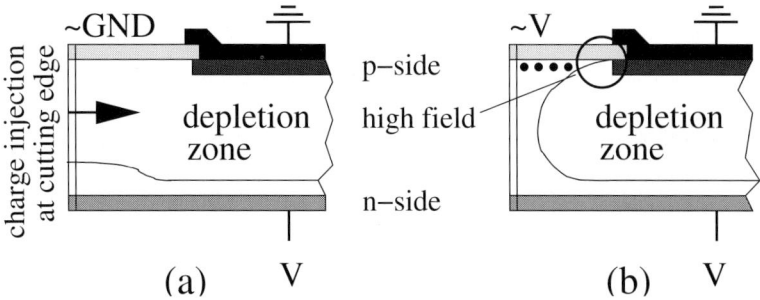

Fig. 2.35. Two problematic situations at the sensor edge. (**a**) The depletion zones reaches the cutting edge and induces a large leakage current. (**b**) The electron accumulation layer adjusts to the backside potential and leads to a high electric field and subsequent breakdown (Drawn after [4])

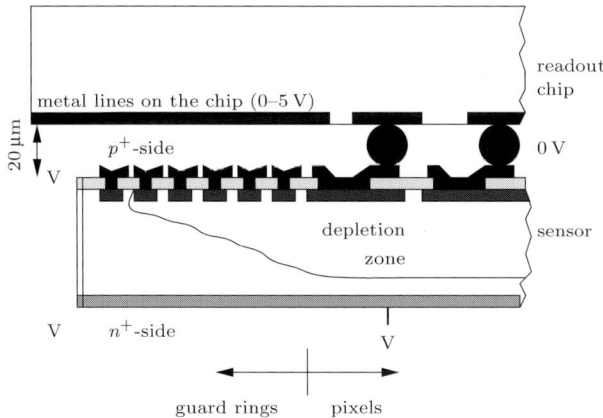

Fig. 2.36. Depletion region below guard rings and potential of surfaces and edges

drop between the rings can be influenced by their spacing and metal overlap. A metal overlap directed outward reduces the electric field at the implant edge. This is a technique used in power electronics since the late 1960s [43]. However it also reduces the punch-through voltage between two neighboring rings if it covers too much of the gap. Therefore these overlaps are usually kept relatively small as indicated in Fig. 2.37. A field plate directed toward the sensitive region suppresses the punch-through hole current and increases the punch-through voltage [126, 127]. This can be understood in terms of a p-MOSFET with the source as the outer of two rings and the drain as the inner ring. If the gate, a metal field plate as shown in Fig. 2.37 on the left, is connected to the outer ring acting as source, the transistor will always be in the switched-off state independently from the state of irradiation. Such

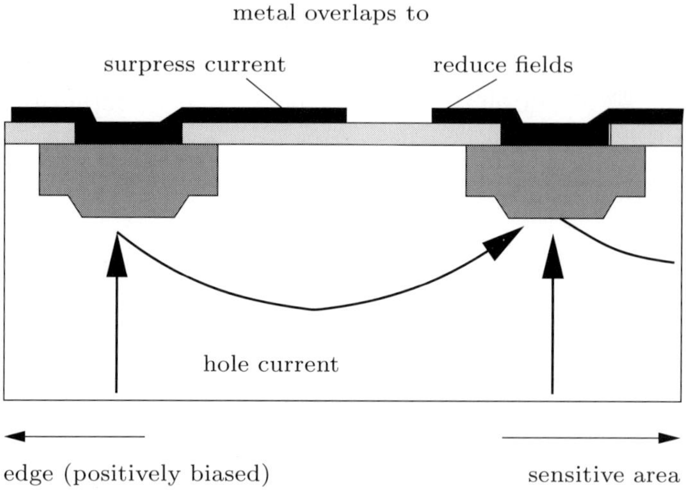

Fig. 2.37. Schematic cross section of a possible guard ring design with metal overlaps according to [126, 127]

geometries are therefore preferred over designs mainly working with outward-directed field plates [127].

The number of guard rings necessary depends on the maximal bias voltage targeted. Structures using between five and more than ten rings have successfully been used [127–129]. If a uniform potential drop between the rings is aimed for, the spacing has to increase from the inner to the outer regions. Although there is always a requirement that the edge region should be small in order to minimize dead material, a useful rule of thumb is to foresee a distance of roughly three times the wafer thickness between the cutting edge and the sensitive region.

It has to be mentioned that a reliable edge termination can also be reached with only two guard rings, one very close to the edge on the backside potential and one very close to the sensitive region on the ground. Both rings have to be connected via a highly resitive (but not perfectly isolating) passivation covering the whole oxide surface in-between to guarantee a well-defined and gentle potential drop. However, such a passivation is not offered by most sensor vendors and therefore multiguard ring structures are widely used.

Although with multiguard rings a breakdown voltage in the kilovolt range can be reached, the maximum operation voltage may be limited by system considerations. In most pixel modules the readout chips stick out above the sensor edge to provide space for wire bond connections. As the spacing between chip and sensor is of the order of 10–20 µm a high voltage difference between the chip (close to ground potential) and the sensor edge (on high potential) might become critical. The breakdown voltage in air is about

1.2 V/μm [130] and can thus easily be exceeded. Sparking between sensor and chip would be destructive and has to be prevented.

2.5.2.4 Biasing Structure

The production of detector modules is a complex procedure that requires a large number of critical steps (see Chap. 4). Especially for large experiments which need thousands of modules, this assembly procedure has to be optimized for good yield. In order to achieve this optimization it is crucial to be able to test all components before and after each assembly step. As the sensor and the readout chip are the most expensive components of a pixel module they both should be tested before joining them by the bump-bonding process.

The easiest method to test sensors is by measuring an *IV*-curve as already small damages in the space charge region lead to an increase of the leakage current. An *IV*-curve with a realistic potential distribution in the sensor can only be measured if all pixels are grounded. It is impossible to contact a significant fraction of the roughly 50,000 pixel cells in a highly segmented sensor with a probe card. The use of conducting rubber is also ruled out by the small size (10–20 μm) of the bump openings in the passivation layer.

One way to solve this problem is to integrate a biasing structure onto the sensor itself. Figure 2.38 shows mask drawings of two such implementations. An implanted line, the "bias grid," is placed between every second pixel column passing close to the pixel implants (Fig. 2.38a). If a bias voltage is applied to this line, the potential of the pixels will be fixed via the punch-through mechanism from the grid to the pixel implant (see Sect. 2.2.7). An

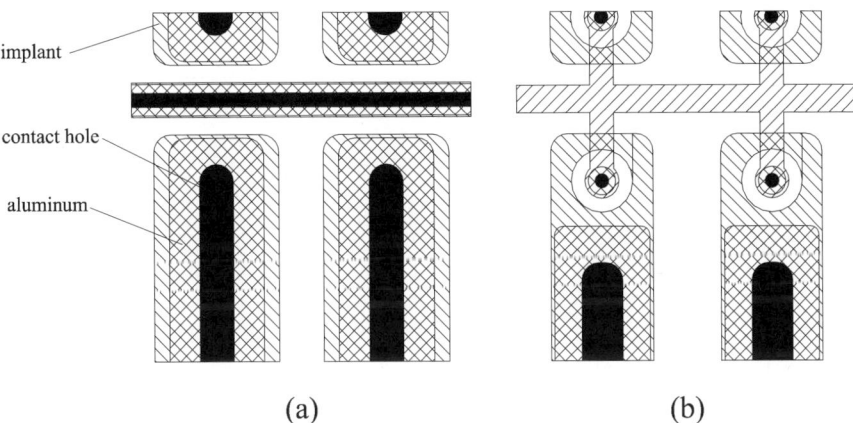

Fig. 2.38. Two possible implementations of a bias grid. (**a**) A simple punch-through structure with minimum requirements to the production process. (**b**) Optimized for minimal loss of sensitive area requiring a smaller feature size and an oxide free of pin holes

IV-measurement with all pixels biased is possible using only two probe needles. One needle has to be placed outside the guard rings, e.g. on the scribe line's metallization, and the other on the grid. After connecting the sensor to the readout chips, the pixels are biased via the bump bonds and the punch-through grid is out of function. For this reason no additional noise has been observed in pixel detectors using this feature contrary to what is known for punch-through biased strip detectors after irradiation [71, 72]. In the case of accidentally missing bump-bond connections the bias grid holds the unconnected pixel cells close to ground potential and prevents destructive discharging to the very closely spaced electronics.

However, the biasing line itself will collect some charge, which will be lost from the signal. It is thus mandatory to minimize the area covered by the bias structure.

The implementation shown in Fig. 2.38a is the simplest biasing possibility. There are no small structures that could approach the limits of the technology and all metal areas are underlaid by an implant. Even pinholes in the oxide would not degrade the performance. In the design shown in Fig. 2.38b the area affecting the charge collection is minimized. The implantation of the bias structure is shrinked to a small *bias dot* placed inside the pixel area. Surrounding the bias dot with the pixel implant guarantees that the signal generated in the region between the pixels is not lost to the grid. The bus in between the pixels is just a metal line and does not influence the charge collection. However such a design makes more stringent demands to the fabrication process: The bias dot is small and has minimum spacing to the pixel implant. The contact hole has also minimal dimension. Furthermore, the area of the narrow aluminum line is sensitive to pinholes in the oxide.

Up to now all pixel detectors with p^+- in-n sensors were fabricated without bias grid [14, 131, 132] and the sensors were fed into the module production either untested or, as the case in the DELPHI experiment, with the guard current measured only. The latter method is capable of detecting "obviously problematic" sensor wafers. However about 13% of the modules produced for the DELPHI pixel detector had to be rejected due to problems which might be connected to the sensor [133]. How much this number could have been reduced by more advanced sensor tests is not clear. The detector systems built with p^+-in-n sensors are rather small and the yield of the sensors is, due to the simplicity of the single-sided process, high enough to justify this approach. As the sensors can be operated underdepleted and will, due to the limited radiation dose, not undergo type inversion, the most frequent damage, namely backside scratches, is not lethal. Finally the operation voltages are much below 100 V and therefore the risk of breakdown, e.g. at unconnected pixels, is generally low. However the implementation of a bias grid can still be recommended.

2.5.2.5 Use and Limitations of p$^+$-in-n Sensors

Due to their simplicity, p$^+$-in-n sensors are the first choice for most detector applications, especially those where radiation damage is not an issue [14, 131, 132]. In these detectors high-resistive silicon is used and operation voltage is always much below 100 V. However, if one wants to exploit the limits of radiation hardness of this sensor type as done by silicon strip detectors, e.g., in CDF (layer00) [134], ATLAS [135], or CMS [136], one needs operation voltages of 300–500 V. The maximum bias voltage is not only limited by the sensor, but also by the module design. In pixel detectors the small distance between the sensor edge and the readout chip (see Fig. 2.36) is of special concern. However, this problem is purely technical and might be solved, e.g., by adding a dielectric layer on top of the sensor that can stand several hundred volts. The MCM-D technique discussed in Sect. 6.2.1 seems to be a good candidate for such a solution. If such a solution is found to be applicable, operation of p$^+$-in-n sensors can be extended up to a fluence of $(2-3) \times 10^{14}$ n$_{eq}$/cm^2.

As already mentioned in Sect. 2.5.1, n$^+$ in p sensors, which are also single sided and therefore potentially cheap, might be used up to even higher particle fluences. They have two main advantages:

- They do not show type inversion and can be operated partially depleted during their whole lifetime.
- They collect electrons which are, due to their higher mobility, less prone to trapping. Trapping is the factor limiting the operation of any silicon sensor in a very high radiation environment.

However this approach has not yet been seriously evaluated and in the experiments requiring a high level of radiation hardness n$^+$- in-n sensors have so far been preferred.

2.5.3 n$^+$ in n: Solution for Applications in a High-Radiation Environment

The large experiments at the LHC, ATLAS, and CMS place their pixel systems very close to the interaction point of the collider and therefore are the first experiments which must assure that the pixel system stays operational up to hadron fluences of $(6-10) \times 10^{14}$ n$_{eq}$/cm^2. After such a high fluence the full depletion voltage may reach or exceed 1000 V (see Sect. 2.4.1). The many problems arising at such high voltages (large dark currents, cable isolation, power supplies, etc.) impose a strongly underdepleted operation after irradiation induced type inversion. This is possible following the n$^+$ in n approach. However, it has to be mentioned that the signal provided by the sensor is proportional to the depletion depth and therefore radiation hardness requires high-voltage stability.

The n$^+$-in-n option has also disadvantages which have to be mentioned. The double-sided processing is much more demanding and increases the wafer-processing costs significantly. Moreover, scratches or other kinds of damage on the sensor's backside destroy the junction and, hence, the whole device. Defects on the pixel side directly destroy pixels. If they cause high currents in a limited number of pixels, they could cause problems in the readout chip. In order to achieve a nonzero field at the pixels, unirradiated sensors have to be operated overdepleted and a reduction of the bias in the case of n-side defects is not possible. Therefore a selection of good devices is necessary before they can be bump-bonded to the readout chips. As the most effective test is an *IV*-measurement, a biasing structure is in this case recommendable.

2.5.3.1 InterPixel Isolation

In contrast to the p$^+$-in-n case, where the isolation of the adjacent p$^+$-implants is provided by the omnipresent electron accumulation layer, this same layer would shorten the n$^+$-implants without further precautions. Isolation is usually provided by a p-type boron implant between the pixels, forming a lateral pn-junction. Depending on the dose of this isolation implant the method is called either p-stop or p-spray. Both techniques are discussed separately in the following. As high-voltage operation in n$^+$-in-n sensors is limited by electrical breakdown at the interpixel isolation structure, the electric fields close to the lateral junction are compared for different stages in the life cycle of a sensor.

p-Stop Isolation

In double-sided strip detectors, a common technique is to introduce a high-dose p$^+$-implant in between the strips as shown in Fig. 2.39a. This is obtained introducing an additional photolithographic step. The alignment of this p-stop mask toward the n$^+$-pixels is critical as an overlapping of the two high-dose implants would result in Zener breakdown. The minimum spacing between two n$^+$-implants is therefore limited by the necessary alignment tolerances. An advantage of this technique is that a typical dose of about 10^{14} boron ions/cm^2 will in any case guarantee a good isolation also after the radiation-induced builtdup of a positive surface charge (see Sect. 2.4.2), and the adjustment of the implantation dose is not at all critical. Different geometries will be discussed in the following section. The electric field close to the lateral pn-junction was numerically calculated using a two-dimensional device simulation package [123]. The results of this study are summarized in Fig. 2.40a. The unirradiated device has a very low oxide charge of $N_{Ox} = 2 \times 10^{11}$ cm^{-2}, resulting in a low electric field of about 0.1 MV/cm. The value of the oxide charge increases after a small irradiation dose of a few kilograys to its saturation value of about $N_{Ox} = 3 \times 10^{12}$ cm^{-2} (see

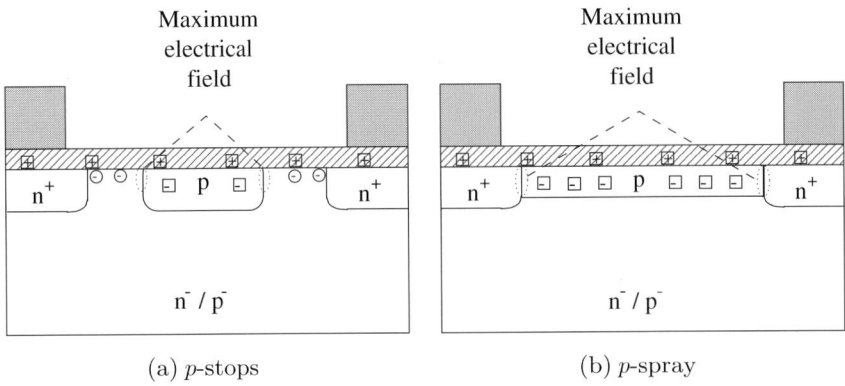

Fig. 2.39. Isolation techniques for adjacent n^+-implants. The type inversion of the bulk material is indicated as well as the fixed (*squares*) and free (*circles*) charge carriers. The field maxima are located at the lateral pn-junctions [123]

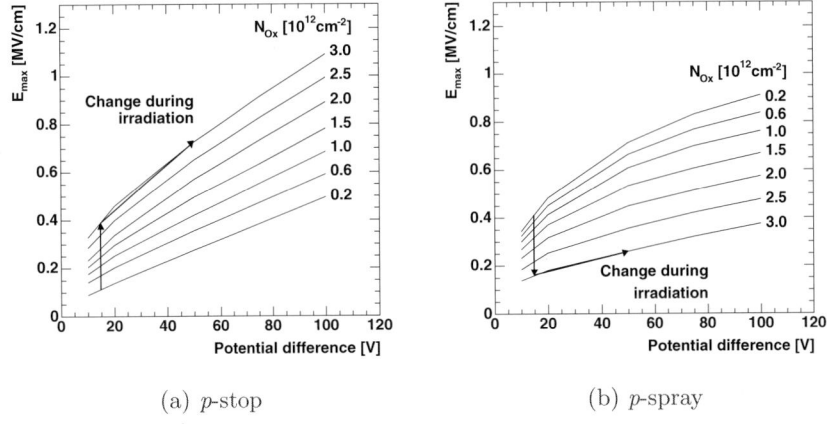

Fig. 2.40. The electric field maximum dependence on the potential difference between the isolating p-layer and the pixel n^+-implant for different values of the oxide charge N_{Ox}. The evolution of the electric field during the lifetime of a detector is indicated by *arrows* [123]

Sect. 2.4.2), which results in an accumulation of electrons close to the surface and subsequently in an increase of the electric field to about 0.4 MV/cm as indicated in Fig. 2.40a. The potential of the p-stop depends on the implant geometry, the backside bias, and the substrate doping. As the latter two quantities are also very high in a highly irradiated sensor, the potential difference between n^+-pixels and p^+-stops increases with ongoing irradiation, leading to an additional increase of the electric field. This means that the breakdown voltage of devices featuring p-stops decreases with irradiation.

p-Spray Isolation

If the dose of the boron implant is matched to the saturation value of the oxide charge which is in the order of 3×10^{12} cm^{-2}, the boron concentration is small enough that an overlap of the boron implant with the pixel n$^+$-implant does not lead to breakdown. The mask separating both implantations can be omitted. The whole surface is then covered by the medium-dose boron implant which is overcompensated at the locations of the high-dose n$^+$-implants as indicated in Fig. 2.39b. Because of the missing mask this technique is called p-spray isolation [123]. The absence of a photolithographic step is an advantage also because it allows narrow spacings between neighboring n$^+$-implants (no alignment tolerances necessary between two different masks). As in the p-stop case, the point of maximal electrical field is at the lateral pn-junction between the isolating boron implant and the n$^+$-pixels as indicated in Fig. 2.39b. The evolution of the electric field during a typical radiation scenario is shown in Fig. 2.40b. The unirradiated device displays the highest electrical field and therefore the lowest breakdown voltage in its lifetime. With the increase of the oxide charge to its saturation value the shallow p-spray layer moves into the depleted state and the electric field decreases as indicated by the arrow in Fig. 2.40b. The lowest electric field is reached when the boron implant matches exactly the saturation value of the oxide charge. However, if the implantation dose is too low, the isolation might not be sufficient. Therefore, one usually chooses an implant dose which is slightly higher than necessary to prevent failure in the case of fluctuations in the production process. With the increase of the effective substrate doping due to irradiation the electric field shows a mild increase also displayed in Fig. 2.40b. This means that the devices have a better high voltage performance after irradiation than before, which is the most attractive feature of the n$^+$-in-n approach.

In order to improve the preradiation high voltage stability of p-spray devices while keeping their good postradiation behavior, the so-called *moderated* p-spray technique [137] has been developed. Here the p-spray implant is performed later in the production process. A topography on the wafer's surface as for example the nitride layer will then be reproduced in the doping profile. A possible pixel layout with a nitride opening between the pixel implants is shown in Fig. 2.41 and the resulting doping profile along the indicated cross section is sketched in Fig. 2.42. The boron dose in the middle of the gap between two pixels can be chosen high enough to ensure interpixel isolation, e.g. twice the expected saturation value of the surface charge. At the same time the boron dose in the surrounding of the lateral pn-junction can be optimized for the best high voltage performance which is reached if the dose is close to the expected saturation value of the surface charge. As the nitride layer is a standard step in the process of many detector vendors this technology, in most cases, does not need an additional photolitographic step. However the freedom to design very small gaps between the pixel implants is lost in the

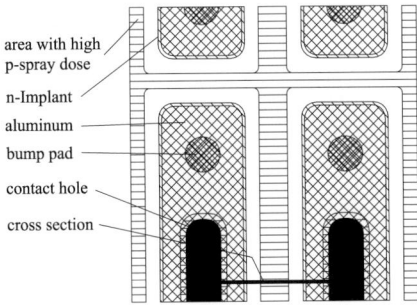 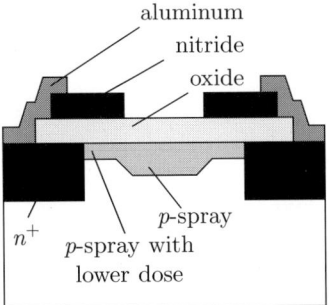

Fig. 2.41. Possible layout of a moderated p-spray design

Fig. 2.42. Schematic cross section of a device with moderated p-spray isolation

moderated p-spray technology as lateral space for the different layers has to be provided.

2.5.3.2 Implant Geometry and Biasing Structure

The choice of the n-side isolation technique has an important impact on the geometry of the pixels and the way a biasing structure can be implemented. The cases for p-spray, moderated p-spray, and p-stops are therefore discussed separately.

Geometries for p-Spray Devices

Thanks to the "missing" mask for the interpixel isolation, the designs of p-spray-isolated n^+-in-n pixels look quite similar to those of p-in-n devices. The design shown in Fig. 2.32 could also be a p-spray-isolated n^+-in-n device if the implant shown on the figure would be of n-type. However, the conducting p-spray layer, which is not shown in the layout drawings and covers the interpixel region, has to be considered. It changes the interpixel capacitance which is made out of different contributions as shown in Fig. 2.43. By increasing the gap width only the contribution C_{nn} is reduced while the serial connection of the two capacitances C_{pn} remains more or less unaffected. A more effective way of reducing the interpixel capacitance would be to interrupt the p-spray layer by introducing floating n^+-implants which would result in a serial connection of more than two capacitances C_{pn} [83]. However, these floating implants turned out to cause a severe loss of signal charge in the interpixel region [66, 138] and therefore this approach has not been followed any further.

All other considerations favor small gaps. As the most important requirement to n^+-in-n sensors is radiation hardness, devices have to be optimized for high-voltage operation. The electric fields have to be as small as possible. Therefore the potential difference between the p-spray layer and the pixel

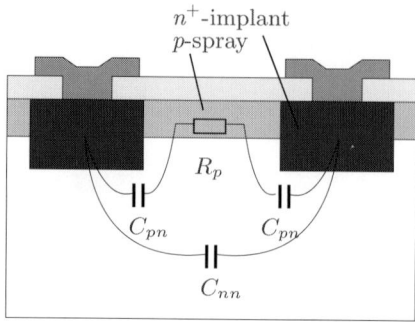

Fig. 2.43. Different components of the interpixel capacitance between two p-spray-isolated pixels

n^+-implant has to be limited as the lateral pn-junction is the most critical region in those devices in terms of electrical breakdown. Furthermore, a very high potential difference between the p-spray and the oxide surface could lead to a conducting inversion layer between pixels. The potential of the p-spray is determined by the largest distance between neighboring n^+-implants which appears in the whole device. These are usually the diagonal gaps in the regions where four pixels join. In order to limit the diagonal distance in the design shown in Fig. 2.32, the gap size between the short edges of the pixels has been reduced. This of course is possible only if the aspect ratio of the pixel cell is much larger than 1.

After irradiation induced type inversion of the bulk material and the subsequent increase of the effective doping concentration, the depletion region starts to grow from the segmented side. Now the same considerations concerning the depletion behavior can be applied as described in Sect. 2.5.2 for unirradiated p^+-in-n detectors also favoring small gaps. However, one cannot extract an exact statement on the largest possible gap and the use of the smallest gap compatible with noise requirements can be recommended.

One advantage of the p-spray technology is the possibility of realizing very small gaps between the n^+-implants. This feature can be used to implement a bias grid in the same way as for p^+-in-n pixels. In an unirradiated sensor all pixels are connected to each other and to the bias grid through the bulk, if the bias voltage applied is below full depletion as indicated in Fig. 2.30(d). If the full depletion voltage is exceeded, all n^+-implants are isolated by the p-spray layer and the leakage current of each pixel has to flow to the bias grid through the depleted bulk via thermionic emission (see Sect. 2.2.7). Due to the narrow spacing (5–6 μm should be possible in most processes) the potential difference between pixel and bias grid will not exceed some volts. Therefore an overdepletion of all pixels is possible and the most frequent failures, like scratches or spikes, can be detected electrically. After bump-bonding the bias grid is out of function as described in the p$^+$- in -n-case since the pixels are biased via the readout chip.

The implementations shown in Fig. 2.38 can be used unchanged in p-spray-isolated n$^+$-in-n sensors. The considerations concerning the requirements on the fabrication process and the bias grid's effect on the charge collection made in the context of p$^+$-in-n detectors are also valid here. The charge collection properties of the two different implementations have been studied in detail [138–140]. The total amount of charge collected versus the impact point of the particle is plotted in Fig. 2.44b for the design shown in Fig. 2.44a. The "hole" visible in the upper right corner corresponds to the implanted bias dot. However some signal charge reaches the preamplifier even if the dot is hit directly. For this reason the probability for detecting a minimum ionizing particle is still very high and was measured to be above 99% for an unirradiated device [139].

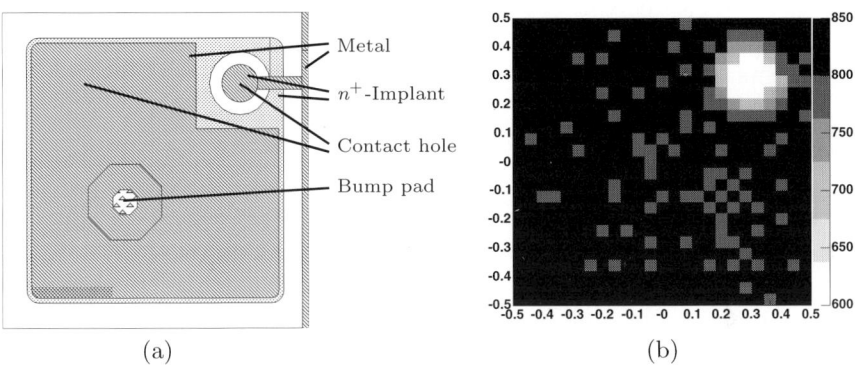

Fig. 2.44. (a) Design of a p-spray isolated pixel. The total pixel dimension is $125 \times 125 \,\mu m^2$. (b) Average of the collected signal charge (in arbitrary units) as a function of the impact position of the particle. The area shown represents one pixel cell. A signal loss of $\approx 30\%$ due to the bias grid is visible in the upper right corner [139]

Geometries for Devices with Moderated p-Spray

In devices featuring moderated p-spray one has to define the regions with the lower (moderated) and the full p-spray dose. In the design shown in Fig. 2.41 the horizontally hatched area marks the openings in the nitride layer where the full p-spray dose is implanted. A certain distance between the pixel's phosphorus implant and the nitride edge has to be kept because of alignment tolerances and to effectively reduce the electric field by the lower boron concentration. In the layout shown in Fig. 2.41 a distance of $5\,\mu m$ was chosen. The nitride opening of course also has a minimum feature size, limiting the minimum possible gap size. In Fig. 2.41 the gap is $20\,\mu m$ between

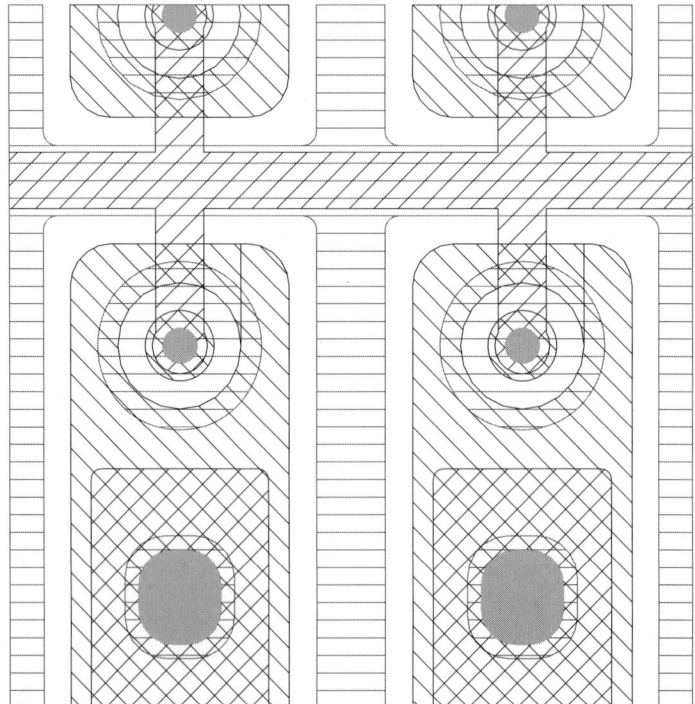

Fig. 2.45. Possible implementation of a bias grid using moderated p-spray isolation. The hatching used is the same as in Fig. 2.41

the long edges and 15 μm between the short edges of the pixels. The purpose of this asymmetry is to keep the diagonal distance small. This is possible only for pixel geometries with a large aspect ratio. For the same reason the corners of the nitride openings are not – or hardly – rounded. After irradiation when the fixed oxide charge has reached its saturation value, it might get close to compensating the moderated part of the boron implantation. This means that the "effective" size of the pixel implant is extended by an electron accumulation layer up to the edge of the nitride opening. So the gap size is larger at the beginning of the operation of the sensor when high-voltage stability is less important than low-noise operation. With irradiation the gap size will decrease, leading to lower electric fields in the irradiated state when high-voltage capability becomes most important [141].

Due to the limitations on the minimum gap between two neighboring implants the design of a bias grid becomes problematic. An implementation of the bias grid according to Fig. 2.38a is not advisable because the large gaps between the bias line and the pixel implant will lead to a large potential drop between them. However, the implementation shown in Fig. 2.38b can be translated into the moderated p-spray design as shown in Fig. 2.45. The gap between the pixel implant and the bias dot is filled completely with the full dose p-spray implant. This is possible because the conducting boron layers between the pixels and in the bias structure are electrically isolated and can have different potentials. In the region between the pixels the potential difference between the p-layer and the pixel implant is higher due to the larger gap but "buffered" by the 5 μm of moderated p-spray. The gap between bias dot and pixel is very small and therefore the potential difference between the p-layer and the pixel implant is much lower. A reduction of the p-spray dose in this area is not necessary. However, this option does not only require a better alignment but is also sensitive to pinholes below the horizontal metal line. Additionally the metal runs above a region where the nitride, which would provide additional safety, is opened. As it is still possible to produce such devices in large quantities with acceptable yield, this design was chosen for the ATLAS and parts of the CMS pixel detector.

Geometries for p-Stop Devices

When designing p-stop isolated devices most attention is paid to the geometry of the p-stops. In the context of strip detectors numerous geometries have been evaluated [142, 143]. In pixel devices "atoll"-like structures are preferred in order to avoid localized problems affecting the whole array [138]. Very good isolation is provided by p-stops; however, in some situations a highly resistive connection between the pixels is desirable, e.g., to perform *IV*-tests of the devices on wafer or to hold unconnected pixels close to ground. Because of the large gaps in p-stop devices the implementation of a punch-through bias grid is not possible. However, a resistor-like connection between neighboring pixels can be implemented by openings in the p-stop implants surrounding every pixel cell. These leave room for a conductive electron accumulation layer. The maximum value acceptable for those resistors is determined by the maximum acceptable voltage drop between pixels and the leakage current, leading to a maximum value of the order of 1 GΩ. The lower limit of the interpixel resistance is given by the requirement not to have a significant signal distribution to the neighbor channels within a typical shaping time of 25 ns. This leads to a lower limit of the resistance of about 1 MΩ. The value of the resistance is dependent on the length and width of resistive path. Different possible resistor geometries have been evaluated [85, 86, 130, 144]. One of the many possibilities is shown in Fig. 2.46a. The pattern of the resistors formed by the electron accumulation layers is sketched in Fig. 2.46b. Each pixel node is attached to an overall "resistive network."

The value of the resistance depends very strongly on the backside voltage as shown in Fig. 2.47. At low voltages the current flows mainly through the undepleted bulk. When the space charge region reaches the n-side, the pixels are separated from each other and the current has to pass through the electron accumulation layer forming the resistor. This leads to an increase of the resistance at bias voltages above 50–60 V (see Fig. 2.47). For devices

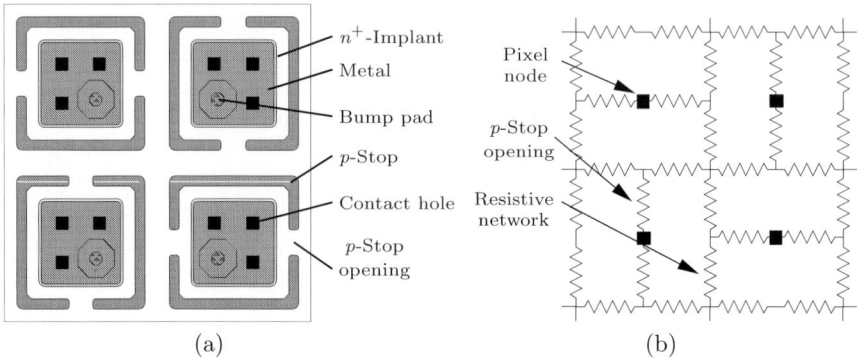

Fig. 2.46. Layout of a pixel with open p-stops (**a**) and the corresponding resistor pattern (**b**) [139]

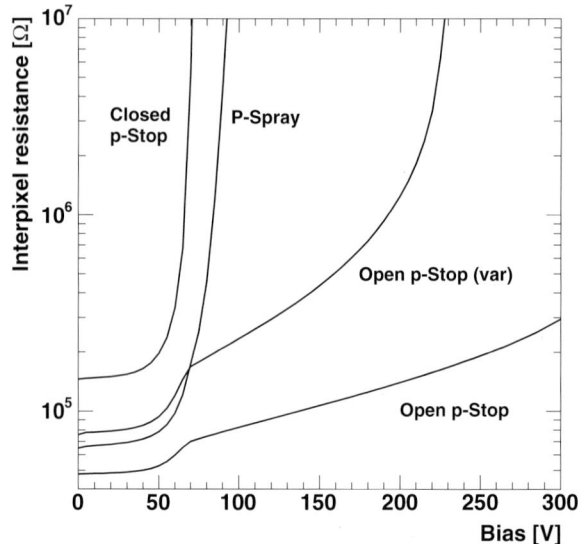

Fig. 2.47. Bias dependence of the interpixel resistance. The design shown in Fig. 2.46a is labeled "open p-stop" [139]

without a resistive connection between pixels (labeled "closed p-stop" and "p-spray") the resistance increases to values far above $1\,\mathrm{G}\Omega$. In the sensors featuring p-stop openings the resistance first grows slowly at further bias increase before increasing very steeply above 200 V for the geometry labeled "p-stop (var)." This is an indication of pinch-off, since with further overdepletion the potential difference between p-stops and n^+-implants increases and a field is growing from the pn-junction between p-stops and bulk also in the lateral direction. The width and height of the "plateau" region between 60 and 200 V depend strongly on the p-stop geometry. In the "open p-stop" design (Fig. 2.46a) it continues without pinch-off up to values much above 300 V. Consequently, it is possible to overdeplete all pixels using two probe needles, one placed on the p-side diode and the other on the n-side guard ring. This allows for a reliable sensor testing.

After irradiation the interpixel resistance saturates at values of several gigaohms almost independently of the design as shown in Fig. 2.48. Although unconnected pixels float to high potentials with respect to their neighbors, no harmful effects have so far been observed. There is no correlation between unconnected pixels (e.g. due to missing bump bonds) and noisy pixels in irradiated samples [85, 86].

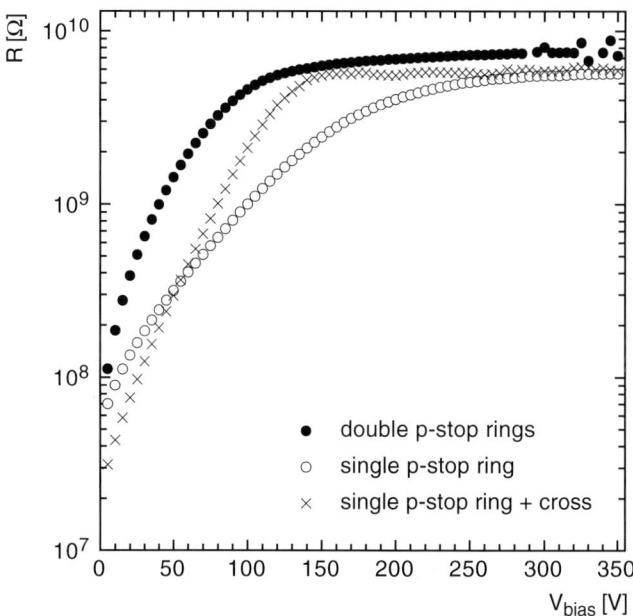

Fig. 2.48. Interpixel resistance of the design shown in Fig. 2.46 after a fluence of $\Phi = 6 \times 10^{14}$ n_{eq}/cm^2 [85, 86]. Resistance of other p-stop insulation designs are also shown

Breakdown in p-stop detectors often occurs at voltages below the several hundred volts required for operation at LHC [145]. At the same time the number of noisy channels starts to increase to unacceptable levels [86, 146]. This is probably due to avalanche breakdown at the p-stop edges and has also been observed in strip detectors [147]. Among the improvements proposed the field plates tested in the context of strip detectors [148, 149] do not appear to

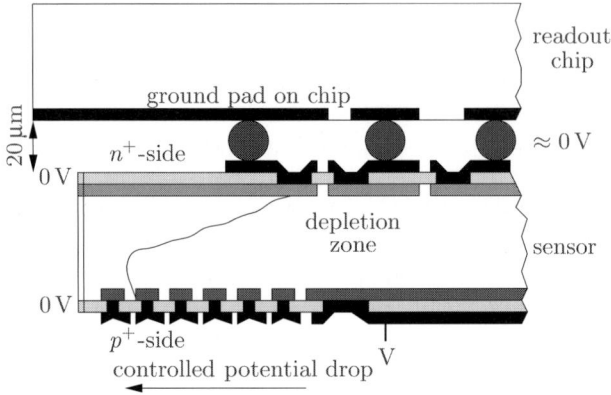

Fig. 2.49. Guard ring concept of an n^+-in-n sensor. The double-sided process is utilized to ground all sensor edges

be suitable for pixels due to lack of space. It has, however, been shown that a reduction of the p-stop implantation dose [86] leads to nice *IV*-curves [139]. The main problem of pixel sensors with open p-stops is that the electron accumulation layer adjusts to the same potential as the n^+-implant and consequently collects signal charge. It seems that in irradiated sensors part of this signal cannot reach the readout node in time, leading to a significant loss of efficiency [139].

2.5.3.3 Guard Rings and p-Side

The p^+-side of the sensor consists of a large area boron implant providing the pn-junction that depletes the detector. This junction must not extend to the scribe line where it would be shorted by the damaged crystal edge. Therefore the p^+-side has also to be segmented requiring a fully double-sided processing of the sensors. This necessary process can also be used to implement the whole guard ring structure on the p^+-side as sketched in Fig. 2.49. The outer region of the n^+-side is covered with a large area n^+-implant connected to a ground pin on the readout chip via bump bonds. The purpose and design of p^+-side guard rings are discussed in Sect. 2.5.2. The main difference is that in n^+-in-n sensors the p^+-side is (negatively) biased and the n^+-side which defines the potential of all sensor edges is grounded. This is an important feature for the module design as one has not to worry about the small air gap between the sensor edge and the readout chip and possible sparking in between.

The design of the large area diode is uncritical. It seems reasonable to extend the p^+-implant some $10\,\mu$m above the sensitive area to provide a homogeneous field also for the edge pixels. The best protection from any mechanical damage is necessary as the sensor is laid on the p^+-side during bump bonding and flip chipping. Scratches reaching the bulk in the diode area will destroy the device. Therefore in most detectors the diode is covered with all layers available: oxide, nitride (if available), aluminum, and passivation (in most cases nitride or oxide). To enable laser studies after flip-chip small holes should be opened in the metallization of every pixel. In small size test sensors where yield is not critical, large openings in the aluminization are very useful for detailed tests with lasers. Contact openings between the aluminum and the p^+-implant can be restricted to small areas, e.g. contact "strips" along the edges. To what extent these design choices improve the yield of large area devices has not been studied so far. However, these guidelines can be followed easily as they influence neither costs nor performance.

2.5.3.4 Limit of Radiation Hardness

The limit of radiation hardness is given by the tolerable reduction of the height of the signal. As this degradation is a steady process it is not possible to give a strict rule when a sensor can be considered unusable. For the readout electronics used in ATLAS and CMS typically a signal of about 5,000 electrons per m.i.p. is required to ensure the particle detection with an efficiency larger than 95%.

The reduction of the signal charge is caused by the increase of the effective space charge and by charge trapping. The former can be compensated up to a certain level by an increase of the operation voltage. Here the limit is set more by practical considerations as it is in principle possible to build detectors with a high-voltage capability exceeding 1 kV. Trapping can be reduced by collecting electrons which are less prone to trapping than holes and by high bias voltages, resulting in short collection times.

For the LHC experiments a radiation hardness up to a fluence of $1 \times 10^{15}\,n_{eq}/cm^2$ has been targeted and reached. In future very high luminosity colliders or a possible LHC upgrade to a luminosity of $10^{35}\,cm^{-2}\,s^{-1}$ will require a radiation hardness of the tracking devices placed at radii below 10 cm up to fluences of $10^{16}\,n_{eq}/cm^2$. Several approaches to improve radiation hardness of silicon devices are followed [150]. The change of the material properties due to irradiation can be influenced by the so-called *defect engineering*. This means that the crystal defects leading to the degradation of performance have to be understood at a microscopic level. Enriching the silicon bulk material with certain impurities, e.g. oxygen, can influence the formation of crystal defects and alter the postradiation material properties. Another approach consists in changing the operating conditions: for example the operating temperature [151] or the polarity of the operation voltage [152]. Other device structures as for example so-called three-dimensional detectors [153] or ultrathin detectors are being considered to minimize the drift path of the signal charge to reduce trapping and to reduce the voltage needed for full depletion as discussed in Sect. 6.2.3. Investigation on other materials like diamond [154] or compound semiconductors which may behave better than silicon under irradiation is also ongoing and will be mentioned in Sect. 2.7.

2.6 Processing of Silicon Wafers

The fabrication technology for silicon detectors is to a high degree derived from the planar technology developed in the field of microelectronics and therefore profits from the large investments made in its technological development. The minimum feature size in silicon detectors of typically some micrometers is quite coarse compared to the submicrometer scales reached in microelectronics today. However, there are special and challenging requirements in sensor processes:

- High purity of detector grade silicon is needed.
- Both sides of the wafer are important for processing.
- Large devices without defects are required.

The conservation of the high purity of the detector grade silicon during the high-temperature steps was considered the most difficult problem in the first attempts of building junction-depleted silicon detectors [155]. Therefore any contamination of the wafer surfaces and the quartz tube in the furnace has to be avoided before and during the high-temperature steps. Furthermore it is useful to minimize the number of high-temperature steps. The unavoidable high-temperature steps should be performed at the lowest temperature possible and as early in the process chain as possible.

In microelectronics all electrical activity takes place in a very thin layer of a few micrometers close to the devices' surface. In silicon sensors the depletion layer extends up to the sensors backside contact. This means that the backside has to be treated with special care also in detectors which do not require a structured backside. Furthermore, at least two processes (backside implantation and metallization) have to be performed on the backside, requiring a special protection of the segmented front side of the sensor. As scratches on either side can destroy the whole device special chucks and wafer handling equipment are necessary.

Last but not least, it has to be mentioned that silicon detectors may have an area ranging from $\approx 10\,\text{cm}^2$ (pixel detectors) up to $\approx 100\,\text{cm}^2$ (strip detectors). The production of such devices with a reasonable yield requires both a careful optimization of the process and a fault-tolerant sensor design.

It is much beyond this scope of this book to discuss the details of semiconductor processing but some basic technology steps will be explained at the level important for the designer. For a detailed presentation one of the numerous textbooks [42, 44] can be consulted.

2.6.1 Production and Cleaning of Silicon

Silicon is isolated from quartzite, which is a relatively pure form of sand (SiO_2), by a reduction with carbon at temperatures above 1400°C. The resulting solid silicon of about 98% purity is treated with hydrochloric acid (HCl) at roughly 300°C to form trichlorosilane ($SiHCl_3$), which has a boiling temperature (32°C) lower than that of the chlorine compounds of unwanted impurities, which allows a distillation. The cleaned trichlorosilane is then transformed back to solid silicon and hydrogen chloride. By this method an impurity concentration of better than 10^{-9} can be reached. This material can be used as starting material for growing large single crystals. Two growing methods are common.

In the *Czochralski* process the silicon is melted and kept together with the desired concentration of dopants in a crucible made of SiO_2. A seed crystal is pulled slowly under rotation from the surface of the liquid as shown in Fig. 2.50. The silicon freezes out at the surface, resulting in a monocrystalline ingot. The diameter of the ingot can be adjusted by the pulling speed. Czochralski-grown silicon contains many unwanted impurities, mostly oxygen atoms originating from the walls of the crucible with a concentration in the order of 10^{18} cm^{-3}. It is therefore usually not used for applications requiring a silicon substrate with a resistivity of more than $10\,\Omega\,\mathrm{cm}$.

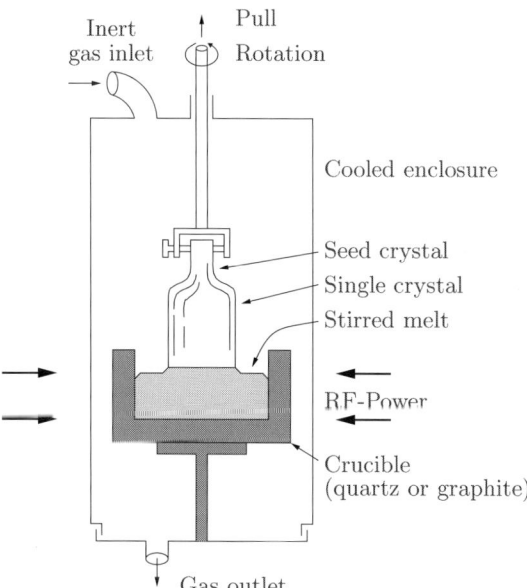

Fig. 2.50. Schematic diagram of a Czochralski crystal puller. The silicon melt is contained in a quartz crucible. A seed crystal is dipped into the melt and pulled out slowly. The diameter of the ingot is defined by the pulling speed

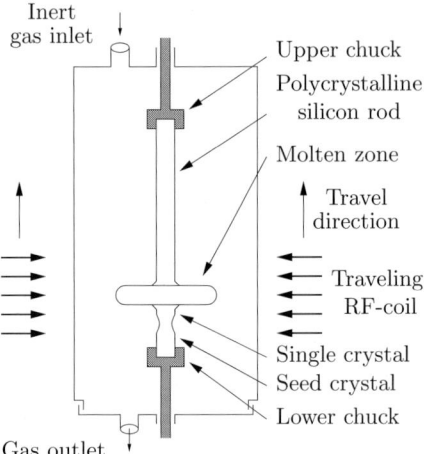

Fig. 2.51. Schematic diagram of a float zone apparatus. A rod of polycrystalline silicon is mounted with a seed crystal between two chucks. A narrow part of the rod is melted by a radio-frequency heater. The heater then moves and the silicon freezes out as a single crystal

For detectors a resistivity of above $1\,\mathrm{k}\Omega\,\mathrm{cm}$ is needed which can be produced using the *float zone* method. A high-purity polysilicon rod is vertically mounted with a seed crystal at the bottom in a quartz envelope evacuated or filled with an inert atmosphere (e.g. argon). A small zone of the rod is melted by a radio-frequency heater. This *float zone* is then moved from the seed crystal through the whole crystal by moving the heater slowly upward (see Fig. 2.51). At the end of the float zone the silicon freezes to a single crystal of the same orientation as the seed. As most impurities are better solvable in the melt than in the crystal they are driven toward the end of the ingot. For a further purification this procedure can be repeated several times up to the desired level.

If a certain substrate doping is targeted, it can be mixed into the atmosphere (e.g. PH_3 or B_2H_6). In the molten zone the gas decomposes because of the heat and the dopant solved in the melt. The substrate doping obtained by this technique shows lateral variations of typically 20–30 % in a 100-mm wafer. For a very high resistive material (above $5\,\mathrm{k}\Omega\,\mathrm{cm}$) the variations are higher. This is tolerable in the pn-diode-like pixel detectors discussed in this chapter. However, devices like silicon drift detectors require a better homogeneity which can be achieved by *neutron transmutation*. Very clean silicon is irradiated with thermal neutrons which are caught by the silicon nucleus to $^{31}_{14}\mathrm{Si}$. This decays with a half-life of 2.6 h in a β^- decay to $^{31}_{15}\mathrm{P}$. As the penetration depth of thermal neutron in silicon is in the order of 1 m the resulting phosphorus doping is very homogeneous.

The single crystal ingots are sawed into wafers whose surfaces are lapped and polished. In this form they are bought by the sensor processing companies.

2.6.2 Thermal Oxidation

The first step in most sensor production processes is to protect the wafer surface with a thin (about 150 nm to 2 µm) layer of SiO_2. This isolating film also provides a good termination of the silicon crystal. The properties of this interface have been studied in great detail in the context of MOS electronics as already mentioned in Sect. 2.2.6.

Silicon oxide is grown by storing the wafer at a temperature between 900 and 1200°C in an oxygen atmosphere. The oxygen from the atmosphere reacts with the silicon to form the oxide. If water vapor is added to the oxide atmosphere, the oxidation process is called *wet* oxidation and reaches growth rates which exceed the ones obtained in *dry* oxidation by about a factor of 5. However, slowly grown dry oxides usually show a superior quality in breakdown stability, pinhole density, interface states, and surface charges. Very thin gate oxides in MOS electronics, e.g., are usually grown dry at very "low" temperatures (around 800°C). The choice of the method is a compromise between cost and performance.

The addition of chlorine during oxidation (in the form of HCl or trichloroethylene) is useful in various aspects. Most important is the formation of volatile compounds with unwanted impurities especially sodium.

The silicon itself is consumed by the oxidation process and the interface between the oxide and the silicon moves about 44% of the oxide thickness into the silicon substrate. As the dopants have a different solubility in the oxide than in the silicon the surface doping concentration is affected by the thermal oxidation. Phosphorus and arsenic are about 10 times less dissolvable in the oxide than in the silicon and therefore accumulate at the surface. The opposite is true for boron which possesses a roughly three times higher solubility in the oxide. Its doping concentration close to the interface is reduced by the thermal oxidation. These *segregation* effects are especially important for the adjustment and fine-tuning of the threshold voltage of MOS transistors. In sensor production segregation is usually unimportant but has to be kept in mind when setting up the process.

2.6.3 Layer Deposition

Silicon dioxide and other materials (e.g. polysilicon and silicon nitride) can also be deposited onto the wafer from the gas phase. The process gases are led into the furnace and react on the surface of the wafer (and all other surfaces).

Chemical vapor deposited (CVD) oxides do not reach the good electrical properties of thermally grown oxides. They are used as covering insulation layers, e.g. on top of the metallization. A low temperature oxide suited to cover aluminum (melting point: 660°C) can be obtained with silane (SiH_4) and oxygen (or nitric oxides like "laughing gas," N_2O). An electrically more stable oxide can be obtained with tetraethylorthosilicate ($Si(OC_2H_5)_4$) which decomposes at about 700°C and forms a silicon oxide layer. Due to this high temperature it cannot be used to cover aluminum and is mainly used to cover polysilicon layers.

In silicon strip detectors capacitors composed of a sandwich of thermal oxide and silicon nitride are commonly used. Nitride (Si_3N_4) can be deposited with dichlorosilane ($SiCl_2H_2$) and ammonia (NH_3) at about 750°C. As the production process for pixel sensors is often derived from strip detectors, nitride layers are also used in pixel sensors, especially in connection with the moderated p-spray isolation technique (see Sect. 2.5.3).

Polysilicon is used in bias resistors of strip detectors (high resistive) or as gate material in MOS transistors or charge coupled devices (low resistive). It can be deposited by pyrolyzing silane at about 600°C. As pixel detectors are DC-coupled, there is no need for a polysilicon layer.

2.6.4 Photolithographic Steps

In order to structure the wafer surface photolithography is used. For that purpose photoresist is spun onto the wafer surface. This is done by mounting the wafer onto a rotating chuck. The resin is dropped in the center of the wafer and distributed by its fast rotation. A typical thickness is about 1 µm, but for lift-off processes or high-energy implantations a higher thickness of several micrometers is required. The resist has then to be dried by a prebake step around 100°C.

The structure is copied by exposing the photosensitive resist through a mask, usually a chrome pattern on a glass substrate. For a good spatial resolution a short wavelength in the UV range is used. The mask can be pressed onto the wafer (contact exposure) resulting in a good resolution, but including the risk to pollute or even damage the mask. In sensor fabrication where no resolution in the submicrometer range is required the proximity illumination with a well-defined distance between the mask and the wafer of about 10–20 µm can be used instead. Projective exposure in which the pattern is transferred to the wafer by an optical system also leads to very good results but requires large investments.

During the development of the resist either the illuminated areas (positive resist, more common) or the nonilluminated areas (negative resist) are removed and therefore exposed to the following process step, either etching or implantation.

The alignment of the mask with respect to a previous photolithographic step is done via special alignment marks. As the sensors might be structured on both sides an alignment of both sides with respect to each other is necessary. This is done either by looking through the wafer with infrared light or aligning the wafer with respect to a reference image. An alignment precision of better than 10 μm can easily be achieved.

2.6.5 Etching

Etching is used to copy the structure of the photoresist into the underlying layers. The two most important properties of the process are its isotropy and its selectivity. In an isotropic etching process the material is removed in all directions. This leads to an underetching of the mask of the order of the resist thickness as illustrated in Fig. 2.52. This fact has to be taken into account in the mask design. The selectivity describes to what extent the process removes only the material desired and stops at the underlying layer. For example hydrofluoric acid (HF) etches SiO_2 about 100 times faster than silicon and is therefore commonly used to open contact holes in the oxide.

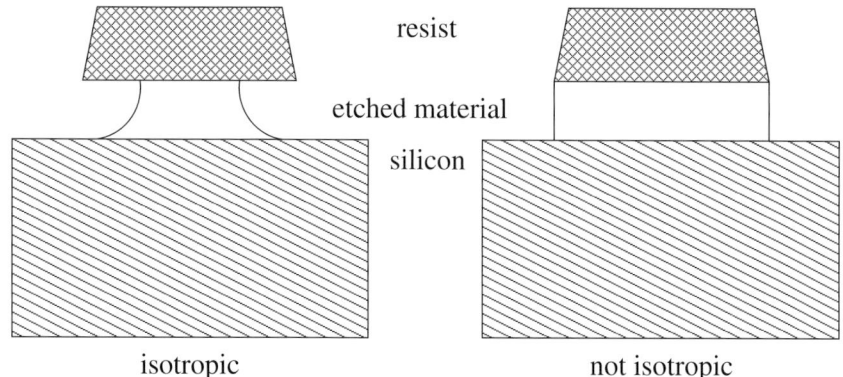

Fig. 2.52. Profiles after isotropic and anisotropic etching process

Wet etching is most commonly used in sensor processing. It requires lower investments than other methods, and allows a very high selectivity and therefore a high grade of process safety. The underetching due to the isotropy of most wet chemical etching processes is not a problem due to the relatively coarse minimum feature size in sensors.

Dry plasma etching, the most widely spread etching method in microelectronics, provides a high degree of anisotropy and therefore allows small structures but lower selectivity. However, the plasma also induces radiation damage in the oxides, another reason for its rare occurrence in sensor technology.

2.6.6 Doping

Diffusion and implantation are the two methods to locally introduce dopants into the silicon's surface. Boron is used for p-type doping, and phosphorus and arsenic for n-type.

2.6.6.1 Diffusion

During diffusion the dopants enter the wafer at a temperature of typically 800–1200°C at open surfaces. As photoresist cannot be heated to such high temperatures the areas which should not be doped have to be protected with an oxide layer. The wafers are placed in a furnace rinsed with an inert carrier gas (N_2, Ar). To this gas the dopant is added either directly as gas (B_2H_6, PH_3, AsH_3) or by leading it through a bubbler containing a liquid form of the dopant (e.g. BBr_3, $POCl_3$). Due to the gradient of the concentration the dopant diffuses into the silicon, leading to a doping profile with its maximum at the surface. The depth of the profile depends on the duration, temperature, and surface concentration and can reach several micrometers. The dopant also diffuses laterally in the silicon. This underdiffusion is about 0.8 times the depth of the profile. It has to be mentioned that the handling of the doping gases is delicate as they are flammable as well as toxic.

A diffusion process very particular for radiation sensors is the production of diffusion oxygenated float zone silicon (DOFZ) developed by the ROSE collaboration [100, 108]. The aim is to obtain a uniform oxygen concentration of the order of several 10^{17} cm^{-3} which slows down the degradation of the material parameters under irradiation as mentioned in Sect. 2.4.1.4. This is achieved by growing a thick (several micrometer) thermal oxide layer on the wafer surface and storing the wafers for a long time (up to 48 h) above 1100°C in an inert atmosphere.

2.6.6.2 Implantation

In the implantation the doping atoms are ionized, accelerated, and shot directly into the silicon wafer. Although this method requires an expensive equipment, it is widely used for the reasons discussed below.

Implantation is performed at room temperature allowing photoresist to be used for masking the areas that are not doped. Its thickness has to be adjusted to the penetration depth of the ions. In the case of self-aligning processes, an implantation can also be masked by a polysilicon or oxide layer.

The implantation dose can be precisely measured, which is important for the reproducibility of the process. The penetration depth of the ions and hence the shape of the doping profile can be adjusted by choosing the energy of the implantation ions. The range of the most common doping elements boron, phosphorus, and arsenic is plotted in Fig. 2.53. The given depth does not take into account the crystal structure of the silicon. The periodical

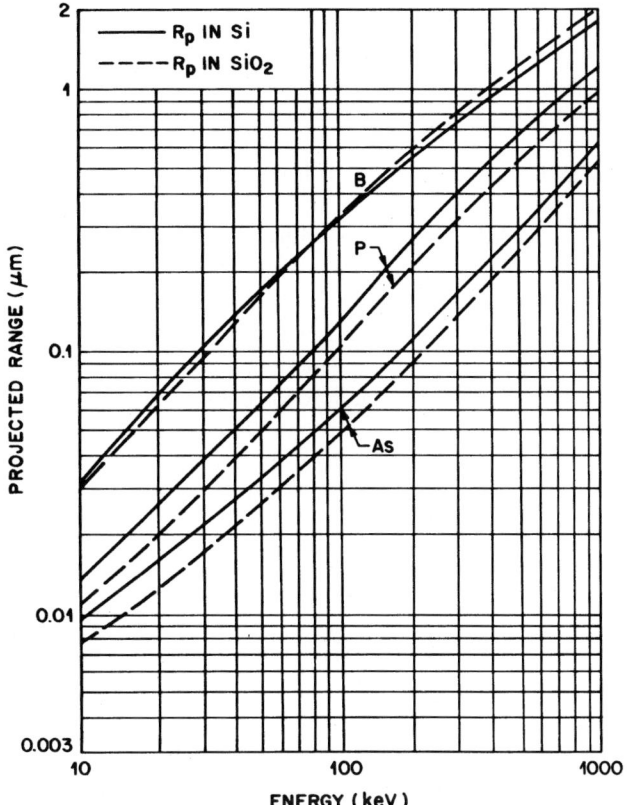

Fig. 2.53. Penetration depth of typical dopants in (amorphous) silicon and silicon oxide [156]

placement of silicon atoms leads to channels into which the doping atoms can be shot. The range of doping atoms following such a channel can be much larger. In order to prevent channeling the wafers are tilted by about 7° with respect to the $\langle 1,0,0 \rangle$-plane of the crystal. For typical energies the ion range is below 1 µm.

When the doping atoms are stopped they are usually not on regular places in the crystal lattice and are not electrically active. In addition the crystal is damaged by the implantation. Very high dose implantation, like the source-drain implant of MOS transistors, can even lead to a total amorphization of the doped region. Therefore every implantation is followed by a thermal treatment for dopant activation and annealing. This annealing also results in a diffusion of the dopants and can be used for "driving" in too shallow junctions.

2.6.7 Metallization

The metallization is used to provide a low resistivity connection between several devices on the same silicon substrate or to form bond pads. The most common metal used is aluminum because of its low resistance and its good adhesion on silicon oxide. A typical thickness is 1 µm. Aluminum can either be evaporated or sputtered. A good ohmic contact (without Schottky barrier) to the silicon is provided only if the silicon is highly doped at the contact areas. Furthermore a sintering process is necessary.

Silicon is very well dissolvable in aluminum with a saturation concentration of a few percent. There is therefore the risk that the silicon might dissolve into the aluminum during annealing and aluminum spikes might grow into the silicon and eventually short shallow junctions. This is especially the case if large aluminum areas are connected to the silicon by a small contact via. Spiking is more likely in $\langle 1, 0, 0 \rangle$ silicon and therefore $\langle 1, 1, 1 \rangle$ material was originally preferred for silicon sensors. Nowadays the targets used for aluminum sputtering are already saturated silicon suppressing spiking to a high degree.

2.6.8 Example of a Process Sequence

In this section a process sequence for the production of a simple p^+-in-n pixel sensor (see Sect. 2.5.2) is discussed. This particular process was originally developed for strip detectors [157] and is schematically shown in Fig. 2.54. It contains only five photolithographic steps including the passivation. The nitride could in principle be omitted reducing the number of mask steps to four. Of course this sequence represents only one of many possibilities to manufacture a pixel sensor. Each vendor adjusts its process to the given equipment and to other (often more important) products produced in the same line.

A special feature of the sample process presented is that there is only one thermal oxidation at the very beginning. The advantage is that the quality of this oxide is very high. On the other hand this implies that the doping has to be done through the oxide by implantation. This also limits the maximum thickness of the oxide as the energy of the implantation is limited. For this reason some producers, which prefer a thicker (about 1 µm) oxide, remove this layer at the doped area and regrow an oxide after implantation or diffusion.

2.6.8.1 Thermal Oxidation

After cleaning the wafers and removing all natural oxide layers with hydrofluoric acid (HF) a high-quality thermal oxide is grown. The thickness has to be small enough to allow implantation and large enough to display the necessary electrical strength (Fig. 2.54-1).

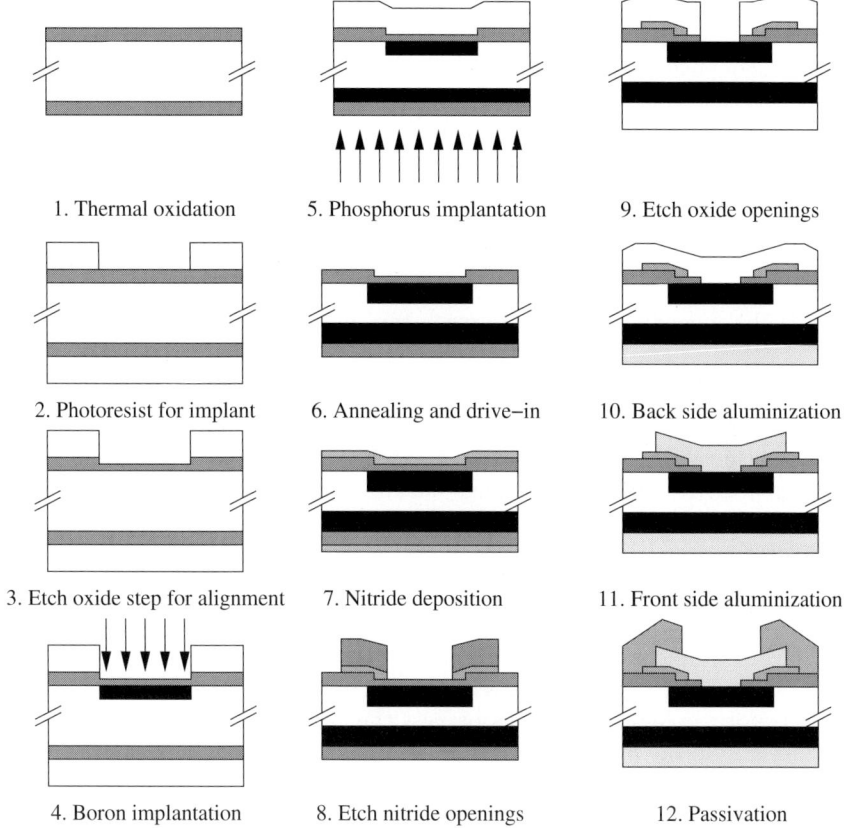

Fig. 2.54. Possible process flow for the production of a p^+-in-n sensor. The dimensions are not to scale

2.6.8.2 p-side Implantation

Photoresist is spun onto both surfaces. On the backside the resist is only a protection, while the front side is exposed with a mask (Fig. 2.54-2). With this mask a small step is etched into the oxide. This step is easily visible and is used to align all further masks (Fig. 2.54-3). Then the boron implant forming the pixel's pn-junction is performed through the same mask (Fig. 2.54-4).

2.6.8.3 n-Side Implantation

Then all photoresist is removed and a new resin is applied on the front side to protect the surface during the phosphorus implantation which forms the ohmic backside contact (Fig. 2.54-5). After removing all resin again the implants are annealed and driven-in by a high-temperature step under inert atmosphere (Fig. 2.54-6).

2.6.8.4 Nitride Deposition and Structuring

Nitride is deposited at the same time on both sides of the wafer (Fig. 2.54-7). For structuring one can use either a plasma etching process or phosphoric acid. In order to avoid radiation damage to the oxide most vendors use a wet etching process for opening the nitride. As the photoresist will be etched by phosphoric acid an additional layer of low temperature oxide (which is not etched by phosphoric acid) is deposited on top of the nitride (not shown in Fig. 2.54). This oxide is conventionally structured with photoresist exposed through a mask and etched in an HF bath. Then the nitride can be wet chemically opened and the oxide mask can be removed.

2.6.8.5 Contact Holes in the Oxide

Now the contact holes in the oxide are etched with a photolithographic step and HF (Fig. 2.54-9). The openings have to be smaller than the openings in the nitride which is not etched by the HF.

2.6.8.6 Metallization

As (in this sample process) the aluminum is sputtered onto the wafer, the deposition has to be done successively on both sides. As the temperature of the wafer is not too high during this process the "unused" side can be protected with resist. On the front side the aluminum is wet chemically structured. For obtaining a good electrical contact to the silicon a sintering step is necessary. In order to reduce the surface currents this step can be done in hydrogen atmosphere.

2.6.8.7 Passivation

The passivation is a protection against mechanical damage and chemical contamination. Most robust and dense is silicon nitride. However it is brittle and causes mechanical stress. Therefore some vendors use polyimide or low temperature silicon oxide. For pixel detectors which will be bump-bonded to the readout electronics, the passivation is a crucial step. Some bump-bonding processes require a very dense passivation to protect the sensor from the chemicals involved in the process. Further the adhesion of the bump metals has to be guaranteed (Fig. 2.54-12).

In this state the sensors are delivered to the customers. After an electrical test on wafer level the bumps are deposited. Then the sensors can be cut out of the wafers and be fed into the module assembly procedure.

2.7 Detector Materials Other Than Silicon

For most applications silicon is the material best suited for particle or radiation detection. The advantages of silicon are as follows:

(a) Its relatively low density and low radiation length X_0 which is particularly important for particle physics while for X-ray detectors this constitutes a disadvantage.
(b) The fact that large silicon crystals can be grown as a substrate material without large inhomogeneities or impurities resulting in large charge carrier lifetimes of $\approx 100\,\mu s$.
(c) The fact that silicon can be doped with specifically tailored doping profiles allowing to shape the potential inside the detector for best charge collection.
(d) Its low price and good availability. Its industrial fabrication and processing has been matured over decades.

Other materials than silicon become attractive as detectors when different requirements overrule these advantages. Table 2.2 lists some basic properties of various semiconductors also including some potentially interesting materials which are at present not yet used as detectors.

Table 2.2. Characteristic properties of some semiconductors

Semi conductor	Density (g/cm^3)	Band gap (eV)	Intrinsic carrier concentration (cm^{-3})	Average Z	$w_{e,h}$ (eV)	Mobility $(cm^2/(Vs))$ e	h	Carrier lifetime
Si	2.3	1.12	1.45×10^{10}	14	3.61	1,415	480	$\sim 250\,\mu s$
Ge	5.3	0.66	2.4×10^{13}	32	2.96	3,900	1,800	$250\,\mu s$
GaAs	5.4	1.42	1.8×10^6	32	4.35	8,800	320	1–10 ns
CdTe	6.1	1.44	10^7	50	4.43	1,050	100	0.1–2 μs
CdZnTe	5.8	~ 1.6	10^7	49.1	4.6	$\sim 1,000$	50–80	$\sim \mu s$
CdS	4.8	2.42		48/16	6.3	340	50	
HgI$_2$	6.3	2.13		62	4.2	100	4	$\sim \mu s$
InAs	5.7	0.36		49/33		33,000	460	
InP	4.8	1.35	1.3×10^7	49/15		4,600	150	
ZnS	4.1	3.68		30/16	8.23	165	5	
PbS	7.6	0.41		82/16		6,000	4,000	
Diamond	3.5	5.48	$<10^3$	6	13.1	1,800	1,200	~ 1 ns

Germanium is superior to silicon with respect to the obtainable energy resolution due to the small energy band gap of 0.7 eV. For the same reason it must however be cooled to very low temperatures to reduce thermally generated leakage current. As a detector, germanium has found applications in nuclear physics and for near infrared photon detection, and–due to its high photon absorption probability–also for X-ray measurements.

As position-sensitive devices, in particular pixel detectors, high-Z semiconductor compound materials are preferred over Ge, most notably GaAs and CdTe/CdZnTe which have been studied for high-energy-physics experiments (GaAs) and for medical imaging. Both materials are attractive for X-ray detection because of their high photoabsorption cross section ($\sigma_\text{photo} \sim Z^4$ for photon energies below 50 keV). Finally, as a very radiation hard detector material with low Z and low radiation length, diamond, which in fact classifies as an insulator, has been investigated in recent years. The following sections address pixel detector developments which make use of these materials.

2.7.1 Gallium Arsenide

Gallium arsenide, like all two-component semiconductors, has impurities in the order of 10^{15} cm^{-3} leading to a carrier lifetime of only 10 ns as compared to 250 µs in silicon. The intrinsic carrier density of undoped GaAs is already small, $n_\text{intr} \simeq 10^6$ cm^{-3}, and comparable to the carrier density in the depletion region of a typical silicon detector. Undoped GaAs therefore is – in the language of Si devices – already depleted. As a direct semiconductor with high electron mobility GaAs is widely used for very high speed electronics circuits and for photonics devices. As particle physics detectors, GaAs sensors have been studied in the late 1990s in the context of its potential radiation hardness due to the large band gap of 1.43 eV as compared to silicon, thus suppressing the reverse bias leakage current [158]. The high density of impurity states (Fig. 2.55), however, causes the mentioned problem of a short carrier lifetime and incomplete charge collection due to trapping.

The existence of the EL2$^+$ deep level mid-gap defects [159] (see Fig. 2.55), caused by arsenic atoms occupying gallium-lattice positions acting as donors,

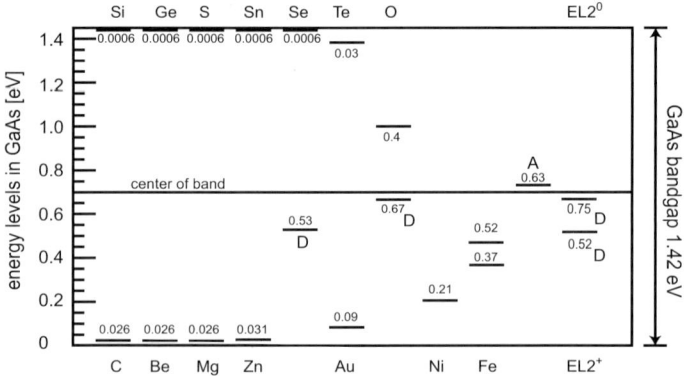

Fig. 2.55. Energy levels of various impurities in the band gap of GaAs. The EL2$^+$ levels act as deep donors (After [40])

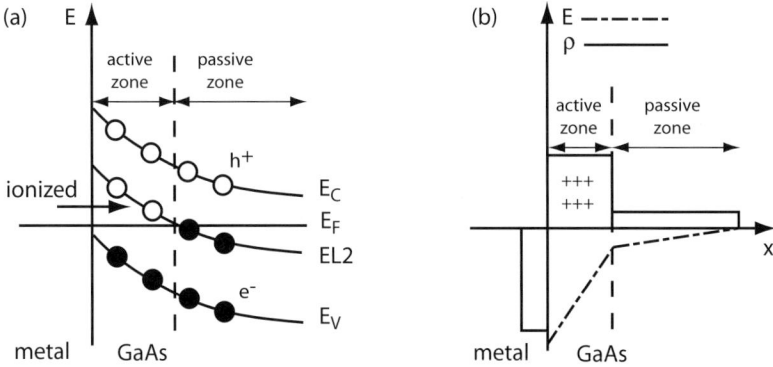

Fig. 2.56. (a) Simplified band diagram of a GaAs Schottky diode. (b) The resulting space charge density and electric field dependence inside the substrate material

introduces field regions of different strengths inside the substrate.[9] GaAs detectors are usually operated as Schottky devices. The upward bending of the energy bands results in regions where the EL2$^+$ levels lie above the Fermi level E_F and are thus ionized, and in regions where they lie below E_F. This is illustrated in Fig. 2.56. The resulting space charge density leads to high and low electric field regions [160], which in turn lead to high and low charge collection efficiencies or *active* and *dead* zones [161].

The quality of GaAs detectors is characterized by the so-called *charge collection efficiency* (CCE) defined as the collected charge on the electrode divided by the deposited charge, or alternatively by the *charge collection distance*

$$d_{CC} = \mu\tau E, \qquad (2.57)$$

where μ and τ are mobility and lifetime, respectively, and E is the applied electric field. Charge collection efficiencies of 90–100 % and charge collection distances of more than 250 µm have been reached [158].

GaAs has interested high energy physicist for reasons of its potential radiation resistance. The radiation hardness of semiconductors strongly depends on the type of radiation (electrons, gamma-rays, or hadrons) and its energy. The radiation-induced damage effects are parametrized by the so-called NIEL-factors (nonionizing energy loss in keV/cm) and thermally stable defects. For the use in high-energy-physics detectors at the LHC the abundance of energetic (>50 MeV) hadrons however renders GaAs (NIEL factor ≈20 keV/cm) less suited than oxygenated silicon (≈2 keV/cm), [102]. For gamma and electron irradiation, and for low-energy protons or neutrons, GaAs is in general more radiation-resistant than Si and hence is often used in military applications.

---

[9] The opposite defect (Ga atom taking an As lattice place) acts as a shallow acceptor

Hybrid GaAs pixel detectors have also been developed in the context of X-ray detection with counting pixel electronics [162, 163] (see also Chap. 5). A 300-μm-thick GaAs sensor with $433 \times 50\,\mu\text{m}^2$ pixels was bump-bonded to a counting pixel chip [164] with hit counting in every pixel cell. The measured CCE was about 80%.

2.7.2 CdTe and CdZnTe

The II–VI compound semiconductors CdTe and $\text{Cd}_{1-x}\text{Zn}_x\text{Te}$ have been studied and used for radiation detection mainly because of their high Z ($Z_{\text{Cd}} = 48$, $Z_{\text{Te}} = 52$, $Z_{\text{Zn}} = 30$) and correspondingly large photon absorption cross section and good energy resolution. The band gap increases from 1.44 eV for pure CdTe to 2.2 eV for ZnTe. In use are pure CdTe detectors operated as ohmic and also as Schottky devices and ohmic CdZnTe detectors with the Zn fraction x between 0.08 and 0.15 which corresponds to a band gap of ≈1.6 eV. The number of produced electron–hole pairs in CdTe is comparatively large, about 22,000 from a 100-keV energy input (e.g. from a γ-ray). The theoretical energy resolution without electronic noise is thus about 500 eV (at an energy of 100 keV) assuming a Fano factor of 0.15. A feature inherent to CdTe and CdZnTe detectors is the low hole mobility (100 cm^2/Vs) and the small mobility–lifetime product for both carrier types:

$$(\mu\tau)_e \approx 10^{-3}\,\text{cm}^2/\text{V}, \tag{2.58a}$$

$$(\mu\tau)_h \approx 10^{-4}\,\text{cm}^2/\text{V}. \tag{2.58b}$$

The intrinsic carrier concentration n_{intr} is of the order of 10^7, i.e. not too different from that of GaAs. Depletion is thus obtained without additional doping; however, chlorine as a dopant with donor character is used to compensate crystal defects by forming Cl-vacancy (Cd) complexes. While in GaAs the charge collection behavior is largely influenced by the EL2-ionization leading to high and low electric field regions in the substrate, in CdTe the shape of the electric field depends on the filling state of traps and hence the time the sensor is exposed to radiation. The nonuniform shape of the field also depends on the choice of contacts, i.e. operation as ohmic or Schottky devices. CdTe and CdZnTe detectors may suffer from incomplete charge collection [165]. In particular the poor hole charge transport results in position-dependent charge collection properties and thus in tails in the measured energy spectra. Another sometimes reported deficiency is the poor homogeneity of CdTe devices [166].

A fairly recent development [167, 168] is the fabrication of CdTe Schottky devices, using chlorine-doped CdTe (p-type) with a low work function metal such as indium as anode material. This device has lower leakage currents than ohmic detectors and allows higher bias voltages to be applied, leading to almost full charge collection [169]. The specific resistance of Schottky CdTe devices is typically in the order of $10^{10}\,\Omega\,\text{cm}$. Schottky-contacted devices suffer however from polarization effects [167], regions of negative space charge which change the electric field inside the sensor. Polarization effects have been observed in chlorine-doped CdTe:Cl Schottky devices [167] and are hence most likely due to charge injection at the Schottky contacts, leading to a time-dependent field distribution inside the device.

Pixel detectors with CdTe substrate material are being developed for X-ray astronomy [169, 170] and as X-ray photon counting devices [171, 172] (see Chap. 5). The bumping process employed is Au-stud bumping. Complete four-chip modules with dimensions of 1.3 cm × 1.3 cm have been used for imaging.

2.7.3 Diamond

Carbon, crystallized in the diamond lattice, is in fact an excellent insulator with a resistivity of 10^{13}–$10^{16}\,\Omega$ m owing to its large band gap of 5.5 ev. Nevertheless, the fabrication of artificial diamond, first as polycrystalline CVD devices [173–177] and recently also as mono-crystals [178], has initiated R&D to use this material for particle detection [179]. The reason for this interest is the apparent radiation hardness of diamond even for radiation doses in excess of those experienced at LHC. The present limits after which a decrease in the charge collection distance is observed are 5×10^{14} neutrons/cm^2, 2×10^{15} pions/cm^2, and 5×10^{15} protons/cm^2 [180, 181]. For electron radiation diamond has been shown to sustain doses of up to 1 MGy [182]. Diamond has, as indirect insulator/semiconductor with a large band gap, a close to zero intrinsic charge carrier concentration resulting in a high resistivity. Diamond detectors can therefore be operated at room temperature even after intense radiation exposure without significant leakage currents. The undoped material does not form rectifying junctions when put in contact with a metal. A typical electrode metallization uses a Ti/W layer and then Au as the contact to the electrode. The device acts as a charge-sensitive parallel plate detector filled with a dielectric. The electric field inside the detector should be in principle that of a parallel plate capacitor filled with an insulator. However, CVD diamond forms a grain structure (see below) in which charge carriers are likely to be trapped at the boundaries, thus superimposing a polarization field on top of the externally applied field. As a result systematic, position-dependent shifts in the space reconstruction of charged tracks occur.

Energy deposits by impinging radiation create one electron–hole pair per 13.1 eV; thus almost a factor of 4 less charge is expected than in silicon. While the mobilities of electrons and holes have values well above $1{,}000\,\text{cm}^2/(\text{Vs})$ the carrier lifetime is very short (a few nanoseconds). The challenge for detector quality devices is to fabricate CVD diamond with charge collection distances in excess of 200–300 µm.

CVD diamond is produced using a low-pressure, low-temperature chemical vapor deposition process. The growth starts from a nucleation by absorption of carbon from the gas phase on a seed substrate (silicon or metals).

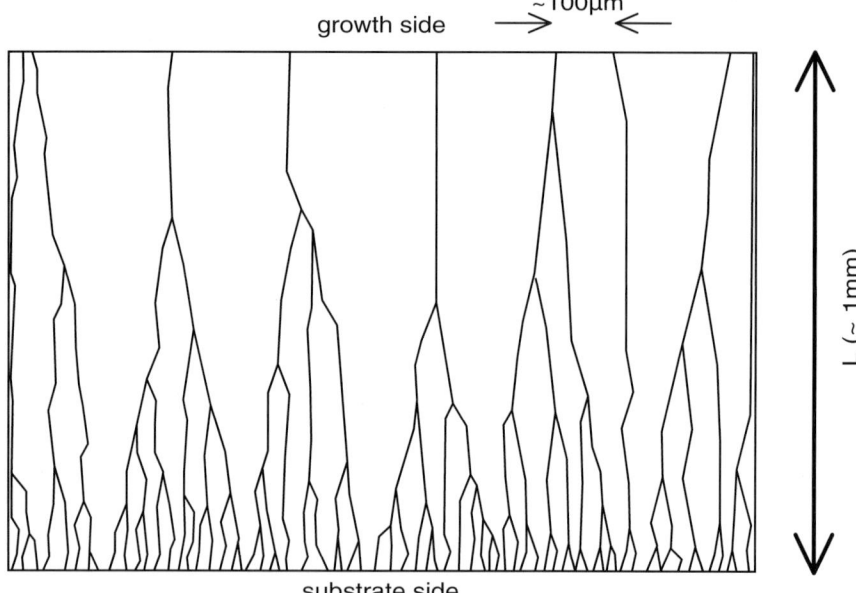

Fig. 2.57. Grain structure as it occurs in CVD diamond material due to the growing process

Fig. 2.58. Electron microphotographs showing the columnar growth structure of CVD diamond (*left*, scale 10 μm) and the grain structure at the growth side (*middle*, scale 100 μm) and at the substrate side (*right*, scale 2 μm) [183]

The diamond (poly)crystals are grown to thicknesses of several hundred micrometers up to ≈1 mm. This growth process causes a grain structure in the material with larger (typically ≈100 μm) structure sizes at the growth side and smaller at the substrate side (see Figs. 2.57 and 2.58). The final detector is lapped down from the substrate side to a total thickness of typically 450 μm. CVD diamond detector-grade material has been grown in wafer sizes as large as 12 cm in diameter [177] which is shown in Fig. 2.59. The dots are Cr/Au metallizations on both sides of the wafer, enabling one to apply voltage and characterize the charge collection properties of the growth. Very recently [178] small size monocrystalline diamond has also been produced

2.7 Detector Materials Other Than Silicon   127

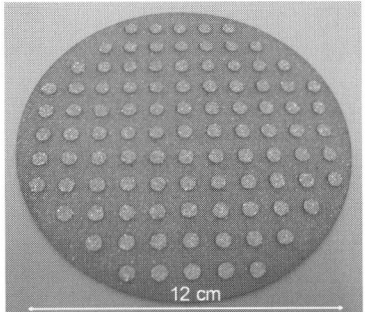 

Fig. 2.59. (*Left*) A 12-cm diameter CVD-diamond wafer. The visible dots are metal contacts for wafer growth characterization (Courtesy of Element Six). (*Right*): A diamond pixel detector module ($2.44 \times 6.34 \, \text{mm}^2$) using ATLAS pixel front-end and readout electronics. The thickness of the sensor is 800 μm

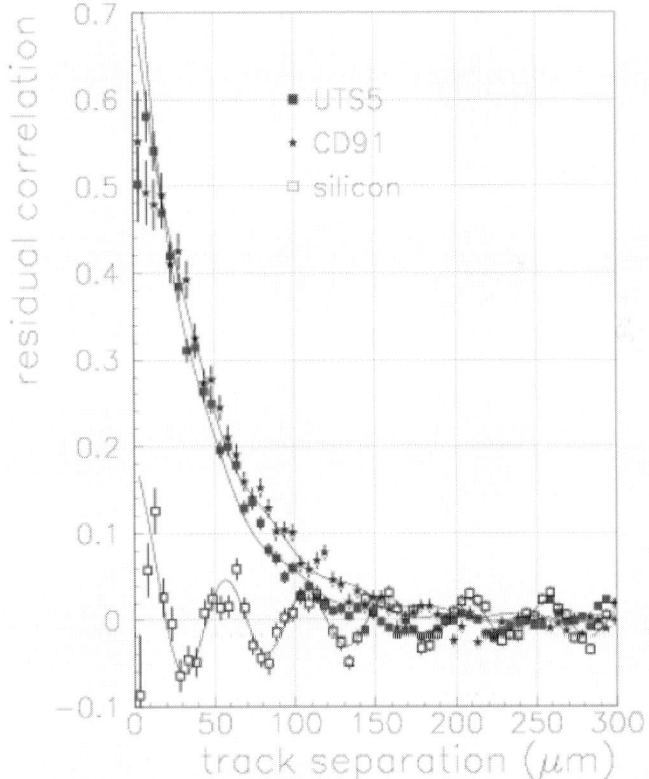

Fig. 2.60. Correlation of the residuals of two track events as a function of the track separation for silicon and two different CVD-diamond pixel detectors (see text), compared with simulations (*solid lines*) [189]. The oscillatory form of the curves is from beating effects due to the pixel pattern

for the first time which does not suffer from the grain-induced deficiencies of CVD-diamond detectors. The production of large detector samples has however not yet been achieved.

The signal from a diamond detector increases during irradiation. This feature [184], is called "pumping" or "priming" effect. It is due to very long trapping times of the trap centers which typically are located deep in the band gap such that thermal emission is very unlikely. Once filled with signal charges the centers are unable to trap further charges. This then leads to larger charge collection distances and hence to larger signals by typically a factor of ∼1.5–1.8. Also optical pumping by UV light has been shown to be possible [185].

Hybrid pixel detectors with CVD diamond sensors have been built using pixel front end chips developed for the ATLAS pixel detector [186, 191] and characterized in the beam [188, 191]. The measurements clearly show the effects of the grain structure leading to systematic shifts in the charge collection [189]. Figure 2.60 displays the correlation of the residuals of two track events, detected by diamond and silicon pixel detectors, respectively, for which the "true" hit positions are given by the extrapolation from a reference beam telescope. With decreasing distance of the hits from the two tracks the correlation becomes stronger for diamond, while for silicon it is roughly unchanged. This effect is naturally explained by the grain structure in CVD diamond, which, for a track separation closer than the grain scale, leads to the same detector response for either track. The comparison of data and simulations (solid lines) is shown in Fig. 2.60. The oscillatory form of the curves is from beating effects due to the pixel pattern. The grain scale is of the order of 150 μm.

3 The Front-End Electronics

Abstract. One of the biggest challenges in the construction of a pixel detector is the design of suited readout chips with several thousand electronic channels. These pixel chips contain a regular arrangement of pixel unit cells, each one serving one sensor pixel connected through a bump-bond connection and, in addition, on-chip circuitry by which hit information is transported to the bottom of each chip from where it can be transmitted out. The pixel unit cells must be as small as possible because their area dictates the sensor pixel size and therefore the spatial resolution. While dissipating typically well below 100 μW, every pixel circuit must provide a low noise amplification of the sensor charge, hit discrimination, and a readout architecture adapted to the application. A sufficient speed of the analog chain and the digital section, a well-defined threshold, and possibly a high-radiation hardness must be guaranteed. This chapter discusses the various requirements and presents solutions to the above-mentioned challenges. Various existing readout architectures for experiments in particle physics and biomedicine are described.

3.1 Introduction

The huge number of channels in pixel systems can only be addressed with highly integrated custom-designed electronics circuits. Only the access to suited chip manufacturing technologies has therefore made the development of pixel systems possible. Early designs had to use technologies with gate lengths of ≈3 μm so that only a few dozen transistors could fit into pixel cells of 330 × 330 μm^2 size [190]. With shrinking technology feature sizes, more complicated functions (in particular in the parameter tuning and in the data processing sections) could be integrated into pixels with significantly smaller area. The access to deep submicron technologies (DSM) with feature sizes of 0.25 μm and below has made possible pixel cells of 55 × 55 μm^2 containing several hundred transistors [192] each. The increasing design possibilities had to satisfy more and more demanding requirements. In the analog part, higher leakage currents and lower signal charges due to radiation damage in the sensor had to be coped with. The analog section had to become significantly faster due to the higher interaction rates in the new experiments (40 MHz at LHC). On the other hand, the power dissipation allowed per pixel decreased because the cooling of electronics and sensor becomes difficult in the low-mass support structures required in particle physics. The digital readout architectures had to process much higher data volumes and so new concepts had to be implemented.

The first operational chips and systems were developed for experiments in particle physics using silicon as the sensor material [191, 193]. In these applications possible hits in the pixels occur at precisely known moments, namely shortly after the collision of a particle beam with a target or with another beam of opposite direction. A trigger signal from other detector components is often available after a fixed time interval, the *latency*, to select interesting events for readout. The generation of the trigger signal typically requires a few microseconds which corresponds to ≈100 interactions at the LHC. Pixel chips for particle physics applications therefore perform an on-chip zero suppression. They buffer the hits until the trigger signal arrives and send only a fraction of the hits to the data acquisition (DAQ). Several different architectures which have been implemented to achieve this goal are presented in Sect. 3.4. Some recent chips [194, 195] send out the zero-suppressed hit data immediately as fast as possible so that the pixel information can contribute to the trigger.

Pixel detectors are also promising devices for medical applications (X-ray imaging, autoradiography, X-ray tomography, Positron Emission Tomography; see Sect. 5.3). Several research groups have built chips which count the number of hits in every pixel [164, 196, 197]. Dual threshold chips with two counters in every pixel [198] allow an energy discrimination for every single hit. This should lead to a contrast enhancement in the image.

3.1.1 Generic Pixel Chip

Although the existing pixel chips use different geometries, readout philosophies, and analog circuits, several building blocks and properties are common to the major part of the designs. As illustrated in Fig. 3.1, the chips can be divided into an *active area* which contains a repetitive matrix of nearly identical rectangular or square pixels and the *chip periphery* from where the active part is controlled, where data is buffered and global functions common to all pixels are located. The wire bond pads for the connection are located only at the lower edge so that the chips can be abutted side by side on a module as indicated in Fig. 3.1. The gap between the chips is made as small as possible and so the size of the underlying sensor pixels must be increased only slightly (see Sect. 5.2.1). The minimum gap size is dictated by the tolerances in chip dicing and by the space required during the flipping procedure.

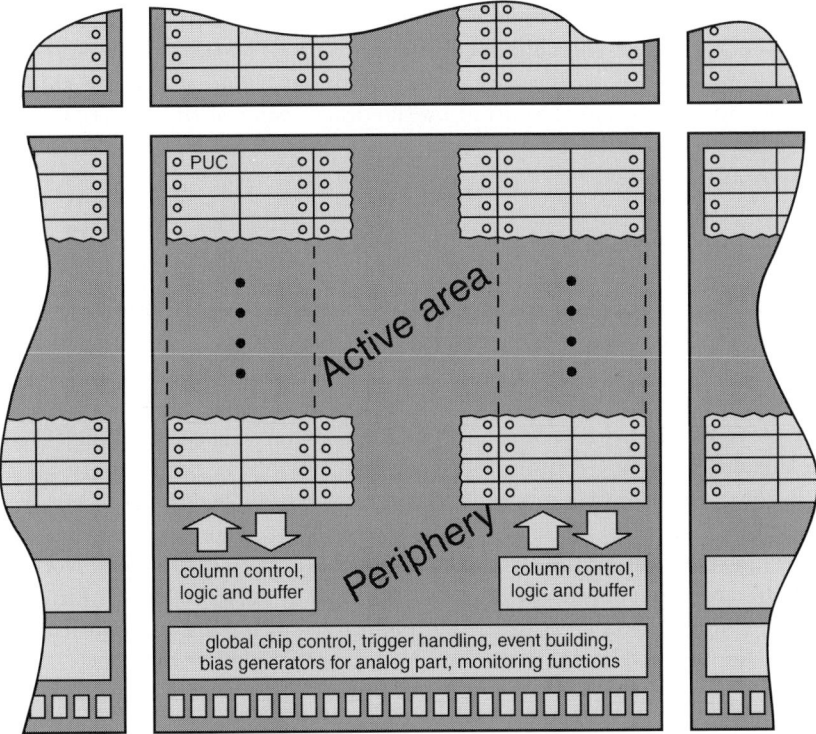

Fig. 3.1. Geometry of a generic pixel chip. The arrangement of several adjacent chips on a module is indicated

3.1.1.1 Active Area

The pixel unit cells (commonly denoted as PUCs) in the active area usually have the same area as the corresponding sensor pixel.[1] They are often grouped in columns; i.e., power, bias, and control signals and the output data flow are routed vertically and only very few signals run horizontally. This column-based approach is advantageous, in particular for rectangular pixels because less area in the PUC is occupied by the bus signals. This is important when only few metal layers are available for routing. Two columns are often grouped together to a "column pair" in order to share circuitry between pixels and to reduce cross talk between the digital and the analog sections. If, for instance, large fractions of the readout section are used by the pixels in two adjacent columns, a mirrored arrangement with the analog parts and the bump pad at the outside and the common readout in the middle of the column pair (see

---

[1]Size and geometry of the pixels on the sensor and the chip can be different if a routing layer is introduced in between, as for instance in the MCM-D approach [199] described in Sect. 6.2.1

Fig. 3.1) is advantageous. The column-based layout with all signals coming from the bottom of the chip requires no circuitry at the side or above the active area and so the distance between the active area and the chip cut edge can be as small as permitted by the design rules of the technology used (typically 50–100 μm). This is important to allow a denser placement of the chips on the module.

3.1.1.2 Chip Periphery

The bottom part of the chip is usually divided into repetitive blocks to interface to the columns, a global control and bias section, and the wire bond pads. An analog test pulse generator is often included on the chip to inject known charges into the pixels. The interfaces to the columns distribute bias signals for the analog sections in the PUCs, they provide buffered digital control signals, and they contain receivers and possibly buffer memory for the data sent down from the pixels. The circuitry to buffer data until reception of a trigger signal can be fairly complicated and space-consuming. The global control part is responsible for the communication with the outside world. In order to save wire bond pads, a serial protocol is very often used for downloading configuration data to the chip and to send hit information from the chip to the DAQ. The configuration data may include the following:

- Bias settings for the analog part which are often generated with on-chip digital-to-analog converters (DACs) as currents or voltages
- A global threshold value and threshold trim values which are written into every PUC
- Mask patterns to disable defective pixels or to switch off complete columns
- Bits to switch between different readout modes, to enable test features, to inject digital test patterns, etc.

The signals going to and coming from the chip which are active during data taking might cross couple into the very sensitive amplifier inputs. Low swing differential signals are therefore often used, so that the number of required wire bond pads is increased. Independent multiple pads are usually used for analog and digital supply voltages to decrease ohmic and inductive losses, and to add redundancy.

3.1.1.3 Examples of Chip Geometries

The size and number of PUCs and the total active area of some existing chips are collected in Table 3.1[2]. The total chip size is a compromise between a good fill factor (requiring large chips) and the production yield (penalizing large chips). The more recent chips use technologies with a gate length of

---

[2] According to [200], the 3 μm self-aligned CMOS (SACMOS) process used for several designs has the density of a typical 1.5 μm process

3.1 Introduction   133

Table 3.1. Geometrical properties of some pixel chips

Experiment or name of chip	PUC (μm^2)	Colums × rows	Active area (mm^2)	Technology (μm)	Ref.
OMEGA1	200 × 200	12 × 9	2.4 × 1.8	3 (SACMOS)	[26]
OMEGA2	500 × 75	16 × 64	8 × 4.8	3 (SACMOS)	[200]
DELPHI	330 × 330	24 × 24	7.9 × 7.9	3 (SACMOS)	[190]
LHC1/OMEGA3	500 × 50	16 × 128	8 × 6.4	1 (SACMOS)	[201]
ATLAS	400 × 50	18 × 160	7.2 × 8	0.25	[202]
CMS 0.8 μm	150 × 150	52 × 53	7.8 × 7.9	0.8	[203]
CMS 0.25 μm	150 × 100	52 × 80	7.8 × 8	0.25	[204]
BTeV, FPIX1	400 × 50	18 × 160	7.2 × 8	0.5	[194]
ALICE/LHCB	425 × 50	32 × 256	13.6 × 12.8	0.25	[205]
MPEC1.0	433 × 50	12 × 63	5.2 × 3.1	0.8	[164]
MPEC2.1	200 × 200	32 × 32	6.4 × 6.4	0.8	[206]
XPAD	330 × 330	24 × 25	7.9 × 8.2	0.8	[207]
PILATUS	217 × 217	48 × 85	9.6 × 17	0.8	[197]
MEDIPIX1	170 × 170	64 × 64	10.9 × 10.9	1 (SACMOS)	[196]
MEDIPIX2	55 × 55	256 × 256	14.1 × 14.1	0.25	[192]

0.25 μm and up to six metal layers for routing so that the pixels can be made smaller. These technologies should still provide a good yield for chips with an area of more than 1 cm^2, which is a common size for commercial chips fabricated in such technologies. Figure 3.2 shows a photograph of a single pixel chip mounted on a printed circuit board for testing.

3.1.2 Simple Sensor Model

A sensor pixel can be modeled by the capacitances to the backside (C_{backside}) and to the first (or more) neighbors (C_{interpix}) as illustrated in Fig. 3.3a (see also Sect. 2.3.4). Assuming that the neighbor pixels are held at constant potential by the connected amplifiers (which is not perfectly true), the capacitances can be summed up leading to the effective detector capacitance C_{det}, which is a crucial quantity for circuit analysis. C_{det} is the central element of the simple sensor circuit model shown in Fig. 3.3b. A constant current source is added to model the fraction of the sensor leakage current flowing into the central pixel. Electron–hole pairs created in the sensor volume are drifting to the electrodes under the influence of the electric field generated by the

134   3 The Front-End Electronics

Fig. 3.2. Photograph of a pixel chip with 18×160 pixels. The regular active area in a column-based structure, the periphery with data buffers at the bottom of column pairs, and the global control circuitry with the wire bond pads at the bottom can be clearly distinguished. A match is shown for size comparison

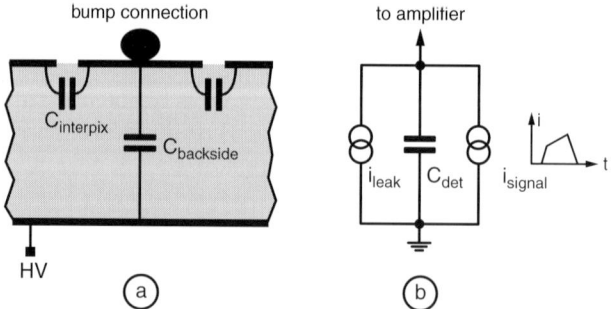

Fig. 3.3. Capacitances in a (**a**) sensor and (**b**) simple equivalent circuit with leakage current source and signal source

3.1 Introduction   135

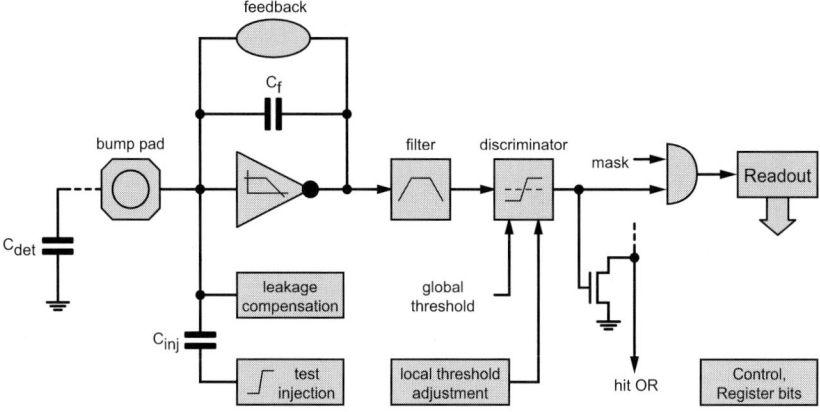

Fig. 3.4. Components of a generic PUC

bias voltage. The signal induced on the pixel and its neighbors already *during* the charge motion is modeled by a time-dependent current source. The exact temporal shape of this current signal depends on many factors like the position of the charge deposition, the sensor material properties (mobilities, trapping), the bias voltage, and the pixel geometry (small pixel effect; see Sect. 2.2.3). A total drift time of ≈ 10 ns is often used to model a 300-μm-thick silicon sensors in which a minimum ionizing particle (see Sect. 2.2.2.1) deposits a total charge of ≈ 4 fC. The polarity of the signal is determined by the type of charge carriers collected on the pixel. For sensor materials with a diode junction like silicon, the voltages must be chosen such that the diode is reversely biased. For other "ohmic" materials like CdTe or diamond, negative input signals are more common due to the higher mobility of electrons. This requires a negative backside bias voltage with respect to the pixels. Typical pixels have capacitances in the order of 100 fF and leakage currents of the order of 10 pA before irradiation.

3.1.3 Generic PUC

Several circuit elements are nearly always present in the elementary PUCs in the active part of the chip. The common circuit blocks shown in Fig. 3.4 are therefore briefly discussed in the following sections. Figure 3.5 shows part of the layout of a PUC as an example.

3.1.3.1 The Bump Pad

A small square or octagonal bump pad is used for the connection to the sensor pixel. The size of the opening in the chip passivation layer depends on the available space and on the bump technology used and ranges from

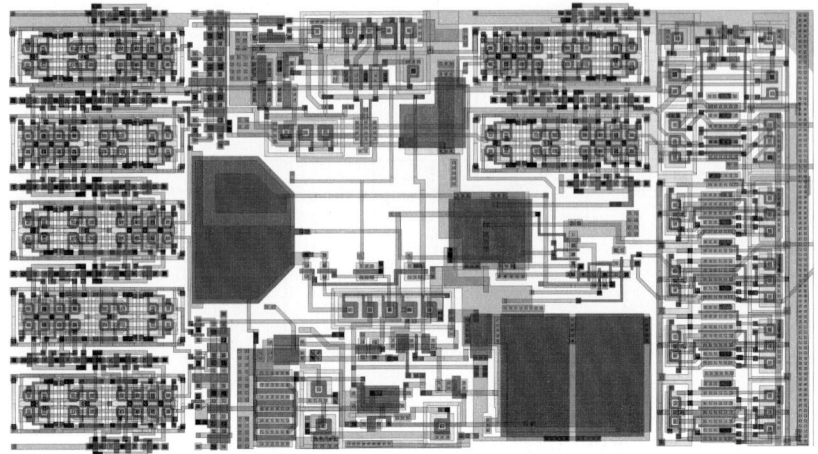

Fig. 3.5. Layout of analog part and control section of the PUC of the ATLAS FEI Chip in 0.25 μm technology. The height of the shown part of the PUC is 50 μm, and the width is ≈100 μm. Only the lowermost two metal layers, the polysilicon layer, and contacts are shown for clarity

12 μm (ATLAS) to 50 μm (MPEC2). The passivation layer must not be too thick for some bump-bonding techniques. The bump pad being connected to the input of the amplifier is a very sensitive node and care must be taken in the layout to shield it toward the underlying circuitry and the substrate. An elegant possibility is to use the capacitance between the pad and the metal shield underneath to implement the feedback and the injection capacitors.

3.1.3.2 The Charge-Sensitive Preamplifier

An inverting amplifier with a feedback capacitor C_f converts an input charge Q_in to a voltage. An infinite gain would keep the input at a perfect virtual ground and the output voltage step in this ideal case is $\Delta U_\mathrm{out} = -Q_\mathrm{in}/C_\mathrm{f}$. If the gain $-v_0$ is finite, however, a small residual voltage

$$\Delta U_\mathrm{in} = \frac{Q_\mathrm{in}}{C_\mathrm{in} + (1 + v_0)\, C_\mathrm{f}} \qquad (3.1)$$

remains at the input. The effective input capacitance

$$C_\mathrm{eff} = (1 + v_0)\, C_\mathrm{f} \approx v_0 C_\mathrm{f} \qquad (3.2)$$

of the charge-sensitive configuration acts in parallel to C_in, which is the sum of the detector capacitance C_det, the preamplifier input capacitance C_amp, and parasitic contributions C_stray. The output voltage step for an amplifier with a finite gain $-v_0$ hence is

$$\Delta U_{\text{out}} = -\frac{Q_{\text{in}}}{C_{\text{f}}} \times \frac{1}{1 + \frac{1}{v_0} + \frac{C_{\text{in}}}{v_0 C_{\text{f}}}} \qquad (3.3)$$

and so a significant fraction of the amplitude is lost when C_{in} approaches $v_0 C_{\text{f}}$. It is therefore usually required that the effective input capacitance be significantly larger than the detector capacitance, i.e.

$$C_{\text{eff}} \approx v_0 C_{\text{f}} \gg C_{\text{in}} = C_{\text{det}} + C_{\text{amp}} + C_{\text{stray}}. \qquad (3.4)$$

This condition also keeps the voltage step at the input ΔU_{in} small so that cross coupling to neighboring pixels via the interpixel capacitance C_{interpix} is reduced.

The rise time of the preamplifier output signal for an instantaneous input charge depends on its gain–bandwidth product and on the closed loop gain, which is roughly set by the ratio of the detector and the feedback capacitor. A large detector capacitance or a small feedback capacitance leads to a slower rise time. The rise time is of course limited by the duration and the exact shape of the signal induced during charge collection, in particular if slow sensor materials or low bias voltages are used.

The choice of the value of C_{f} is therefore a compromise between charge gain, input impedance, speed, stability, and matching (gain uniformity between channels). Values of 4–30 fF are used in most pixel front-end chips.[3] A typical inverting gain of a few hundred leads to an effective input capacitance of a few picoFarads which is an order of magnitude larger than typical sensor capacitances.

The preamplifier is one of the most crucial parts in the PUC. It must provide an inverting gain of well above 100 with a sufficient bandwidth. Its power consumption must be kept very low in most applications (typically 50 μW in particle physics) in order to limit the heat dissipated in the active area. As the noise of the preamplifier is a crucial factor for the performance of the PUC, both the frequency-independent white noise as well as the $1/f$-noise must be minimized for a given detector capacitance and shaper characteristics. The input transistor is usually the dominant noise contribution and so its type (NMOS, PMOS, bipolar transistor), its size, and the biasing condition must be chosen carefully (see Sect. 3.3.6). Another design goal is a good immunity of the amplifier to fluctuations in the supply voltage which could be generated by changes in chip activity and cross coupling from the digital to the analog part. A high power supply rejection ratio in the frequency range of interest is therefore desirable. It can be achieved, for instance, using differential topologies.

---

[3]Chips for the readout of strip detectors must use much higher values of C_{f} because of the larger detector capacitance

3.1.3.3 The Feedback Circuit

A feedback circuit is required to define the DC-operation point of the charge-sensitive preamplifier and to remove signal charges from the input node (or from C_f after the dynamic response of the amplifier) so that the preamplifier output voltage returns to its initial value. The discharge should be slow if further filtering is used as shown in Fig. 3.4, otherwise the pulse shape after the filter is degraded by the falling edge of the preamplifier output signal (see Fig. 3.7). The discharge must be fast enough, on the other hand, to avoid saturation or nonlinearities of the preamplifier at the maximum allowed hit rate. If the discriminator is DC-coupled to the preamplifier output, a "pileup" of pulses leads to unwanted threshold shifts.

The discharge of a feedback capacitor of $C_f = 10$ fF in $\tau = 1$ µs with a resistive element requires the very large resistance of $R_f = \tau/C_f = 100$ MΩ which is difficult to implement on a chip. MOS transistors operated in the linear region for instance, which are commonly used in chips for the readout of strip detectors, cannot be used easily because the drain saturation voltage becomes very small at the required very low gate voltages. Possible solutions to this problem are presented in Sect. 3.3.2.

In a circuit where no separate filter is present and the preamplifier output is directly used for hit discrimination, the discharge must be completed before the next signal arrives. A fast discharge is required in this case. This can lead to a reduction in the peak amplitude if the discharge starts before the signal has reached its maximum due to limited rise time of the amplifier. This "shaping loss" is illustrated in Fig. 3.6a, b.

Several chips use a constant current discharge which can be implemented very easily. This leads to saw-tooth-shaped output signals (Fig. 3.6b, d) with a width proportional to the input charge. This property can be used to determine the deposited charge by measuring the width of the discriminator output pulse, the "time over threshold" (ToT). A constant current feedback is less suited for high-rate applications due to the increasing dead time for large signal amplitudes.

In some rare cases large charges deposited in the sensor by heavy particles, curled tracks, or pulse height fluctuations can saturate the amplifier for a relatively long time, in particular if constant current feedback is used. Some designs therefore introduce an additional nonlinear element (e.g. a diode) in the feedback which provides a high discharge current for anomalously high output voltages.

3.1.3.4 The Leakage Compensation Circuit

The sensor pixels are usually DC-coupled to the preamplifier inputs (this eliminates the need for biasing structures and AC-coupling capacitors on every sensor pixel) so that the leakage current I_{leak} of up to several 10 nA after irradiation must be sunk/sourced by the pixel circuit. Without any

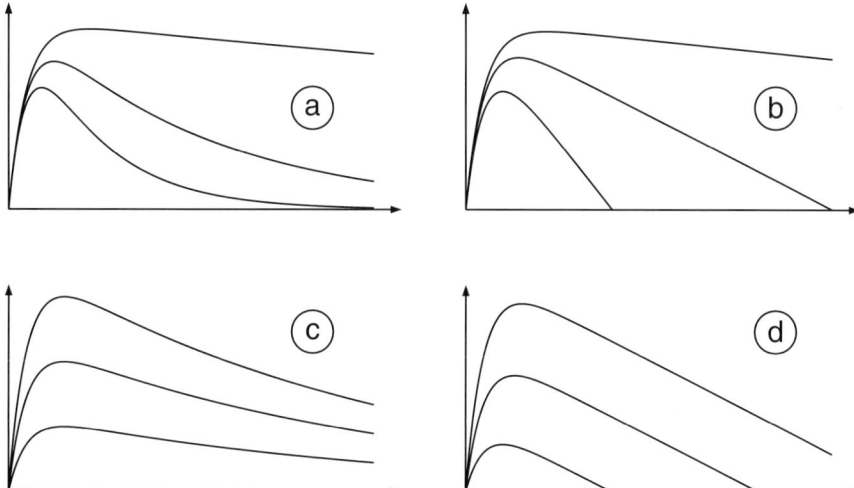

Fig. 3.6. Preamplifier signals (amplitude vs. time) for resistive feedback *(left)* and for constant current feedback *(right)*. The upper graphs (**a**) and (**b**) show a variation in the feedback time constant, and the lower graphs (**c**) and (**d**) show different input charges

further measure, this current would have to flow through the feedback circuit. It would introduce a DC-offset at the preamplifier output of $R_F I_{leak}$ for the case of a purely resistive feedback. For $I_{leak} = 10$ nA and $R_F = 100$ MΩ this leads to $\Delta V \approx 1$ V, a value which would bring typical circuits out of their operation regime. Even much smaller offsets would lead to significant threshold shifts in schemes where the discriminator is DC-coupled to the preamplifier.

A compensation circuit to sink all or a significant fraction of the leakage current is therefore often implemented. A simple solution is the subtraction of a fixed current which is determined, for instance, by measuring the leakage in one or more "dummy" pixels [193]. More sophisticated designs use a current source in every pixel which is regulated such that the preamplifier output reaches a certain average DC-level [208]. Because signal current and leakage current can only be distinguished by their time structure, a very long time constant is involved in this solution. A different, very simple solution is provided by a constant current feedback which is able, due to its nonlinear nature, to provide leakage compensation and the discharge of the feedback capacitor simultaneously [209]. Various feedback circuits are presented in Sect. 3.3.2.

Most leakage compensation circuits can only sink (or source) current and so the polarity of the leakage current, and hence of the signal, is fixed. These pixel chips can then be used for one polarity type of readout only.

3.1.3.5 The Shaper

A band-pass filter (commonly referred to as *shaper*) is often included to explicitly limit the bandwidth of the preamplifier output signal. This is beneficial for a reduction of high- and low-frequency noise contributions introduced in particular by the sensor leakage current and by the input device. Special filter functions can be used to achieve ultimate noise performance for a particular noise spectrum. These filters are difficult to implement with few device components, however, and so a sequence of N simple high-pass (CR) and M low-pass (RC) stages is often used. The noise reduction by such CR^N–RC^M-shapers is discussed in Sect. 3.3.6.

In the time domain, higher order filters lead to shorter pulses for a given peaking time (see Fig. 3.30 in Sect. 3.3.6.1). This makes them useful for high-rate applications where the baseline must be restored as quickly as possible in order to be ready for the next signal pulse. The shaper outputs a unipolar signal only when its input is a step function (with a finite rise time). The preamplifier output, however, has to return to its baseline after some time in order to avoid saturation. This signal decay leads to an undershoot in the shaper response as illustrated in Fig. 3.7c, d. This problem can be addressed by a technique often referred to as "pole-zero cancellation".

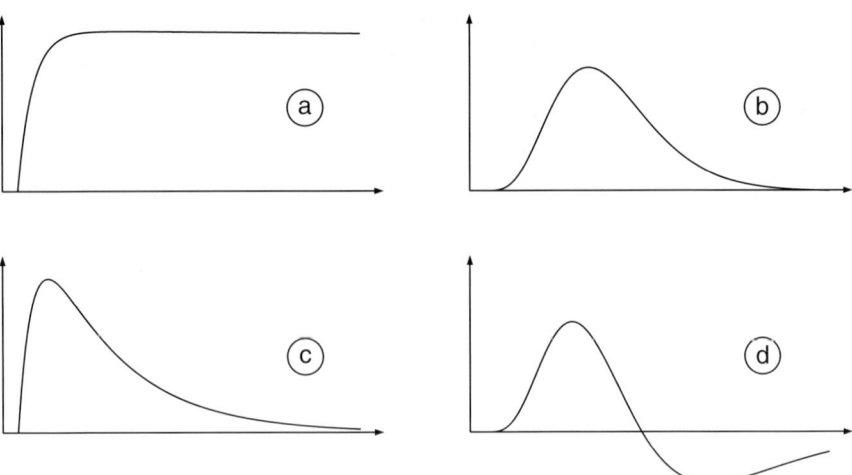

Fig. 3.7. Pulse shapes (amplitude vs. time) after a preamplifier with finite rise time *(left)* and after a CR–RC^4-shaper *(right)*. For a slow discharge of the feedback capacitor (**a**), the corresponding shaper output (**b**) is nearly unipolar. For fast discharge (**c**), however, the shaper output (**d**) has an undershoot

3.1.3.6 The Discriminator and Threshold Trim

Hits with a sufficiently large input charge are detected by a discriminator which compares the shaper output to a threshold value which is distributed globally to all pixels. The threshold is set as low as possible in order to maximize the detection efficiency but not too low, on the other hand, to keep the rate of noise hits at an acceptable level. Variations of the threshold of the individual pixels caused by transistor mismatch, voltage drops or preamplifier gain variations can lead to an increased noise hit rate or to a reduced sensitivity. It is therefore common practice to include a local threshold fine-tuning in every pixel to compensate for these variations. A well-defined threshold is particularly important in applications where the discriminator picks out hits from a continuous amplitude spectrum. In chips for X-ray detection for instance, two discriminators with different thresholds are used to distinguish between high- and low-energetic X-rays [198]. The possible contrast enhancement depends very much on the precision of the amplitude cuts provided by the discriminators and so a precise threshold trimming is required.

The local threshold adjustment is most often implemented by simple digital to analog converters (DACs) with 3–7 bit resolution [195, 210]. This solution requires several digital registers in every pixel to store the trim values. A more compact solution is the dynamic storage of a voltage on a capacitor [206]. This allows a continuous trimming over a wide range with high precision at the expense of the need for regular refresh operations.

The response time of the discriminator (in combination with the rise time of the preamplifier) is crucial in applications where the arrival time of a signal must be detected with high precision. This is for instance the case in particle physics experiments at the LHC where particle interactions occur every 25 ns. The "time walk" curve in Fig. 3.8 shows that hits with high amplitudes lead to a fast response. The discriminator needs much longer, however, to detect hits with amplitudes just above the threshold Q_{Thr}. A $\Delta T < 25$ ns wide time window (some jitter must be allowed for other system components)

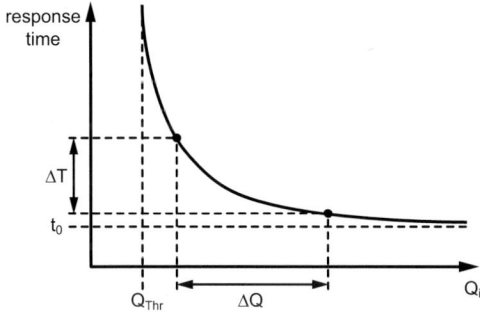

Fig. 3.8. Response time of a discriminator as a function of the input charge

can be selected on this curve by adjusting external system delays. This fixes an interval ΔQ of charges which are detected "in-time." The lowest amplitude which can still be associated with the correct interaction (the "in-time threshold") is higher than the threshold of the discriminator.

3.1.3.7 Test Charge Injection

In order to verify the correct operation of the PUCs already on the wafer level (when no sensor is connected), the controlled injection of known charges into the preamplifier inputs is crucial. This is easily accomplished by applying a known voltage step to a well-defined calibration capacitor C_{inj}. The relative matching of medium-size metal–metal capacitors with a capacitance of a few femtoFarads across a chip can be on the percent level so that the charge error can be made small. The voltage step can be supplied externally to the chip or generated in a "chopper" circuit in the chip periphery. It must be distributed to all pixels with a low-resistive bus so that the rise time degradation due to RC-effects is small. In order to inject only into a selected set of pixels (most readout concepts cannot cope with too many simultaneous hits) a switch in the PUC can disconnect the injection capacitor from the injection bus. The switch should be designed carefully because the parasitic drain–source capacitance in the open state can still lead to spurious charge injection. An alternative to a global voltage step (requiring a fairly powerful chopper circuit) is the local generation of the step in every pixel starting from static voltage levels distributed to the PUCs.

A drawback of the capacitive injection is the difficulty to generate several consecutive charge signals. This would require staircase signals which are more difficult to generate and which are limited in amplitude. Another approach is therefore to steer a known current I_{inj} into the preamplifier input during a known time interval T_{inj}. The injected charge is $Q_{\text{inj}} = I_{\text{inj}} T_{\text{inj}}$, and so a current of 400 nA is required to accumulate a charge of 4 fC in a time interval of 10 ns. The injection can be accomplished easily with a simple differential pair. The transistors used to generate I_{inj} must be sized carefully to guarantee an adequate matching between the different channels.

3.1.3.8 Control and Test Circuitry

Most chips generate a fast hit-OR from the PUCs in one column. This signal can be used to test the analog part independently from the readout or to provide a fast timing signal. Some column readouts are started when a hit in the column is flagged by the hit-OR. Chips requiring a trigger signal can use the hit-OR to automatically produce a trigger whenever a signal is detected somewhere on the chip. This "self-triggered" mode is very useful for module testing with a radioactive source where the events occur at unknown times.

Depending on the area available in the layout, various control bits are used to disable ("mask") the readout of individual PUCs (for instance because a broken front-end or a bad sensor pixel generates an excessive noise hit rate), to disable the fast OR, to turn on a digital test mode, or to switch off the preamplifier completely.

The control bits, the bit for test injection, and those for the threshold trim circuitry are stored in static registers in the pixels. They must be loaded with data during the chip initialization phase. x–y-Decoding schemes or shift registers to address the pixels have been used. A read back feature is useful to increase the testability. As with all registers on the chip, single event upset (i.e. the flipping of bits due to charges deposited on the storage node by ionizing particles) must be addressed for instance with the DICE cell discussed in Sect. 3.2.2.

3.1.3.9 The Readout

Various concepts for the further processing of the hit information have been proposed. The details of the readout architecture depend strongly on the target application. Two important classes are:

- Chips which buffer all incoming hits for a short time interval until a trigger signal selects a subset for readout and
- Chips which count the number of hits in every pixel

They are discussed in some detail in Sect. 3.4.

3.1.3.10 Extensions to the PUC

The existing pixel chip implementations add application-specific blocks to the generic PUC. These can be a second discriminator to sort hits according to their amplitude [192, 198], multiple discriminators to implement a low resolution ADC [194], an individual gain correction [210] or an analog sample and hold for a high-resolution readout of the signal amplitude [203].

3.1.4 Module Controller Chips

The hit information of the individual front-end chips must be sent to the data acquisition (DAQ) for further processing and storage. Serial links carrying digital or analog [195] signals are normally used in order to reduce the number of cables. If the bandwidth of the links is not exhausted by the data rate of a single front-end chip, the number of cables can be further decreased by merging the hit information already on the pixel module. This task is often accomplished by a separate "module control chip," the MCC [211]. As illustrated in Fig. 3.9, the MCC receives hit information from several pixel chips. A star topology is often preferred for its higher bandwidth and fault tolerance. The MCC buffers the hits until all pixel chips have delivered the data belonging to a given event and outputs a compact data block to the

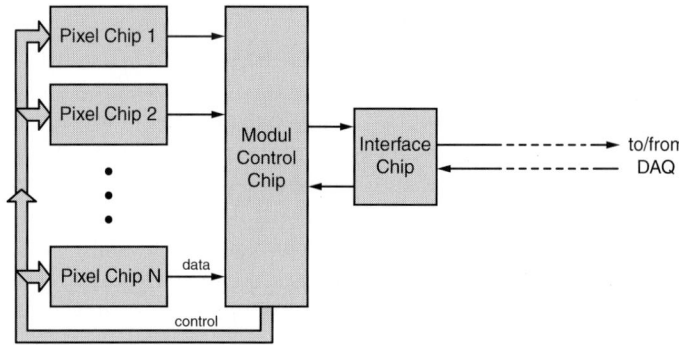

Fig. 3.9. Typical components of a pixel module with an MCC. The MCC receives data from the front-end chips, merges them, and sends them to the data acquisition. It is also used to configure the pixel chips and to provide timing and trigger information. Interface chips can be used, for instance, to serve an optical link

DAQ. It must provide means to handle various error conditions like defect pixel chips, buffer overflows, bad check sums, etc. The MCC can also be used to distribute timing and control signals and configuration data to the pixel chips. The token bit manager chip [212] of the CMS pixel system provides, among other functions, a start signal for the daisy chain readout of the front-end chip.

The short local connections between pixel chips and MCC can use single-ended CMOS, differential CMOS, or differential low swing signals (LVDS) depending on how fast signals are and on how sensitive the environment is to cross coupling. LVDS is often preferred at the expense of twice the number of signal traces and more power dissipation in the LVDS drivers due to the required termination resistors. The long distance between the module and the DAQ often uses optical links which provide high speed, low cross talk and no risk of ground loops. The steering of the laser diode or LED for the outgoing data and the detection of the incoming timing and control signals are sometimes done with separate interface chips.

The architecture of an MCC is very experiment-specific. MCCs are therefore not further discussed in this book.

3.2 Design Aspects

This chapter discusses various aspects of the design of pixel chips. Section 3.2.1 lists some typical specifications which guide the design. The radiation hardness of the circuits (i.e. the long-term stability when ionizing and nonionizing particles irradiate the device) and the ability to cope with local charge depositions which are large enough to flip storage nodes (single event upset, "SEU") are discussed in Sect. 3.2.2. The digital activity on the chip can easily inject spurious signals into the analog section. This cross talk can significantly degrade the analog performance (noise, lowest possible threshold). It is addressed in Sect. 3.2.3.

3.2.1 Typical Specifications

Several partially conflicting requirements must be met by the readout electronics. Most of them concern the analog pixel part which consists of a low-noise charge-sensitive preamplifier, a discriminator with adjustable threshold (two discriminators in a dual discriminator scheme) and very often a test charge injection circuit. The most important requirements are discussed in this section.

3.2.1.1 Pixel Size and Geometry, and Spatial Resolution

A small pixel size is the prime requirement to achieve a high spatial resolution. While the effective pixel *pitch* must be identical for sensor and electronics in both the x- and the y-direction in order to achieve a one-to-one connectivity through the bump bonds, the exact geometries can be different. It is for instance possible to connect an array of hexagonal sensor pixels to a chip with rectangular pixels with a side ratio of $x/y = 2\sqrt{3}/3 \approx 1.155$ as shown in Fig. 3.10. While rectangular sensor pixels seem to be the most natural choice, a hexagonal geometry has the advantage that at most three pixels touch in the corners. The deposited signal charge is therefore shared between three pixels only and so more charge is left per pixel as compared to rectangles. The same advantage is obtained if square pixels are "bricked"; i.e., every second row is shifted by half a pitch (see Sect. 2.3.6). It can also be shown easily that the theoretical resolution of such geometries with a purely binary readout is slightly better than for chessboard square pixels. Hexagonal pixels have probably not been used widely so far because of the slightly more

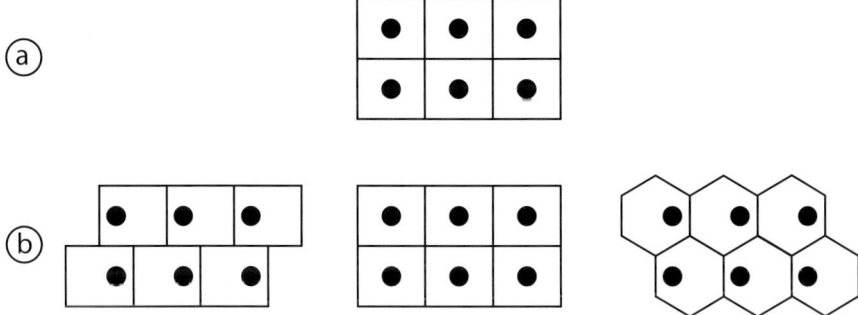

Fig. 3.10. A readout chip with rectangular pixels with side ratio $x/y = 2\sqrt{3}/3$ (**a**) can be mated with a sensor with bricked, rectangular, or hexagonal pixels (**b**)

complicated implementation and coordinate reconstruction. Table 3.1 lists the geometries of some existing pixel readout chips. Rectangular pixels are used in applications where asymmetric resolutions in x and y are required, for instance in particle physics where tracks are bent in a magnetic field. The measurement of the curvature of the tracks in the plane perpendicular to the field is particularly important. From the chip layout point of view rectangular pixels simplify the distribution of signals if the busses run in the "short" direction. If technologies with few metal layers are used, this can be a significant advantage over square pixels of the same area. The rectangular geometry has the drawback that the capacitance between adjacent pixels is relatively large which can lead to a noise increase and to increased cross talk between pixels.

The exact bump pad location within the pixel is sometimes chosen differently for odd and even rows or columns, the advantage being for instance that the analog sections of adjacent pixels are located next to each other on the readout chip (see Fig. 3.1). Power and control signals can be shared between the pixels and the cross coupling between analog and digital parts is reduced. This can be important if few routing and shielding possibilities are offered by the technology.

The spatial resolution for point-like hits (i.e. when the charge clouds by electron and hole diffusion in the sensor are much smaller than the pixel pitch) is $a/\sqrt{12}$, the rms of a box distribution of width a, a being the pixel pitch (see Sect. 2.3.6). Hits close to the pixel edge have the largest errors in this case. When charge diffusion spreads the hits over two adjacent pixels, the "double hits" occurring at the pixel edges have small reconstruction errors. The resolution is further increased if a pulse height information, i.e. the amount of charge deposited in the pixels of a hit cluster, is available. This general behavior is regularly observed in test beam measurements where the position reconstructed from the pixel data is compared to the true impact position measured by a high-resolution reference detector, for instance a silicon strip detector. Examples of such measurements are presented in Sect. 2.3.6.

3.2.1.2 Threshold and Threshold Variations

The thresholds in the many pixel elements on one chip are not perfectly identical. They exhibit random variations due to component mismatch (production fluctuations of doping concentrations, oxide thickness, geometrical size, etc.) and possibly systematic fluctuations due to voltage drops along the column or component mismatch from mirrored layouts. Furthermore, the effective threshold varies as a function of the capacitance connected to the preamplifier input. This is due to the finite gain of the preamplifier which leads to an amplitude loss for increasing input capacitance (see (3.3)). This situation typically occurs at the side and at the top of the chip where the sensor pixels are made larger to fill the area between adjacent chips (refer to Sect. 5.2.1).

The local threshold trim possibility described in Sect. 3.1.3 has two main goals. On one hand, few pixels with particularly low or high thresholds should be brought close to the nominal value. This requires a large trim step size in order to extend the reachable range. On the other hand, the width of the threshold distribution of the majority of the pixels should be narrowed. This requires a small trim step. The optimal trim step is a compromise between these conflicting goals. It is therefore very much desirable to be able to adjust the trim step size dynamically (at least on the chip level) during the threshold trimming procedure.

The thresholds have an rms spread of σ_{thr} after trimming. Pixels with particularly low thresholds are susceptible to noise hits, while pixels with high thresholds may be less efficient. As an example for the estimation of the minimum required threshold, a pixel system at LHC with 250-μm-thick silicon sensors is considered. The signal charge collected from a minimum ionizing particle after several years of operation and corresponding irradiation degrades to $\approx 12,000$ e$^-$ due to a reduction of the depletion depth and to charge trapping (see Sect. 2.4.1). This charge is often spread over two adjacent pixels so that only $\approx 6,000$ e$^-$ may be seen per pixel. Taking into account statistical fluctuations of the signal charge (see Sect. 2.2.2), a threshold of $Q_{\text{thr}} \leq 3,000$ e$^-$ is typically required for good efficiency after degradation of the sensor due to irradiation.

The lower limit for the average threshold is set by the fluctuations σ_{thr} and by the pixel noise σ_{noise}. The noise hit rate is kept negligible if

$$Q_{\text{thr}} \gtrsim 5 - 6 \times \sqrt{\sigma_{\text{thr}}^2 + \sigma_{\text{noise}}^2} \tag{3.5}$$

The threshold dispersion should therefore be comparable to the noise. A typical conservative design goal is $\sigma_{\text{thr}} \approx \sigma_{\text{noise}} \approx 200$ e$^-$.

A specification of the threshold dispersion after trimming can be necessary, for instance, in systems which select certain signal amplitudes (see Sect. 3.4.3), but very often it is only required that the noise hit rate be kept low while setting the lowest feasible threshold. This goal could be achieved with a threshold trimming mechanism which observes the noise hit rate and increases the threshold until the noise hit rate is sufficiently low. This approach would lead to minimal thresholds albeit at unknown absolute values.

3.2.1.3 Noise

The noise of a channel can be defined as the root mean square (rms) of the voltage fluctuation at the end of the analog processing chain divided by the gain (in volts per coulomb). The *equivalent noise charge* at the input,

$$\text{ENC} := \frac{\text{noise output voltage (rms)}}{\text{signal output voltage for an input charge of one electron}}, \tag{3.6}$$

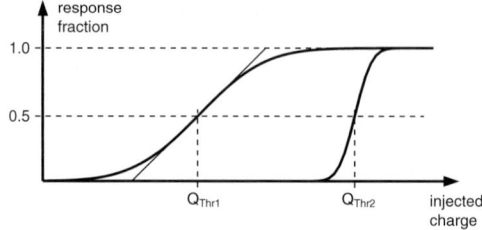

Fig. 3.11. Response of a discriminator (fraction of injections of a given charge firing the discriminator) for a channel with a high noise at a low threshold and for a channel with a low noise at a higher threshold

is commonly quoted in units of electrons as the figure of merit. In binary systems, the analog signal cannot be measured directly and so the ENC is determined from the response of the discriminator to multiple charge injections with increasing amplitude. Figure 3.11 illustrates this kind of measurement for noise with an assumed Gaussian distribution. The threshold is defined as the charge where 50% of the injections fire the discriminator. The slope s at this point can be used to determine the noise:

$$\text{ENC} = \frac{1}{\sqrt{2\pi}} \frac{1}{s}. \tag{3.7}$$

For quick estimations one can determine the interval from ≈5 to ≈95% which corresponds to ≈ 3.3 ENC.

The noise is mainly determined by the total input capacitance and by the speed of the amplifier. It can often be reduced by increasing the transconductance of the input device (see Sect. 3.3.6) if the bandwidth of the system is kept constant.

As described in the previous section, the noise must be kept significantly lower than the hit threshold to limit the noise hit rate to an acceptable level. This is particularly difficult for the edge and corner pixels where the noise is higher due to the increased capacitance. A noise in the order or below the threshold dispersion level of $\sigma_{\text{Thr}} \approx 200$ e$^-$ is usually required. In systems with an analog readout of the hit amplitude an even lower noise figure may be beneficial to increase the spectroscopic resolution or to improve the spatial resolution obtained by interpolation.

The calculations in Sect. 3.3.6 show that such a noise can be achieved fairly easily with an idealized amplifier. Several other sources of noise, in particular cross talk within the chip and noise on the supply lines and on the substrate, voltage drops, etc., make it a design challenge, however, to reach the theoretical value on a large chip when many pixels are active simultaneously.

3.2.1.4 Input Impedance and Cross Talk

A charge Q deposited on a single pixel can induce parasitic signals on neighboring pixels due to the interpixel capacitance. This "cross talk" can lead to an increased fraction of double and triple hits and may have to be taken into account in the position reconstruction algorithms. The approximate magnitude of this effect is calculated in this section for cross coupling along one dimension only as it is the case for elongated rectangular pixels where the interpixel capacitance on the long edge dominates. Figure 3.12a shows a simplified cross section of an (infinitely wide) sensor with backside capacitances C_{back} and interpixel capacitances C_{C}. Next-to-next neighbor capacitances etc. are not considered in this approximation. Every pixel is connected to a charge-sensitive preamplifier with an effective input capacitance C_{eff} (see (3.4)) which is usually designed to be much larger than the total pixel capacitance in order to "pull" all the deposited charge into the amplifier. The capacitance C_{G} of every node to ground is therefore much larger than the pixel capacitances:

$$C_{\text{G}} \approx C_{\text{eff}}, \qquad C_{\text{G}} \gg C_{\text{back}}, \qquad C_{\text{G}} \gg C_{\text{C}} \tag{3.8}$$

The deposition of a charge Q on pixel 0 in the lumped arrangement of Fig. 3.12b is analyzed. The infinite capacitor chain to the right consisting of $C_{\text{C}}, C_{\text{G}}, C_{\text{C}}, \ldots$ can be replaced by a single equivalent capacitor C_{X} as illustrated in Fig. 3.12c. Its value can be determined from the observation that a C_{C}–C_{G} pair followed by C_{X} must have the effective capacitance C_{X} due to the infinite arrangement. This is illustrated in Fig. 3.13. C_{X} must therefore be equal to the series connection of C_{C} and $C_{\text{X}} + C_{\text{G}}$. This leads to

$$C_{\text{X}} = \frac{C_{\text{G}}}{2}(W-1) \quad \text{with} \quad W = \sqrt{1 + 4\frac{C_{\text{C}}}{C_{\text{G}}}}. \tag{3.9}$$

The charge Q deposited on pixel 0 with the total capacitance $C_{\text{G}} + 2C_{\text{X}}$ leads to a voltage $U_0 = Q/(C_{\text{G}} + 2C_{\text{X}})$ at that node. The charge Q_0 on C_{G} of this

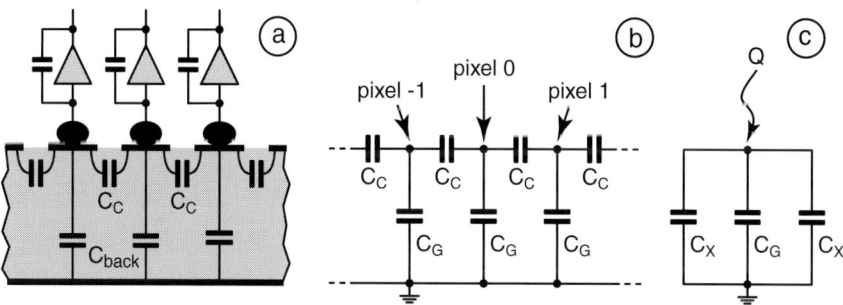

Fig. 3.12. (a) Model used for the estimation of cross talk to neighboring pixels, (b) lumped model of capacitances (b), and (c) left and right capacitances merged to C_{X}

pixel is $U_0 C_G$, and so the corresponding charge fraction q_0 is

$$q_0 := \frac{U_0 C_G}{Q} = \frac{C_G}{C_G + 2C_X} = \frac{1}{W}. \qquad (3.10)$$

The remaining charge is located on the right (and left) C_X:

$$q_x = \frac{1}{2}(1 - q_0) = \frac{W-1}{2W}. \qquad (3.11)$$

C_X can be replaced by the equivalent circuit of Fig. 3.13 for the calculation of the charge on C_G of pixel 1. This results in

$$q_1 = q_x \frac{C_G}{C_G + C_X} = \frac{1}{W}\frac{W-1}{W+1}, \qquad (3.12)$$

and, more generally,

$$q_i = \frac{1}{W}\left(\frac{W-1}{W+1}\right)^{|i|}. \qquad (3.13)$$

The q_i sum up to 1 for $i := -\infty, \ldots, \infty$ as expected. The q_i basically represent the charge fractions seen by the preamplifiers because C_G is dominated by C_{eff}.

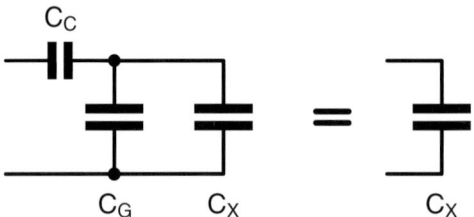

Fig. 3.13. Adding another C_C–C_G pair in front of C_X must again give C_X in an infinite arrangement

The cross talk can be quantified by the fraction of charge seen on the first neighbors, i.e. q_1. Using the approximation (3.8) leads to $W \approx 1 + 2C_C/C_G$ and so (3.12) becomes, to the first order,

$$q_1 \approx \frac{C_C}{C_G} \approx \frac{C_C}{C_{\text{eff}}}. \qquad (3.14)$$

A realistic cross-talk value can be estimated as follows: The total capacitance of the rectangular pixels considered here is dominated by C_C, and so $C_C \approx C_{\text{det}}/2$. A typical preamplifier has an effective input capacitance which is significantly larger than the pixel capacitance, say, $C_{\text{eff}} \approx 25 C_{\text{det}}$. This leads to a cross talk q_1 of 2%, a value which is acceptable in most applications.

In reality, the cross-talk issue is very complicated because the transient signals induced on neighboring pixels by the drifting charges must also be taken into account.

3.2.1.5 Speed

Depending on the application, the speed requirement for the preamplifier–shaper–discriminator chain can be different. At least two aspects must be considered.

The first important figure of merit is the *timing precision* with which the arrival time of a hit is determined. This is important for instance in particle physics experiments at the LHC where hits must be associated with one particular bunch crossing with a precision of better than 25 ns. Although constant delay contributions can, in principle, be calibrated out, the time resolution is improved by a short delay between the charge deposition and the reaction of the discriminator. A short preamplifier rise time, a high shaper bandwidth (or no shaper at all), and a fast discriminator are therefore desirable. This generally requires more current in the corresponding circuit blocks. The response time is a function of the input charge and so the "time walk" curve, as illustrated in Fig. 3.8, can be used to characterize the behavior. Charge depositions just above the threshold generate only a small discriminator overdrive so that the response becomes slower. If the amplitude of a hit is known, the measured time can, in principle, be corrected for on the basis of the known time walk characteristic. This task can already be performed on the front-end chip [210]. Note that a good timing precision can also be obtained with a "slow" preamplifier using special techniques like the deconvolution of the analog pulse shape [213] or the detection of the zero crossing of an appropriately shaped pulse.

A second important characteristic is the maximum possible *hit rate* which is limited by the time required to process one hit. This "dead time" ΔT can be dominated by the analog section but it may also be determined by the time required in the following readout circuitry to process the hit. A very complex behavior is introduced when the readout speed depends on buffers being filled, on the average occupancy, etc., and so Monte Carlo simulations are required to estimate the fraction of lost hits in a realistic environment.

In order to determine the fraction of hits lost due to a dead time ΔT with random events at an average rate of r hits per second, two different (idealized) cases can be distinguished:

When a hit occurring during the dead time interval *extends* the dead time by another ΔT, no hit pair with an arrival time difference smaller than ΔT can be detected. The loss fraction can therefore be found by integrating the normalized probability density of time intervals t between consecutive hits

$$p(t) = r\, e^{-r t} \tag{3.15}$$

from zero to up to ΔT. This leads to the loss fraction for *extended dead time*,

$$\mathrm{loss}_{\mathrm{ext}}(r, \Delta T) = \int_0^{\Delta T} p(t) dt = 1 - e^{-r \Delta T}, \tag{3.16}$$

which is $\approx r\Delta T$ for $\Delta T \ll 1/r$. Nearly all hits are lost when $\Delta T \gg 1/r$ because the dead time is permanently extended by new hits. An example for an extended dead time is the recovery time required in the preamplifier to discharge the feedback capacitor.

The situation is slightly different when hits occurring during the dead time do *not* extend the dead time. After a dead time of length ΔT, the detector finds a new hit after a time $1/r$, on average, and so the loss fraction for *nonextended dead time* is

$$\text{loss}_{\text{nonext}}(r, \Delta T) = \frac{\Delta T}{\Delta T + 1/r} = \frac{r\Delta T}{1 + r\Delta T}. \tag{3.17}$$

This is $\approx r\Delta T$ for $\Delta T \ll 1/r$, as before. The losses for the nonextended case are always smaller than for the extended case because the system is still able to detect a new hit after the fixed dead time even if the rate is high. An example for a nonextended dead time can be the digital hit processing.

A low dead time is, for instance, important in counting pixel chips where rates of up to 10 MHz per pixel can be required. The dead time often depends on the amplitude of a hit. Furthermore, the threshold for the following hit is clearly a function of the time delay as the shaper output must come back to its initial value. Higher order shapers (see Fig. 3.30) and a careful cancellation of the undershoot due to the feedback circuit (see Sect. 3.1.3) are important in such applications.

3.2.1.6 Power Consumption

The maximally allowed power dissipation in the pixel chip also depends on the application. The front-end chips can be cooled relatively easily in systems for X-ray detection because these consist of only one sensor layer in which the particles are absorbed. Components used for cooling underneath the front-end chips, hence, do not degrade the system performance. In detector assemblies in particles physics, on the contrary, several layers of sensors with front-end chips are usually used to measure several points of the track of penetrating particles (see Sect. 4.1). The flight path of such particles is influenced by multiple scattering in the material which therefore must be kept to a minimum. The cooling (see Sect. 4.5.2) is simplified if less power must be taken out of the system. As the dissipation of the front-end chips is usually the dominant heat source (other sources are the module control chip, the interface chips, the power supply cables, and the sensor itself), their power consumption must be reduced. A typical figure for the power used in particle physics applications is ≈ 50 μW/PUC. The total current drawn by preamplifier, shaper, discriminator, readout, etc., should therefore not exceed 25 μA if a supply voltage of 2 V is used.

This limitation constitutes a challenge to achieve the analog performance goals for noise and in particular for speed. In order to obtain an impression of the significance of the problem, note for instance that a single D-flip-flop implemented in a 0.25 μm technology using enclosed NMOS devices for radiation hardness (see Sect. 3.2.2) and clocked at a frequency of 40 MHz at a supply of 2 V already draws an average current of ≈10 μA. A signal trace running across a pixel of 100- μm height has a capacitance of the order of 20 fF in a 0.25 μm technology and so a power of 3 μW is dissipated when the trace is clocked at 40 MHz at a supply of 2 V. Another source of power dissipation in the digital circuitry can be short circuit currents between power and ground during the switching of CMOS inverters or gates. The transconductance of the transistors offered by modern technologies is high, and so short circuit currents of several tens of microAmperes are common. The duration of the short circuit currents must be minimized by keeping the rise time of all digital signals very fast. This makes the design delicate and can cause increased cross coupling to the analog section. Possible remedies to this problem are discussed in Sect. 3.2.3.

3.2.1.7 Analog Charge Measurement

For monitoring purposes and in order to improve the spatial resolution, the measurement of the signal charge with a modest resolution is desirable. An amplitude information with a resolution of only a few bits can lead to a noticeable improvement in the spatial resolution in some applications. A better amplitude precision often does not significantly improve the spatial resolution [13, 214].

A straightforward approach is the sampling of the peak amplitude on a storage capacitor and the subsequent readout of the stored voltage (or charge) through a multiplexer. This approach has been implemented, for instance, in the CMS pixel readout [195] described in Sect. 3.4.2.

The BTeV readout chip uses a 2 bit-FADC (flash analog-to-digital converter) to get a very low resolution pulse height measurement [194].

The analog signal processing can be avoided by measuring the width of the discriminator output signal. This ToT is an almost linear function of the charge if a constant current feedback is used. The added complexity is small if the rising and the falling edge of the discriminator output are measured in units of the system (bunch crossing) clock. The quality of the analog information depends on the uniformity of the preamplifier/discriminator chain and on the maximum duration allowed for the hit signal (long signals introduce dead time). A resolution of ≈4 bits can be easily achieved. The ATLAS pixel chip described in Sect. 3.4.2 implements such a ToT readout.

3.2.1.8 Summary

Some typical specifications of readout chips for pixel detectors in particle physics are summarized in Table 3.2.

Table 3.2. Typical specifications for pixel chips used in particle physics

Quantity	Specification
Pixel area	2,500–40,000 μm^2
Noise (ENC)	<200 e$^-$
Threshold dispersion	<200 e$^-$
Power dissipation per pixel	<50 μW
Sensor leakage current tolerance	0–100 nA
Hit time measurement	<25 ns

3.2.2 Radiation-Tolerant Design

Pixel detectors are often used to detect ionizing radiation and highly energetic photons which traverse both the sensor and the readout chips. The MOS devices on the chips can therefore be exposed to high levels of radiation. When low-energetic radiation is detected (e.g. synchrotron radiation of some 10 keV), a significant fraction is absorbed in the sensor and so the readout chip is partially shielded. In particle physics experiments at the LHC, fluences of 10^{15} n$_{eq}$/cm^2 are expected during 10 years of operation. This corresponds to a radiation dose of 500 kGy.

The dominant effect of ionizing radiation on CMOS electronics is the shift of device thresholds mostly due to the accumulation of positively charged holes in the gate and field oxide [215]. It has been known for a long time that the electrons can tunnel out of the oxide from a thin surface layer and consequently it was observed that the upcharging is significantly reduced when the oxide becomes as thin as ≈12 nm [32]. Modern "deep submicron" (DSM) CMOS technologies with gate lengths of 0.35 μm and below have even thinner gate oxides. The thin oxides and the high-quality processing steps employed lead to devices which show nearly no more threshold shifts for fluences of 300 kGy and above [216]. Parasitic NMOS transistors with thick field oxide can, however, develop thresholds low enough to open up parasitic leakage current paths. They must therefore be avoided. The current path can be interrupted by p$^+$ guard rings [217], so that a design with enclosed NMOS devices separated by guard rings can make the chip radiation-tolerant. This has been successfully verified for several modern DSM technologies. No special measures need to be taken for PMOS transistors because the thresholds of parasitic devices in an n-type substrate or in an n-well increase rather than decrease with irradiation. An alternative geometry suited to implement NMOS devices with small W/L ratios has been suggested in [218]. Figure 3.14 shows schematically a radiation-tolerant layout for one or two series connected NMOS transistors.

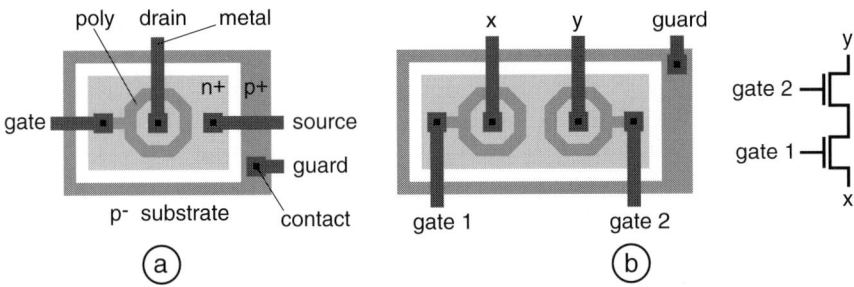

Fig. 3.14. Schematic layout of a (**a**) radiation-tolerant annular NMOS with guard ring and (**b**) series connection of two devices

In the particle physics community the very encouraging results of [216] have triggered a migration from the so far used specialized radiation-hard technologies to DSM technologies. The expected high yield for large area chips, the very high integration density, and the many levels of routing metal render them particularly well suited for pixel chips.

The design of mixed mode circuits using DSM technologies with radiation-tolerant design techniques, i.e. annular NMOS devices with guard rings, has several consequences some of which are mentioned in the following:

- The *device models* provided by the vendor are usually not suited for devices with enclosed geometry. Test structures must therefore be prototyped in order to extract parameters for reliable circuit models. Among the quantities to determine are the effective W/L ratio, output conductance, capacitances, noise, and matching, which has been measured to be worse for enclosed devices than for rectangular layouts [216].
- The W/L *ratio* of NMOS devices with enclosed layouts cannot be decreased below ≈ 2 [216] because the effective width of the device increases with increasing length. This large W/L ratio can be a serious limitation in analog design. High-quality NMOS current sources are difficult to implement and their noise contributions must be taken into account carefully. As a further consequence, NMOS devices are often operated in weak inversion at the currents used in pixel analog sections.
- The *capacitances and output conductances* of the inner and outer terminals of the enclosed devices as well as the gate overlap capacitances are different. This must be taken into account in the circuit models. The association of one or the other terminal to drain and source is an additional degree of freedom in the design. The inner terminal is normally used as the drain because of its significantly smaller capacitance. The asymmetry can cause problems, for instance, of charge injection in MOS switches.
- The *series connection of NMOS devices* must follow the rule that n$^+$ regions on different potentials must be separated by a gate or by a p$^+$ guard ring. Two devices connected in series are possible within one guard

ring when the inner terminals are used as illustrated in Fig. 3.14b. The intermediate node has a very high parasitic capacitance in this case due to the large area of the n$^+$ region. More than two devices require additional guard rings. Serially connected NMOS devices should therefore be avoided if possible, for instance by replacing CMOS NAND gates by NOR-type logic with parallel NMOS devices.
- The *input capacitance* of CMOS logic gates is large, in particular if the W/L ratio of the PMOS devices is matched to the mostly unnecessarily large W/L value of the enclosed NMOS devices. This leads to a significantly increased dynamic power consumption as compared to "minimum-size" logic. Furthermore, the short circuit current in such gates during the input signal transition is large (up to 200 μA for a "matched" inverter operated at 2 V) and so *all* signal transitions must be very fast in order to avoid large supply current spikes.

3.2.2.1 Single Event Upset

Strongly ionizing particles (in particular slow heavy ions) can deposit very large charges in a very small volume of silicon. If the charge happens to be deposited on a storage node (DRAM or SRAM cell, latch, flip-flop), this node can flip such that the stored information gets corrupted. If such a "single event upset" (SEU) occurs in a state bit of a state machine, a wrong behavior with possibly serious consequences may follow. If the state machine enters a "forbidden" state, a permanent lockup may completely block the system. (State machines should therefore be designed such that they recover from all possible states.) The rate for SEU events depends very much on the circuit schematic and on the physical layout and on weather the bit flips from 0 to 1 or from 1 to 0. Shift registers can be more sensitive when being clocked as compared to the static state. The sensitivity of a circuit to SEU is often quoted as a cross section. Measured values for various flip-flop designs for instance range from 10^{-12} to 10^{-16} cm^2 [219–222]. The cross section decreases rapidly with smaller energy depositions and becomes negligible below a certain threshold energy.

Standard techniques to address SEU are the use of error correcting logic (e.g. Hamming codes), for instance, in RAMs or self-correcting triple redundancy cells for individual bits. These solutions require significant additional hardware so that they can be used only when really needed. In some applications it may be sufficient to detect an SEU by a checksum mechanism (e.g. a parity bit) and react externally by a reconfiguration of the circuit. A fairly simple and efficient approach to reduce SEU in a standard SRAM cell consisting of two cross-coupled inverters is the addition of a capacitor between the two storage nodes as illustrated in Fig. 3.15. The capacitor slows down the cell and so short transients on the storage nodes cannot easily flip the cell. A reduction of the SEU cross section by 2 orders of magnitude has been reported [221](see also Fig. 5.4).

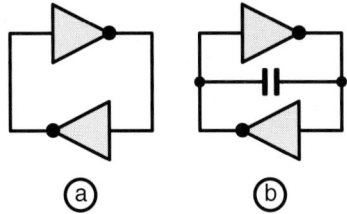

Fig. 3.15. Standard (**a**) and SEU-protected SRAM cell (**b**)

The ingenious DICE cell [223] drawn schematically in Fig. 3.16 uses a different concept to address the problem. The information (and its inverse) is stored on 2+2 independent storage nodes which are cross-coupled in such a way that a temporary flip of one node due to an SEU does not permanently flip the cell. The circuit is very simple and compact. The layout must assure that corresponding storage nodes are not flipped simultaneously by the same charge deposition.

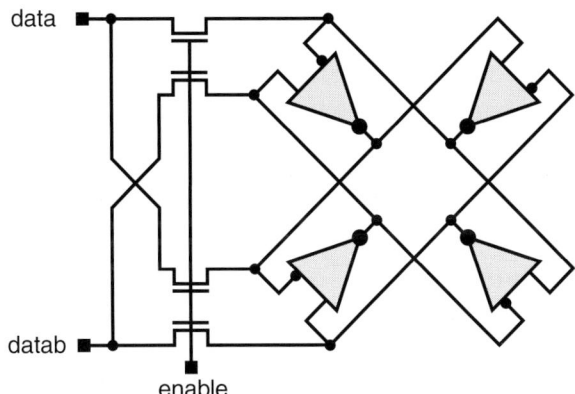

Fig. 3.16. SEU-tolerant DICE storage cell, redrawn from [223] to stress the symmetry. The *triangular symbol* is an inverter consisting of two transistors with separate inputs for the gate of the NMOS and the PMOS transistor

3.2.3 Cross Talk

Pixel electronics is particularly affected by cross talk from the digital to the analog section because these are densely packed in the active area. The intrinsic noise and the achievable charge thresholds of the analog section are low and so already very small disturbances degrade the overall performance. A voltage step of 1 V, for instance, injected through an extremely small parasitic capacitance of only 1 fF generates a cross-talk charge of 1 fC or 6,250 electrons, which is much more than the noise and typical thresholds. Crosstalk signals can be injected capacitively (across the chip, via the substrate or via the sensor) or through spikes in supply or bias signals. Several precautions can be taken to address the problem.

3.2.3.1 Shielding

A metal *shield* should cover the chip so that signal swings do not couple capacitively to the sensor. A capacitance of $1.8\,\mathrm{aF}/\mu\mathrm{m}^2$ between chip and sensor has been measured [194]. This value depends only on the distance between sensor and chip, i.e. on the height of the final bump connections. A hermetic shield often requires more than one metal layer due to technology limitations in the individual layers (maximum metal density, slots in metal). The shield must be connected to a "clean" net because voltage spikes on the large area would be detrimental. The connection of the shield must have low resistance and low inductance because its large area is itself prone to pickup from the signals below.

The *input pad* is the most sensitive node due to its relatively large area. It must be well shielded. The integration of the feedback and test charge injection capacitors in the metal stack underneath the pad can be a solution (see Sect. 3.3.2).

Guard rings are often used around the analog parts. They may be less effective in DSM technologies, however, because the substrate resistance can be very low, allowing an efficient signal path underneath the guard implantations.

3.2.3.2 Power Supplies

Separate power supplies for analog and digital parts and possibly also for large buffers and CMOS output pads are commonly used. Low inductance supply traces and multiple bond pads are standard techniques.

On-chip voltage regulators can be used to provide clean voltages and to compensate for voltage drops on the supply cables. The CMS pixel chip, for instance, uses four regulators for two analog and two digital supplies [203].

The *injection of switching noise* from the digital supplies into the substrate can be reduced if a separate net is used for the bulk and possibly the well connections [224].

Spikes on the digital supply are generated by the charging and discharging of capacitive loads during signal transitions, but also by the short circuit current during transitions, which can be quite important when enclosed devices with large W/L are used to achieve radiation hardness. While slow rise times would decrease the magnitude of capacitive spikes, the DC-currents would become unacceptable if ratioed CMOS logic (i.e. with equally strong PMOS and NMOS devices) were used. Other logic families should therefore be considered:

- Small PMOS devices (i.e. not matched to the large W/L of the NMOS) reduce the input capacitance and the short circuit current which lasts, however, longer due to the slow signal rise times. This approach requires very careful design because rise and fall times and propagation delays are very asymmetric. The noise immunity is reduced because the switching thresholds are low.
- The addition of a current limitation (current source) into the CMOS logic reduces the short circuit current and the rise times [225]. The limited drive strength must be individually adapted to the capacitive load and so a very careful extraction of all parasitics is necessary. The underestimation of a load can limit the operation speed of the whole chip. Programmable drive strength in crucial places are conceivable.
- Differential full swing logic with adjacent signal pairs produces balanced emissions (toward shield and substrate), but the supply spikes are even increased, because more lines must be driven. The differential voltage swing is twice the supply voltage and so a large charge is stored in the capacitance between the two traces.
- Differential logic with a reduced signal swing supplied with a constant current is a very attractive candidate because spikes are small and emissions are low. The limited drive strength, however, requires a careful design, as mentioned above. The total power dissipation of such a logic is not necessarily higher than for CMOS, because the dynamic contribution is significantly reduced by the low signal swing.

It may be sufficient to use low noise logic at particularly critical places, like in the PUCs.

A *reduction of the switching noise* is achieved by avoiding the simultaneous flipping of signals with large loads (busses), for instance by using Gray-encoded counters. Less (more optimized) logic produces less spikes.

Periodic signals may be less dangerous than intermittent activity because their effect on the analog part is constant.

3.2.3.3 Further Remarks

A *high power supply rejection* of the analog section reduces the sensitivity to spikes on the supply. This can be achieved, for instance, with a differential preamplifier design [225] at the expense of an increased analog power dissipation. Note that the power supply rejection of the circuit must still be significant in the signal frequency band. The second input of the differential amplifier is an extremely sensitive node and must be treated with care.

Decoupling capacitors can be added locally on the chip. Although the achievable total capacitance is not very large (tens of nanoFarads), they can be very efficient because of the small inductances of the local connections. A possible yield problem introduced by the large area of the thin gate oxide can be overcome by using thin metal traces for the connection acting as fuses and by a special circuitry to switch off defective devices [226]. These techniques are commonly used in industrial designs. They lead to an area overhead of roughly a factor of 2, depending on how sophisticated the test circuitry is. A large chip capacitance in conjunction with the wire bond inductance can lead to oscillations. These are very difficult to predict because the damping elements are difficult to model.

The *input/output signals* going to and coming from the chip during data taking should use low swing signals, slew rate limited signals, current outputs, or differential signaling (LVDS).

3.2.4 Testability and Ease of Operation

3.2.4.1 Testability

The large number of channels on pixel chips makes it very time-consuming to perform precise analog measurements for every pixel. These are required, however, to characterize the design and to sort out known good dies before mounting them onto the sensor. It is therefore crucial to foresee internal circuitry to simplify testing. These features can also be used to monitor correct operation of mounted chips.

- The *injection of known test charges* into the preamplifiers must be possible. The simultaneous stimulation of several pixels, preferably in a freely programmable spatial pattern, should be possible so that a realistic signal activity can be generated on the chip. This feature is useful, for instance, to measure weather a significant activity in the digital readout degrades the analog performance by cross talk. Simultaneous injections can also be used to characterize several pixels simultaneously.

    The injection circuitry must be designed very carefully in order not to increase the noise due to the extra components connected to the amplifier input. The test charge is often generated by applying a voltage step to small injection capacitors located in the pixels. This solution guarantees fairly identical charges in all pixels because the matching of on-chip capacitors is usually very good. The value of the injection capacitors must be determined with high precision in order to calculate the absolute amount of injected charge. An elegant method is an on-chip measurement for instance using the very simple charge pump circuit presented in [227]. The calculated charge can be cross calibrated with the charges deposited by monoenergetic γ-rays once a sensor has been connected to the chip.

    Another method to inject charges are current pulses of well-defined amplitude and duration. They can be generated easily by steering a known current to the preamplifier input with a differential pair. This concept allows the generation of multiple successive hits in short time intervals. This is not easily possible with capacitive injection (it requires staircase-shaped signals), but can be a valuable feature in applications where a

high count rate is required. A good matching of the charges injected into different channels is more difficult to guarantee, however, because active MOS devices are involved.
- A *leakage current injection* into every pixel is useful to verify the correct operation of the leakage compensation circuitry.
- The injection of *digital test patterns* can be valuable for quick tests of the digital section. This feature can also be used, for instance, to quickly fill data buffers with known contents and to study overflow conditions etc.
- *Monitoring* of internal DC-levels, of bias currents, supply voltage drops, etc., can be useful. The interesting signals can be connected to few test pads with analog multiplexers so that the external test overhead is small.
- The observation of *analog waveforms* can be accomplished by multiplexers in the pixels which connect selectable buffered signals to a readout bus.
- It should be possible to *verify* the contents of all internal registers.

3.2.4.2 Ease of Operation

The operation and test of chips before their assembly and inside a sensor system can be simplified and accelerated by additional features. For instance, defective pixels which show hits with no signal applied can mask themselves off automatically. A very time-consuming operation is the local fine-tuning of the pixel thresholds. This is usually done by an iteration of threshold scans followed by a change of the trim DAC settings. Some intelligence inside the pixel can be used to search for the DAC setting which corresponds to a 50% hit fraction. The continuous injection of the desired threshold charge is then sufficient to correctly set the thresholds in all participating pixels. Such an "auto-tune" feature has been implemented, for instance, in the ATLAS FEI3 chip [210]. Further self-test features could be implemented to verify buffers, check overflow conditions, and others.

3.3 Analog Signal Processing

This section presents some commonly used circuits for the most important building blocks of the analog part of the PUC (see Fig. 3.4). A low-noise, low-power amplifier fed back by a small capacitance is used to convert the input charge to a voltage. The feedback capacitor must be discharged after a hit by an appropriate circuit. The leakage current of the sensor flows into the pixel input in DC-coupled arrangements. It must therefore be absorbed by a suited compensation circuit. A discriminator is used to convert the analog pulse to a digital hit signal. Some chips use a bandwidth-limiting shaping amplifier between preamplifier and discriminator to reduce the noise and to increase the double hit capability. Preamplifier, feedback, and discriminator with threshold trim are briefly discussed in Sects. 3.3.1 to 3.3.3. A rough calculation of the noise expected at the output of a CMOS charge amplifier with no shaping is presented in Sect. 3.3.5. The effect of a simple shaper is analyzed in some detail in Sect. 3.3.6.

3.3.1 Charge Amplification

A simple single-ended cascoded amplifier is very often used in pixel and strip readout chips [228] owing to its simplicity and current efficiency. In the "direct" or "straight" cascode configuration shown in Fig. 3.17a the input device M_{in} is mainly biased by the current source I_{B1}, which is often operating from an additional, lower supply voltage to save power. The cascode device M_{casc} keeps the drain of M_{in} at a constant potential so that the signal current flows to node v_1 where it generates a voltage signal. The smaller current I_{B2} in this branch makes it easier to achieve a high output impedance of the current source and thus a high DC-gain. A source follower is often added to reduce the capacitive loading of the dominant node v_1 in order to increase the bandwidth. The very popular [193, 229, 230] "folded" arrangement shown in Fig. 3.17b is slightly less current efficient when the same supply voltage is used (bias current I_{B2} does not flow through the input device) but it is better suited for DC-feedback and low supply operation owing to its higher useful signal range. Variations of this basic cell using regulated cascode structures have been used as well.

Both circuits can of course also be implemented with PMOS input devices. The choice to be made is influenced by many factors, some of which are briefly listed here.

- The transconductance of the input device should be high in order to increase the bandwidth or to reduce the channel noise (see Sect. 3.3.5). This favors NMOS devices.

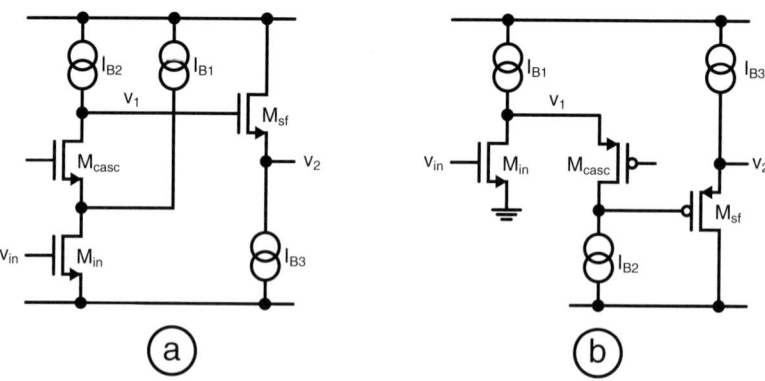

Fig. 3.17. The "direct" cascode (**a**) and the "folded" cascode (**b**) are commonly used single-ended amplifier configurations. They are often decoupled from the following circuits with voltage followers which are also indicated in the circuit diagrams

- The 1/f-noise, however, is usually better for PMOS transistors [231].
- The coupling of substrate noise into the sensitive input node through the input device can be reduced if it is located in a separate well. This can be connected to a "clean" potential.
- When designing for radiation hardness, NMOS devices require special attention in order to avoid leakage after irradiation (see Sect. 3.2.2). One approach is the use of an enclosed geometry for this transistor type. This approach leads to NMOS devices with fairly large transconductances and so current sources with a large output resistance are difficult to implement. The crucial current sources I_{B2} in Fig. 3.17 are therefore better implemented with PMOS devices if a high DC-gain is desired. This fixes the input device type for a given topology. (When using devices with large transconductance for current sources, their noise contribution may be significant.)
- The source followers in Fig. 3.17 have asymmetric (large signal) rise and fall times. The polarity of the input charge signal determines the direction of the leading edge and therefore constraints the design.

The power supply rejection of the amplifier is improved if a differential pair is used [201]. The current required for the same noise or speed is, however, increased. One input of the differential amplifier is connected to the sensor. The second "reference" input is an extremely sensitive node and must be connected to a very clean potential. In the LHC1 chip, this potential is generated internally.

3.3.2 Feedback and Leakage Compensation

The value of the feedback capacitor C_f is chosen as a compromise, among others, between a high gain (small C_f) and a large effective input capacitance (larger C_f) as discussed in Sect. 3.1.3. Typical values range from a few to tens of femtoFarads. All parasitic contributions from other devices, in particular in the feedback circuit, must be carefully considered. Such small capacitors can easily be implemented as metal–metal structures with an area of $<100\,\mu m^2$. Their geometry must be designed carefully in order to reduce additional parasitic capacitances to the input node. One possibility is to use the bump pad as one capacitor plate as illustrated in Fig. 3.18. This also provides a shield for the bump pad. If the value of C_{f1} is not sufficient, another capacitor C_{f2} can be added, its bottom plate being shielded by the injection capacitor in this example.

The charge deposited on the input node must eventually be removed, i.e. the feedback capacitor C_f must be discharged. Some general requirements for the reset circuit are a reasonably fast discharge, linearity, a small noise contribution, and the response to leakage current at the input. Some commonly used approaches are briefly discussed here (see also [232]).

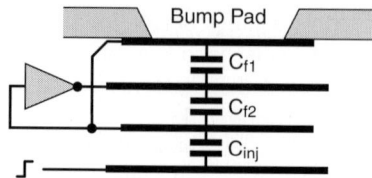

Fig. 3.18. A sandwich structure of metal plates can be used to implement the feedback capacitor while simultaneously shielding the input pad and the preamplifier input from noise sources

3.3.2.1 Resistive Feedback (MOSFET in Linear Region)

If a resistor in parallel to the feedback capacitor is used to discharge for instance $C_f = 5$ fF in $\tau = 100$ ns, a value of $R_f = 20$ MΩ is required. This is too high to be implemented as a passive device, and the associated capacitive parasitics would be prohibitive. A MOSFET operated in the linear region (see Fig. 3.19a) has a channel resistance of $R = [K(W/L)(V_{GS} - V_T)]^{-1}$. For a reasonably long device with $W/L = 0.1$ and a transconductance parameter of $K = 50$ μA/V^2, the required gate overdrive would be only $V_{GS} - V_T = 10$ mV. This small value leads to a very low saturation voltage and so only very small output voltages are possible if a true RC-discharge is desired. Larger output signals show very different behavior depending on the polarity. Sensor leakage current flowing through the feedback changes its behavior [229]. Transistor mismatch can lead to significant channel to channel variations. The simple solution of a MOSFET operated in the linear region works fairly well for AC-coupled strip readout chips where feedback capacitors are larger (required by the larger sensor capacitance) and output swings are smaller. It is difficult to bias in pixel chips, however, and saturates early unless very long devices are used.

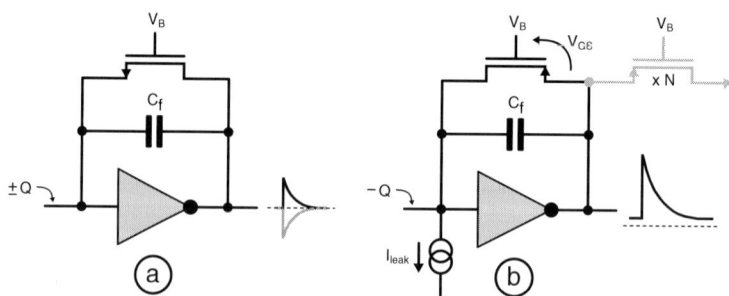

Fig. 3.19. Feedback (**a**) with a long NMOS device operated in the linear region and (**b**) with a PMOS operated in saturation

3.3.2.2 Feedback with MOS in Saturation

The feedback device of Fig. 3.19b can be kept in saturation if a small current (from sensor leakage or sometimes as input bias of the amplifier) always flows. The device type must be chosen such that the source is at the output. This requires the use of a PMOS for negative input charges (and therefore also negative leakage current). The stationary output voltage increases with higher input leakage currents. A negative charge deposition at the input leads to a positive edge at the output and so $|V_{GS}|$ of the PMOS rises. The increasing drain current discharges C_f. Due to the nonlinear relationship between the output voltage swing and the drain current, the shape of the discharge curve depends on the signal amplitude. The discharge time constant is also determined by the characteristics of the feedback transistor, by the value of the feedback capacitor, and by the leakage current. It cannot be adjusted externally. The time integral over the discharge current in the feedback device equals the input charge and so a second, N times wider transistor (shown in gray) can be used as a "current mirror" to provide a precisely N times larger replica of the input charge. Its gate is connected to V_B, the source to the output of the amplifier of Fig. 3.19b and the drain to a voltage equal to the input voltage [233].

3.3.2.3 Constant Current Feedback

The circuit shown in Fig. 3.20a discharges C_f with a nearly constant current and so the triangular pulse shapes of Fig. 3.21 are observed. The simple circuit [194, 209] is, at the same time, able to sink leakage currents I_{leak} much larger than the bias current Ib.

The function of the circuit is derived here for identical devices M_1 and M_2 operating in strong inversion with the usual square law current relationship $I_D = K'(V_{GS} - V_T)^2$ in saturation and $I_D = 2K'V_{DS}(V_{GS} - V_T - \frac{1}{2}V_{DS})$ in the linear region [234]. The abbreviation $K' = \frac{K}{2}\frac{W}{L}$ is used, with K being

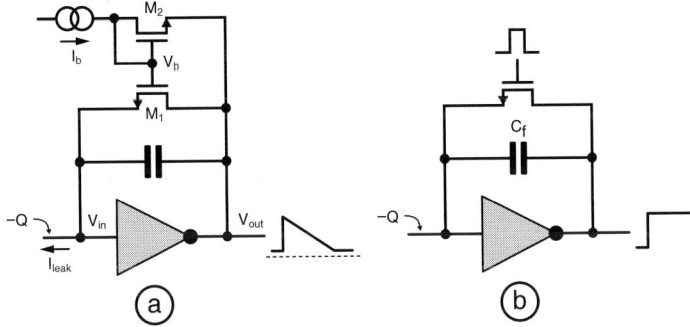

Fig. 3.20. Feedback (**a**) with a constant current source and (**b**) with a MOS switchd

the transconductance parameter and W and L the width and the length of the transistors, respectively. The substrate effect is neglected.

The configuration shown in Fig. 3.20a is suited for negative input charges which produce positive output signals. A leakage current I_{leak} leads to a slightly positive potential at the output so that the source of M_1 is actually at the input. V_{GS} of M_1 can be expressed as

$$V_{GS,1} = V_{GS,2} + V_{DS,1} = V_T + \sqrt{\frac{I_b}{K'}} + V_{DS,1} \qquad (3.18)$$

with I_b flowing through M_2 in saturation. Equation (3.18) shows that M_1 operates in the linear region ($V_{DS,1} < V_{GS,1} - V_T$) when no output signal is present and so

$$I_{leak} = 2K'V_{DS,1}\left(V_{GS,1} - V_T - \frac{1}{2}V_{DS,1}\right). \qquad (3.19)$$

Injecting (3.18) into (3.19) and solving for $V_{DS,1}$ leads to

$$V_{DS,1} = \sqrt{\frac{I_{leak} + I_b}{K'}} - \sqrt{\frac{I_b}{K'}}, \qquad (3.20)$$

and, again using (3.18),

$$V_{GS,1} = V_T + \sqrt{\frac{I_{leak} + I_b}{K'}}. \qquad (3.21)$$

When a negative charge is deposited at the input, the output becomes positive and M_1 goes into saturation. The potential at the input is held nearly constant by the feedback action of the amplifier. $V_{GS,1}$ remains unchanged as long as the voltage at node V_b does not vary (this can be guaranteed by appropriate decoupling). The current through M_1 becomes

$$I_{1,active} = K'(V_{GS,1} - V_T)^2 = I_{leak} + I_b.$$

The feedback in this situation provides the leakage current and exactly I_b to discharge C_f. A very similar result is obtained, surprisingly, if both devices operate in weak inversion. The increase of the feedback current by I_b is independent of the output amplitude and so the feedback capacitor C_f is discharged with a constant slope. This result is confirmed by the measurement in Fig. 3.21a where the feedback current I_b is varied. The output pulse comes back to the baseline after a time

$$T = Q/I_b. \qquad (3.23)$$

This expression remains valid even if the output saturates. Figure 3.21b shows the response of the circuit to increasing input charges for fixed I_b. The width

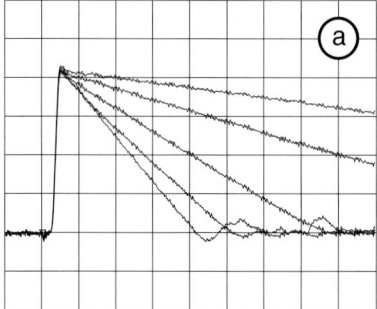 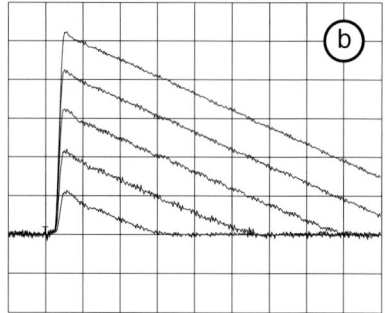

Fig. 3.21. Measured output signals of a charge preamplifier with constant current feedback (200 ns resp. 200 mV per division) [235]. The feedback current is varied in (**a**) from small values for the topmost traces to large values for the fast discharges. The injected charge is varied in (**b**)

of the pulse increases linearly with the injected charge and so a measurement of the width of the discriminated output signal can be used to determine the analog charge deposition. This method is used in the ATLAS FEI pixel chips [202]. Note that the leakage current I_{leak} can be significantly larger than the feedback current I_{b}. The only consequence in this simplified analysis is a DC-shift of the output potential according to (3.20).

3.3.2.4 Reset Switch

A simple MOS switch as shown in Fig. 3.20b can be used in applications where a continuous reset is not needed. This method has been successfully used in one of the first chips for microstrip readout [228, 236, 237]. Input and output of the amplifier are shorted and C_{f} is discharged before a signal arrives. The circuit becomes an ideal integrator once the switch is opened. Regular resets are required in order to avoid saturation of the amplifier output. Design issues are the charge injection from the gate to the input, when the switch is opened, and also the switching noise ("kTC-noise").

3.3.2.5 The Krummenacher Scheme

The circuit in Fig. 3.22 proposed already in 1991 [208] compares the DC-value of the preamplifier output to a reference voltage (which can be the voltage at the input of the amplifier) with the differential pair M_{1a}/M_{1b}. The current in the leakage compensation device M_2 is regulated such that it equals $I_{\text{leak}} + I_{\text{b}}$ in the equilibrium state. When a positive charge is deposited on the input node, the output goes negative as indicated. The complete bias current $2I_{\text{b}}$ is steered through M_{1b} while the current through M_{1a} is turned off. The input node is therefore discharged with a net current of I_{b}, independent of I_{leak}. The current through M_2 is not changed significantly because the large

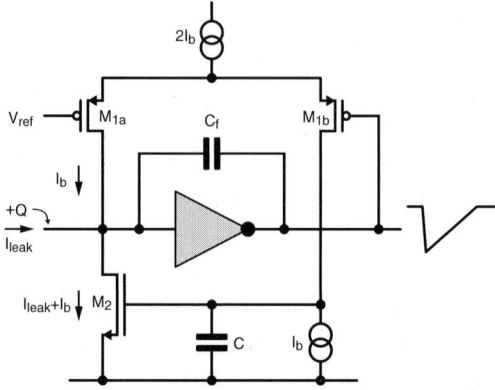

Fig. 3.22. Circuit to adjust the leakage current compensation in M_2 so that the output reaches a given reference level [208]. When a signal arrives, the current through M_{1a} is cut off and the input node is discharged with I_b.

capacitor C keeps the gate voltage nearly constant. The noise contribution of M_2 is reduced by using a long device with a small transconductance. This scheme or one with minor modifications has become very popular and is used in several chips for strip and pixel readout [192, 205, 238].

3.3.3 Hit Discrimination

A discriminator is used to detect a preamplifier (or shaper) output signal above a given threshold. It generates a digital hit signal which is fed to the readout (see Sect. 3.4) and which can also be used, for instance, to sample the analog signal amplitude for a later readout [195]. Important design aspects are power dissipation and speed, layout area, the achievable threshold range, and the homogeneity of thresholds in different pixels. Some proposed implementations are briefly presented here.

3.3.3.1 Differential Pair

A very popular circuit for the discriminator is the differential amplifier because of its simplicity and constant current operation. The threshold can be set, for instance, by generating a voltage offset between the two inputs. The circuit shown in Fig. 3.23a achieves this by pulling a current I_{thr} through a resistor R, leading to an offset RI_{thr}. The actual implementation [206] uses a PMOS device operated in the linear region for R and a weak PMOS current source for I_{thr}. The offset is a linear function of the threshold current and can be set all the way down to zero.

The threshold can be set directly if the dc output level of the driving signal can be controlled, for instance with a feedback circuit of the type shown in Fig. 3.22 where the stationary output level equals V_{ref}.

Another possibility to influence the switching point of the differential amplifier is the injection of an additional current in one of the branches as illustrated very schematically in Fig. 3.23b. This concept can be improved by using a transconductance amplifier with a wider linear range and a current comparator at the output [192].

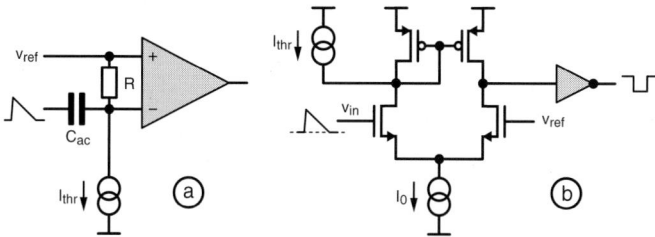

Fig. 3.23. Discriminators using differential amplifiers. The threshold can be set by introducing a DC-offset between the two inputs with the help of a resistor supplied by a current (**a**) or by unbalancing the currents in the two branches (**b**)

3.3.3.2 Cascade of Low Gain Limiter Stages

The task of the discriminator is to amplify a small ("analog") voltage difference to a digital full swing signal. It can therefore be considered as a high gain amplifier. It is well known that fast high gain amplifiers can be obtained by cascading several low gain stages. The switching is further improved by limiting the output swing of the individual stages. This concept has been proposed for pixel detectors in [208].

3.3.3.3 Diode Biassed Inverter

The very simple circuit illustrated in Fig. 3.24 has been used in the PILATUS chip [197]. An offset voltage is introduced between the input and the output of a simple CMOS inverter by pulling a current I_{thr} through a forward biassed pn-diode connecting output and input. The input signal is AC-coupled through a sufficiently large capacitor. The generated threshold is very constant because the properties of the pn-junction are much less subject to process variations than, for instance, MOS threshold voltages. The dependency of the threshold on mismatch in the inverter and in the current source is very small, and so a dispersion of only 120 e^- has been achieved without threshold trim. A drawback of this circuit is the limited range of possible thresholds.

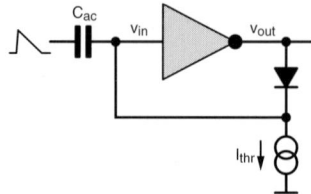

Fig. 3.24. Discriminator of the PILATUS chip. The threshold is set by the voltage across a diode supplied with current I_{thr}. The input signal is AC-coupled through a sufficiently large C_{ac}

3.3.4 Threshold Trim

A narrow threshold distribution without large tails is required in order to guarantee a constant hit efficiency and a low noise hit rate (see also Sect. 3.2.1). A careful design of the discriminator and all circuit components affecting the threshold is therefore necessary. Transistor matching and, possibly, voltage drops in supply and bias lines must be taken into account. If the required maximum threshold dispersion for a given application cannot be guaranteed by design, circuitry to fine-tune the threshold(s) in every single pixel must be included. Two basic approaches for the threshold "trim" have been pursued.

3.3.4.1 Digital Trim

Digital correction values of 3 bit [192, 195, 197, 205] to 7 bit [210, 239] resolution can be stored in every pixel. Voltage- or current-mode DACs are used to convert the binary information to control voltages or currents which influence the threshold of the discriminator(s). The "coarse" threshold is often set by a global control signal generated in the periphery of the chip with a high-resolution DAC (see Fig. 3.4). The fine adjustment in every pixel is a small correction to the global setting. The threshold change per trim step should be adjustable in order to find a good compromise between a wide trim range required to correct largely offset thresholds and a fine step size to achieve a low net dispersion after the trimming procedure. Figure 3.25 shows the threshold dispersion of a FEI2 chip before (a) and after (b) trim. The digital approach provides stable trimming once the correction values have been determined and downloaded to the pixel registers. The area required for the storage cells and the DAC is significant, however, so that the resolution is limited.

3.3.4.2 Analog Trim

Another approach is the dynamic analog storage of a correction voltage on a capacitor. The precision of the adjustment depends only on the precision of the supplied voltage, so that very wide tuning ranges with extremely high

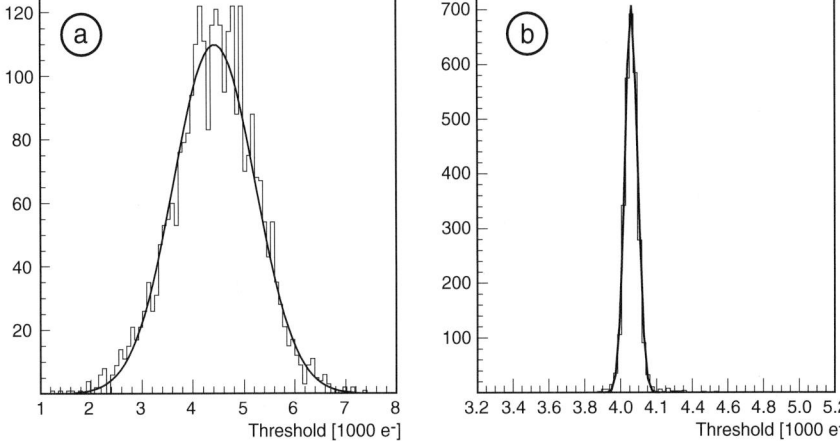

Fig. 3.25. Threshold distribution of the 2,880 pixels of a FEI chip before (**a**) and after (**b**) adjustment. The initial dispersion of $\sigma = 796\,\mathrm{e}^-$ drops to $\sigma = 35.5\,\mathrm{e}^-$

resolution can be achieved. The drawback of this approach is the unavoidable droop of the stored voltage due to leakage. This drift of the thresholds can be acceptable in applications where the chip is operated for short time intervals only, such that the values can be refreshed (like for instance in medical X-ray applications). The concept is less suited for continuous operation. The drift of the thresholds can be kept at a very low level if the leakage from the storage node is reduced. The simple circuit in Fig. 3.26 decreases leakage through parasitic diodes and the turned-off channel of the switch transistor M_2 by keeping the potentials of its drain and bulk terminals equal to the stored voltage. This is achieved by a unity gain buffer with a very small output current and so writing of an external voltage is possible through the open switches M_1 and M_2. Subthreshold leakage can be further reduced by overdriving the switch gates when they are off. A droop of the threshold of $\approx 0.1\,\mathrm{e}^-/\mathrm{s}$ has been measured at room temperature [206]. Figure 3.27 demonstrates the threshold precision which can be obtained with the analog approach. An initial rms dispersion of 327 (233) electrons, (for the two discriminators in every pixel) is reduced to 13 (11) electrons, which is basically the error in the determination of the thresholds.

3.3.5 Noise in a Simple FET Amplifier

This section gives a brief overview of the methods used to determine the noise at the output of the amplifier. A simplified circuit is treated here for illustration to give an idea of the noise values that can be achieved. It is assumed here for simplicity that the input capacitance is dominated by the sensor so that stray capacitances and in particular the input capacitance of

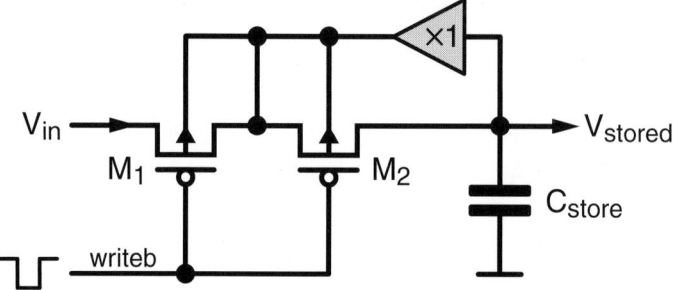

Fig. 3.26. Circuit to reduce leakage from an analog storage node. The bulk node connection and the intermediate node of two serial switches are kept at the same potential as the stored voltage by a very weak unity gain buffer so that currents flowing through leakage elements become small

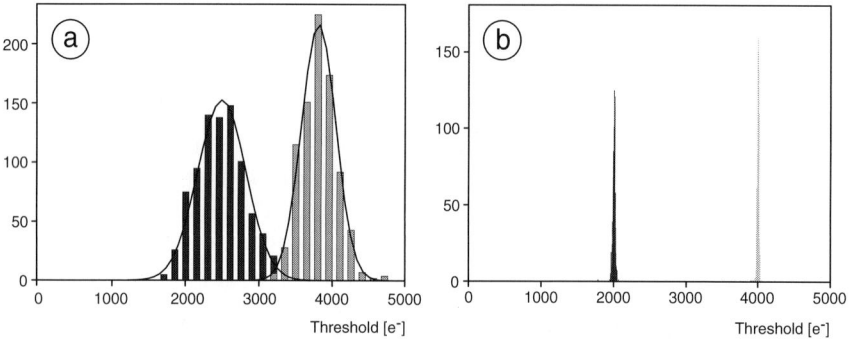

Fig. 3.27. Threshold distribution for the two discriminators in every pixel of the MPEC2.1 chip without sensor before (**a**) and after (**b**) analog trim. The rms of the distributions drop from 327 (233) electrons for the low and high threshold to 13 (11) electrons, respectively

the amplifier itself can be neglected. Only the dominant noise contributions from the sensor leakage current, from thermal and $1/f$-noise in the channel of the input transistor, and from the thermal noise of the feedback resistor are considered for simplicity. A more detailed treatment of a system with preamplifier and shaper is found in Sect. 3.3.6.

3.3.5.1 Transfer Function

Figure 3.28 shows the circuit used for the calculation in this section. The inverting preamplifier Fig. 3.28a consists of a transistor with transconductance g_m driving into a resistor R_o. The capacitance at the preamplifier output node is C_o. This circuit has the frequency-dependent voltage gain

$$v(s) = \frac{v_\mathrm{out}(s)}{v_\mathrm{in}(s)} = -\frac{v_0}{1 + s/\omega 0} , \qquad (3.24)$$

with the complex frequency variable $s = i\omega$ and with the DC-gain and bandwidth constants

$$v_0 := g_m R_o \quad \text{and} \quad \omega_0 := \frac{1}{R_o C_o}. \tag{3.25}$$

The voltage gain of this amplifier is v_0 up to the bandwidth ω_0 from whereon it drops inversely proportional to the frequency until unity gain is reached at the unity gain bandwidth $\omega_{\text{GBW}} = g_m/C_o$.

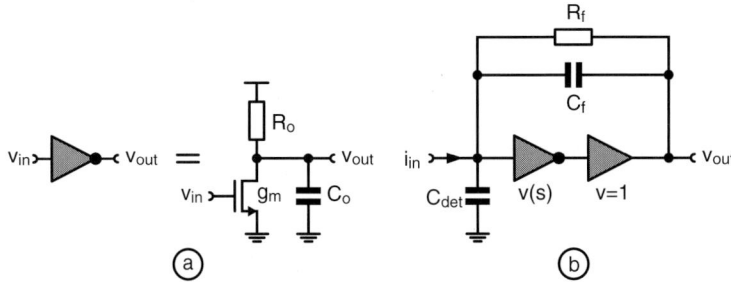

Fig. 3.28. Circuit used for the noise calculation. The preamplifier (**a**) is a simple gain stage with a capacitive load C_o. A unity gain buffer is added in the charge-sensitive amplifier configuration (**b**). The feedback capacitor C_f is discharged with a resistor R_f

The preamplifier is followed by an ideal unity gain buffer (source follower) with a small output impedance in the charge-sensitive configuration shown in Fig. 3.28b. A feedback capacitor C_f is discharged by a large resistor R_f in this example. The buffer is often added to minimize the bandwidth-limiting load at the output of the gain stage. It also simplifies the gain transfer function because a direct signal path through the feedback network is disabled. The feedback capacitor C_f is sometimes connected directly behind the preamplifier to avoid extra phase shifts introduced by the buffer. Summing currents at the input leads to

$$i_{\text{in}} = sC_{\text{det}} v_{\text{in}} + (v_{\text{in}} - v_{\text{out}}) \left(\frac{1}{R_f} + sC_f \right).$$

The elimination of v_{in} by (3.24) results in

$$-\frac{v_{\text{out}}}{i_{\text{in}}} = \frac{v_0}{\frac{1+v_0}{R_f} + s\left[\frac{1}{\omega_0 R_f} + C_{\text{det}} + (1+v_0)C_f\right] + s^2 \frac{C_{\text{det}}+C_f}{\omega_0}}. \tag{3.26}$$

As explained in Sect. 3.1.3.2, the effective input capacitance of the preamplifier should be significantly larger than the sensor capacitance. Furthermore, a realistic amplifier has a gain much larger than unity. C_f should be usually much smaller than C_{det} so as to achieve a high charge gain. With the conditions

$$v_0 C_f \gg C_\text{det} \gg C_f \quad \text{and} \quad v_0 \gg 1$$

and using (3.25), expression (3.26) simplifies to

$$H(s) = -\frac{v_\text{out}}{i_\text{in}} \approx \frac{R_f}{1 + s\left(\frac{C_o}{g_m} + R_f C_f\right) + s^2 \frac{R_f C_\text{det} C_o}{g_m}}. \quad (3.27)$$

This transfer function describes how a current signal of frequency $s = i\omega$ at the input is converted to a voltage signal at the output of the preamplifier. For DC-signals ($s = 0$), the gain is simply R_f.

The term $R_f C_f$ is the discharge time constant τ_f of the feedback capacitor. It should be larger than the rise time of the output signal τ_r in most applications as explained in Sect. 3.1.3. τ_r is the reciprocal of the closed loop bandwidth which is given by the unity gain bandwidth of the amplifier g_m/C_o multiplied by the feedback factor C_f/C_det. It is therefore reasonable to use

$$R_f C_f \gtrsim \tau_r = \frac{C_o}{g_m} \frac{C_\text{det}}{C_f} \gg \frac{C_o}{g_m}$$

so that (3.27) simplifies to

$$H(s) = -\frac{v_\text{out}}{i_\text{in}} \approx \frac{R_f}{1 + as + bs^2} = \frac{R_f}{(1 + s/\omega_l)(1 + s/\omega_h)}, \quad (3.28)$$

with

$$a = R_f C_f \quad \text{and} \quad b = \frac{R_f C_\text{det} C_o}{g_m}. \quad (3.29)$$

The two poles of the transfer function at ω_l and ω_h have been written down explicitly in (3.28). They can be determined from a and b. Simple expressions are obtained if the low-frequency pole ω_l and the high-frequency pole ω_h are widely separated, i.e. $\omega_h \gg \omega_l$, so that

$$\left(1 + \frac{s}{\omega_l}\right)\left(1 + \frac{s}{\omega_h}\right) = 1 + s\left(\frac{1}{\omega_l} + \frac{1}{\omega_h}\right) + \frac{s^2}{\omega_l \omega_h} \approx 1 + \frac{s}{\omega_l} + \frac{s^2}{\omega_l \omega_h}$$

and therefore

$$\omega_l \approx \frac{1}{a} = \frac{1}{R_f C_f} = \frac{1}{\tau_f} \quad \text{and} \quad \omega_h \approx \frac{a}{b} = \frac{g_m}{C_o} \frac{C_f}{C_\text{det}} = \frac{1}{\tau_r}. \quad (3.30)$$

3.3.5.2 Noise from Sensor Leakage

The sensor leakage current I_leak has a white noise spectrum with a spectral density (with units A^2/Hz) of [234]

$$\frac{d\langle i_\text{leak}^2\rangle}{df} = 2qI_\text{leak}. \quad (3.31)$$

This input noise spectrum is filtered by the transfer function (3.28) of the amplifier and so the squared rms noise at the output is obtained by integration of the filtered noise over all frequencies:

$$\langle v_{\text{out}}^2 \rangle_{\text{leak}} = \int_0^\infty |H(s)|^2 \frac{\mathrm{d}\langle i_{\text{leak}}^2 \rangle}{\mathrm{d}f} \, \mathrm{d}f = 2qI_{\text{leak}} R_f^2 \int_0^\infty \left| \frac{1}{1 + as + bs^2} \right|^2 \mathrm{d}f$$

$$= 2qI_{\text{leak}} R_f^2 \frac{1}{4a} = \frac{qI_{\text{leak}}}{2} \frac{R_f}{C_f}. \tag{3.32}$$

The integral can be solved in the complex plane. This noise voltage is usually referred back to the input by division by the output signal q/C_f of a single electron (the shaping loss is small in this example). The resulting ENC characterizes the system by the fluctuation of the input charge (in electrons) required to cause the observed voltage noise at the output. It can be compared directly with the signal charge to determine the signal-to-noise ratio. The ENC caused by a leakage current I_{leak} in this simple system is

$$\text{ENC}_{\text{leak}} = \frac{C_f}{q} \sqrt{\langle v_{\text{out}}^2 \rangle_{\text{leak}}} = \sqrt{\frac{I_{\text{leak}}}{2q} \tau_f} = 56 \, \text{e}^- \times \sqrt{\frac{I_{\text{leak}}}{\text{nA}} \frac{\tau_f}{\mu\text{s}}}. \tag{3.33}$$

Example: A leakage current of $I_l = 5\,\text{nA}$ at a fall time of $\tau_f = 500\,\text{ns}$ leads to an ENC of 88 electrons.

3.3.5.3 Transistor Channel Noise

The white thermal noise in the channel of the input transistor leads to an equivalent white noise voltage source at its gate with a white spectral density of [234]

$$\frac{\mathrm{d}\langle v_{\text{therm}}^2 \rangle}{\mathrm{d}f} = \frac{8}{3} \frac{kT}{g_m}. \tag{3.34}$$

A small signal serial voltage v at the input leads to an input current v/Z_{in} where the lumped input impedance Z_{in} contains the parallel connection of the detector, feedback, stray, and gate capacitance of the input device. This can be verified by explicitly writing down the transfer function for a serial voltage source (see also Sect. 3.3.6.2). Z_{in} is dominated by the sensor capacitance in this example and so the equivalent input noise current is $\langle i_{\text{therm}}^2 \rangle \approx \langle v_{\text{therm}}^2 \rangle |(sC_{\text{det}})^2|$. The output noise becomes

$$\langle v_{\text{out}}^2 \rangle_{\text{therm}} = \int_0^\infty |H(s)|^2 \frac{\mathrm{d}\langle i_{\text{therm}}^2 \rangle}{\mathrm{d}f} \, \mathrm{d}f = \frac{8}{3} \frac{kT}{g_m} C_{\text{det}}^2 R_f^2 \int_0^\infty \left| \frac{s}{1 + as + bs^2} \right|^2 \mathrm{d}f$$

$$= \frac{8}{3} \frac{kT}{g_m} C_{\text{det}}^2 R_f^2 \frac{1}{4ab} = \frac{2}{3} kT \frac{C_{\text{det}}}{C_f C_o}$$

and so

$$\mathrm{ENC}_{\mathrm{therm}} = \frac{C_{\mathrm{f}}}{q}\sqrt{\langle v_{\mathrm{out}}^2\rangle_{\mathrm{therm}}} = \sqrt{\frac{kT}{q}\frac{2C_{\mathrm{det}}}{3q}\frac{C_{\mathrm{f}}}{C_{\mathrm{o}}}}$$

$$= 104\ \mathrm{e}^{-} \times \sqrt{\frac{C_{\mathrm{det}}}{100\ \mathrm{fF}}\frac{C_{\mathrm{f}}}{C_{\mathrm{o}}}}. \tag{3.35}$$

This expression does not contain the transconductance of the input transistor because the decrease in noise for higher g_{m} is cancelled by the increase of the bandwidth. The situation is different when a shaper limits the system bandwidth independently of g_{m}. An increase of the transistor current in that case would lead to a lower noise (see Sect. 3.3.6).

Example: For a detector capacitance of $C_{\mathrm{det}} = 500$ fF, a feedback capacitor $C_{\mathrm{f}} = 5$ fF, and a load capacitor $C_{\mathrm{o}} = 50$ fF at the bandwidth-limiting output node of the amplifier, the ENC due to the transistor channel noise is 74 electrons. With an input device with a transconductance of $g_{\mathrm{m}} = 500\ \mu\mathrm{S}$, the rise time would be $\tau_{\mathrm{r}} = (C_{\mathrm{o}}/g_{\mathrm{m}} \cdot C_{\mathrm{det}}/C_{\mathrm{f}}) = 10$ ns.

3.3.5.4 Transistor $1/f$-Noise

Various expressions are used to describe the $1/f$-noise in MOS devices. For instance, the $1/f$-noise voltage (in $V^2/$ Hz) at the gate of a FET operated in strong inversion can be modeled by the expression [231]

$$\frac{\mathrm{d}\langle v_{1/f}^2\rangle}{\mathrm{d}f} = \frac{K_{\mathrm{f}}}{C_{\mathrm{ox}}WL}\frac{1}{f}, \tag{3.36}$$

with K_{f} being the device- and technology-dependent constant. C_{ox}, W, and L are the gate oxide capacitance per unit area and the (effective) transistor width and length, respectively. This expression is also used in some SPICE implementations for circuit simulation (in PSPICE for instance with switch NLEV = 2 and AF = 1).[4] The noise density depends on the frequency f and is largest at low frequencies. As before, the output noise is

$$\langle v_{\mathrm{out}}^2\rangle_{1/f} = \frac{K_{\mathrm{f}}C_{\mathrm{det}}^2 R_{\mathrm{f}}^2}{C_{\mathrm{ox}}WL}\int_0^\infty \left|\frac{s}{1+as+bs^2}\right|^2 \frac{\mathrm{d}f}{f}$$

$$\approx \frac{K_{\mathrm{f}}C_{\mathrm{det}}^2 R_{\mathrm{f}}^2}{C_{\mathrm{ox}}WL}\frac{a^2}{a^4-b^2}\ln\left(\frac{a^2}{b}\right)$$

$$\approx \frac{K_{\mathrm{f}}}{C_{\mathrm{ox}}WL}\frac{C_{\mathrm{det}}^2}{C_{\mathrm{f}}^2}\ln\left(\tau_{\mathrm{f}}\frac{g_{\mathrm{m}}}{C_{\mathrm{o}}}\frac{C_{\mathrm{f}}}{C_{\mathrm{det}}}\right).$$

---

[4] Another common parametrization of $1/f$-noise uses C_{ox}^2 in the denominator and a coefficient usually named K_{a}.

The approximation of widely separated poles has again been used in this calculation and some approximations have been made. The ENC becomes

$$\text{ENC}_{1/f} \approx \frac{C_{\text{det}}}{q}\sqrt{\frac{K_f}{C_{\text{ox}}WL}}\sqrt{\ln\left(\tau_f \frac{g_m}{C_o}\frac{C_f}{C_{\text{det}}}\right)}. \quad (3.37)$$

Example: The expression under the logarithm is 50 with the values used in the previous section, and so the last term becomes ≈ 2. In a 0.25 μm technology with $C_{\text{ox}} = 6.4$ fF/μm² and $K_f = 33 \times 10^{-25}$ J for an NMOS device [231] and a transistor with $W = 20$ μm and $L = 0.5$ μm, the ENC is

$$\text{ENC}_{1/f} \approx 9 \text{ e}^- \times \frac{C_{\text{det}}}{100 \text{ fF}} \quad \text{(NMOS input device)}. \quad (3.38)$$

The 1/f-noise contribution for the presented example with $C_{\text{det}} = 500$ fF is therefore ≈ 45 electrons. This value can be decreased in particular by using a PMOS input device with a 27 times smaller K_f [231] or, for instance, by increasing the input transistor size.

3.3.5.5 Thermal Noise of the Feedback Resistor

The white thermal noise current

$$\frac{\text{d}\langle i_{Rf}^2\rangle}{\text{d}f} = \frac{4kT}{R_f} \quad (3.39)$$

of the feedback resistor has the same effect on $\langle v_{\text{out}}^2\rangle$ as noise from the sensor leakage current, and so the term qI_{leak} in (3.32) can be simply replaced by $2kT/R_f$, yielding

$$\langle v_{\text{out}}^2\rangle_{Rf} = \frac{kT}{C_f}.$$

The ENC of

$$\text{ENC}_{Rf} = \frac{C_f}{q}\sqrt{\frac{kT}{C_f}} \approx 13 \text{ e}^- \times \sqrt{\frac{C_f}{\text{fF}}} \quad (3.40)$$

is relatively small (29 e⁻ for the values above). It does not depend on the value of the feedback resistor in this particular case, because the increase of the thermal noise for smaller resistor values is cancelled by the simultaneous reduction of the integration time.

3.3.5.6 Summary

The approximative calculations of this section have illustrated some general facts:

- The noise is determined by the frequency-dependent transfer function of the system and by the magnitude of individual primary noise contributions.

- Noise can be reduced by manipulating the transfer function, for instance, by a shaping amplifier as it will be described in the following section.
- The fundamental noise due to sensor leakage current increases in slower systems.
- The contributions from transistor thermal channel noise and from $1/f$-noise become larger with increasing detector capacitance.
- $1/f$-noise is often small compared to other sources. It becomes important, however, if very small output noise must be achieved.
- For pixel detectors, ENC values in the order of 100 electrons can be achieved without too much effort for sensors with several 100 fF capacitance and input devices with moderate g_m.

The effect of a shaper is discussed in some detail in the next section.

3.3.6 Noise in Charge-Sensitive Amplifier/Shaper Combination

The noise at the output of the amplifying chain can be decreased by the reduction of the bandwidth with appropriate filters, commonly referred to as "shapers." The system depicted in Fig. 3.29 consisting of a preamplifier and a "semi-Gaussian" shaper modeled by a combination of N high-pass and M low-pass stages is treated here as a simple example [240]. The effect of a feedback circuit is neglected. An ideal, buffered preamplifier is assumed, a realistic assumption as the shaper is usually the bandwidth-limiting element. The case of a simple CR-RC-shaper is studied in more detail. A MOS input device and a bipolar input transistor are considered for comparison.

3.3.6.1 Pulse Shapes after the Shaper

The output signal of the preamplifier can be approximated by a step function with amplitude $U = Q_\mathrm{in}/C_\mathrm{f}$ if the rise time of the amplifier is small and the discharge time constant $C_\mathrm{f} R_\mathrm{f}$ is large compared to the filter time constant of the shaper. The shaper is described by a Laplace transform $\mathcal{L}_\mathrm{HP} = s\tau(1+$

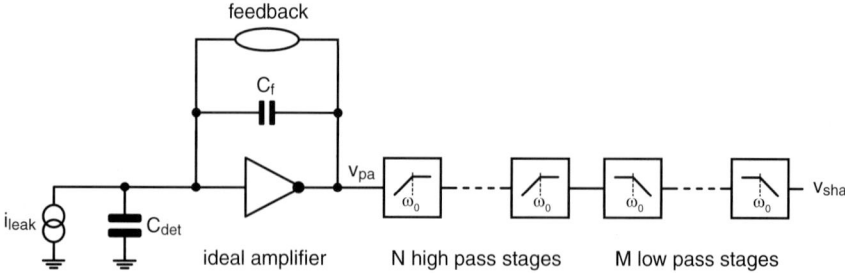

Fig. 3.29. Ideal charge preamplifier followed by N high-pass and M low-pass stages for signal filtering

$s\tau)^{-1}$ for each of the N high-pass stages and by $\mathcal{L}_\text{LP} = (1+s\tau)^{-1}$ for each of the of M low-pass stages, where $s = i\omega$ is the complex frequency variable. The filter time constant $\tau = 1/\omega_0$ is the reciprocal of the corner frequency ω_0. It is assumed identical for all high- and low-pass sections for simplicity. The complete Laplace transform of the output signal of the (N, M)-shaper with a unity step as an input signal (with Laplace transform $\mathcal{L}_\text{Step} = 1/s$) therefore is

$$\mathcal{L}^{(N,M)}(s) = \frac{1}{s}\left(\frac{s\tau}{1+s\tau}\right)^N \left(\frac{1}{1+s\tau}\right)^M = \frac{\tau^N s^{N-1}}{(1+s\tau)^{N+M}}. \tag{3.41}$$

The output signal in the time domain is calculated as the inverse Laplace transform which can be carried out by determining the residuum of the $(N+M)$-fold pole at $s = -1/\tau$.

$$\begin{aligned}
f^{(N,M)}(t) &= \text{Res} \left.\frac{\tau^N s^{N-1} e^{st}}{(1+s\tau)^{N+M}}\right|_{s=-1/\tau} \\
&= \frac{\tau^N}{(N+M-1)!} \lim_{s \to -1/\tau} \frac{d^{N+M-1}}{ds^{N+M-1}} \left[\frac{s^{N-1} e^{st}}{(1+s\tau)^{N+M}}\left(s+\frac{1}{\tau}\right)^{N+M}\right] \\
&= \frac{1}{(N+M-1)!}\left(\frac{t}{\tau}\right)^M \sum_{i=0}^{\infty} \frac{(-t/\tau)^i}{i!} \frac{(M+i+N-1)!}{(M+i)!}. \tag{3.42}
\end{aligned}$$

For a shaper with only one high-pass section ($N = 1$), this simplifies to

$$f^{(1,M)}(t) = \frac{1}{M!}\left(\frac{t}{\tau}\right)^M e^{-t/\tau} \tag{3.43}$$

with peaking time and maximum amplitude of

$$t^{(1,M)}_\text{peak} = M\tau = \frac{M}{\omega_0} \quad \text{and} \quad f^{(1,M)}_\text{max} = \frac{1}{M!}\left(\frac{M}{e}\right)^M. \tag{3.44}$$

For the simple CR–RC-filter with only one low pass ($M = 1$) this becomes

$$f^{(1,1)}(t) = \left(\frac{t}{\tau}\right) e^{-t/\tau} \tag{3.45}$$

with peaking time and maximum amplitude of

$$t^{(1,1)}_\text{peak} = \tau = \frac{1}{\omega_0} \quad \text{and} \quad f^{(1,1)}_\text{max} = \frac{1}{e}. \tag{3.46}$$

Figure 3.30a shows $f^{(1,M)}(t)$ according to (3.43). The shape of the pulses is compared in Fig. 3.30b where all curves are normalized to the same peak amplitude and peaking time. This requires the corner frequencies to be increased

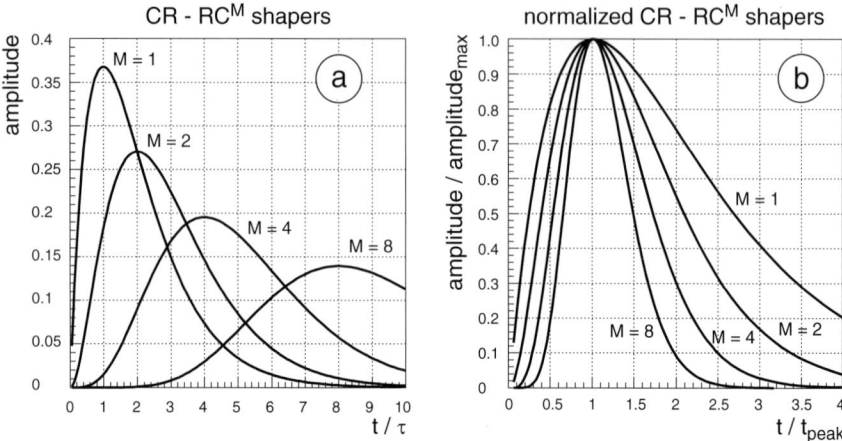

Fig. 3.30. Step responses of $CR\text{-}RC^M$-shapers for $M = 1, 2, 4, 8$ (**a**) and pulse shapes normalized to same peak amplitudes and times (**b**)

by a factor of M. The width of the normalized pulses decreases with increasing M and so higher order shapers are better suited if good double pulse resolution is required. An increase of M above 4 does not shrink the pulse width very much further. For a shaper with two high-pass sections ($N = 2$), expression (3.42) becomes

$$f^{(2,M)}(t) = f^{(1,M)}\left(1 - \frac{t/\tau}{M+1}\right). \tag{3.47}$$

This function crosses zero at $t = (M+1)\tau$ and has a negative undershoot (Fig. 3.31). This behavior is often unwanted and so additional (parasitic) high-pass stages must be avoided.

3.3.6.2 Total Noise

The noise of the amplifier is modeled by a serial noise voltage source and a parallel noise current source at its input as shown in Fig. 3.32. The noise sources are characterized by the frequency spectrum of the (squared) rms voltage $\langle v^2(f) \rangle$ or current $\langle i^2(f) \rangle$. White noise sources have frequency-independent noise while the spectrum of $1/f$-noise sources increases at low frequencies. In the following calculation, the parameterized spectral noise densities

$$\text{Serial noise voltage:} \quad \frac{\mathrm{d}\langle v^2(f) \rangle}{\mathrm{d}f} = V_0 + V_{-1}f^{-1} \tag{3.48a}$$

$$\text{Parallel noise current:} \quad \frac{\mathrm{d}\langle i^2(f) \rangle}{\mathrm{d}f} = I_0 \tag{3.48b}$$

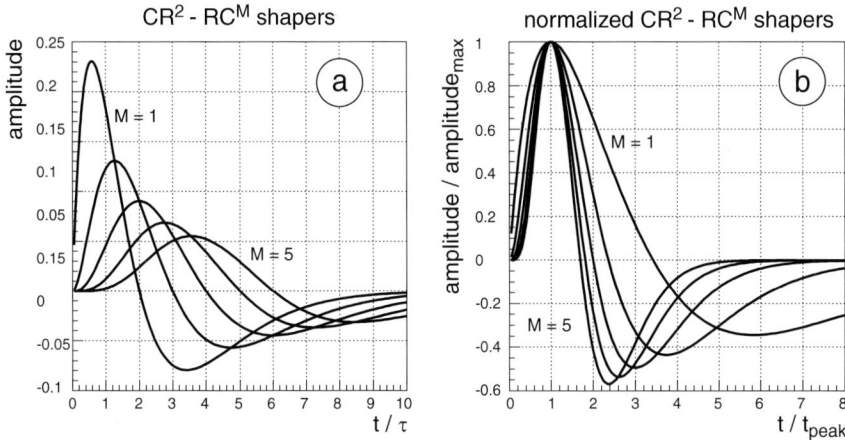

Fig. 3.31. Step responses of CR^2–RC^M-shapers for $M = 1$–5 according to (3.47) (**a**) and normalized to the same peaking amplitude and time (**b**)

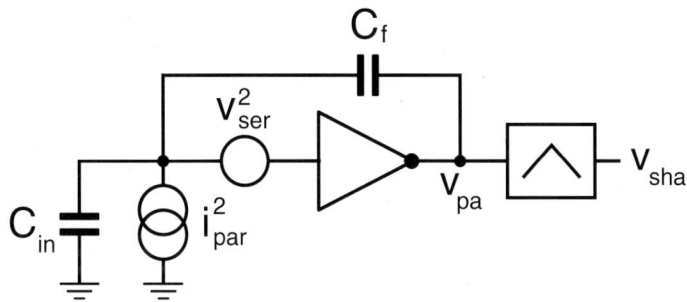

Fig. 3.32. Charge-sensitive preamplifier and shaper with input noise sources and capacitive input load

are used. V_0 and I_0 are the coefficients for the white noise contributions and V_{-1} is for the $1/f$-noise. Specific values for these parameters are given later in (3.57) and (3.59).

The effect of the two noise sources on the preamplifier output voltage can be calculated separately if they are uncorrelated: The parallel noise current must flow through the feedback capacitor assuming a perfect virtual ground at the input of the amplifier so that

$$\text{Parallel noise}: \quad \frac{d\langle v_{\text{pa}}^2(\omega)\rangle}{d\omega} = \frac{d\langle i_{\text{par}}^2(\omega)\rangle}{d\omega}\frac{1}{(\omega C_\text{f})^2} = \frac{I_0}{2\pi}\frac{1}{(\omega C_\text{f})^2}, \quad (3.49)$$

where the spectrum has been expressed as a function of the angular frequency $\omega = 2\pi f$. The serial noise voltage is related to the preamplifier output voltage through the capacitive divider made up of C_f and $C_\text{in} = C_\text{det} + C_\text{parasitic} + C_\text{preamp}$ by

$$v_{\text{pa}}^2 = v_{\text{ser}}^2 \left(\frac{C_{\text{in}} + C_{\text{f}}}{C_{\text{f}}}\right)^2.$$

Under the realistic assumption that $C_{\text{f}} \ll C_{\text{in}}$, the output noise spectrum from this noise contribution becomes

$$\text{Serial noise:} \quad \frac{\text{d}\langle v_{\text{pa}}^2(\omega)\rangle}{\text{d}\omega} \approx \left(V_{-1}\omega^{-1} + \frac{V_0}{2\pi}\right)\left(\frac{C_{\text{in}}}{C_{\text{f}}}\right)^2. \quad (3.50)$$

Equations (3.49) and (3.50) can be combined to

$$\frac{\text{d}\langle v_{\text{pa}}^2(\omega)\rangle}{\text{d}\omega} = \sum_{k=-2}^{0} c_k \omega^k, \quad (3.51)$$

with

$$c_{-2} = \frac{I_0}{2\pi C_{\text{f}}^2}, \quad c_{-1} = V_{-1}\frac{C_{\text{in}}^2}{C_{\text{f}}^2}, \quad \text{and} \quad c_0 = V_0 \frac{C_{\text{in}}^2}{2\pi C_{\text{f}}^2}. \quad (3.52)$$

The preamplifier output $v_{\text{pa}}^2(\omega)$ is filtered by a shaper with N differentiating high-pass stages and M integrating low-pass stages with identical corner frequencies ω_0. The squared transfer function of such a shaper with gain A is given by

$$H_{N,M}^2(\omega) = A^2 \frac{(\omega/\omega_0)^{2N}}{[1+(\omega/\omega_0)^2]^{N+M}}. \quad (3.53)$$

The total squared shaper output voltage becomes

$$\langle v_{\text{sha}}^2 \rangle = \int_0^\infty H_{N,M}^2(\omega)\,\text{d}\langle v_{\text{pa}}^2(\omega)\rangle = \sum_{k=-2}^{0} \int_0^\infty c_k \omega^k H_{N,M}^2(\omega)\,\text{d}\omega$$
$$= \frac{A^2}{2}\frac{1}{\Gamma(N+M)} \sum_{k=-2}^{0} c_k \omega_0^{k+1} \Gamma\left(N+\frac{k+1}{2}\right)\Gamma\left(M-\frac{k+1}{2}\right), \quad (3.54)$$

where Γ is the gamma function with $\Gamma(x+1) = x\Gamma(x)$, $\Gamma(1) = 1$, and $\Gamma(1/2) = \sqrt{\pi}$.

3.3.6.3 Equivalent Noise Charge with a Simple CR–RC-Shaper

For the simple CR–RC-shaper with $N = M = 1$, (3.54) simplifies to

$$\langle v_{\text{sha}}^2 \rangle = A^2 \frac{\pi}{4}\left(\frac{c_{-2}}{\omega_0} + \frac{2}{\pi}c_{-1} + \omega_0 c_0\right). \quad (3.55)$$

This noise voltage at the output must be normalized to a typical signal. If the peak output signal amplitude for an input charge of a single electron

$V_{\max} = \frac{1}{e} A \frac{q}{C_{\mathrm{f}}}$ from (3.46) is used, the equivalent noise charge(ENC) referred back to the input (the "ENC") becomes

$$\mathrm{ENC}^2_{\mathrm{CR-RC}} = \frac{\langle v^2_{\mathrm{sha}}\rangle}{V^2_{\max}} = \frac{e^2}{4q^2}\left(\frac{\tau}{2}I_0 + \frac{1}{2\tau}V_0 C^2_{\mathrm{in}} + 2V_{-1}C^2_{\mathrm{in}}\right), \qquad (3.56)$$

where the definitions from (3.52) have been used and the corner frequency of the bandpass filter ω_0 has been replaced by the reciprocal of the peaking time τ according to (3.46). (Note that this relation is valid only for $N = M = 1$.)

3.3.7 FET Preamplifier

For an amplifier with a FET input stage, the three contributions in (3.56) are

From leakage current I_{leak}: $\qquad I_0 = 2qI_{\mathrm{leak}}$, $\qquad$ (3.57a)

From transistor channel noise: $\qquad V_0 = \frac{8\,kT}{3\,g_{\mathrm{m}}}$, $\qquad$ (3.57b)

From $1/f$-noise: $\qquad V_{-1} = \frac{K_{\mathrm{f}}}{C_{\mathrm{ox}}WL}$. $\qquad$ (3.57c)

For the 0.25 μm technology used in Sect. 3.3.5 with $C_{\mathrm{ox}} = 6.4$ fF/μm^2, $K_{\mathrm{f}} = 33 \times 10^{-25}$ J, and the same NMOS input device with $L = 0.5$ μm and $W = 20$ μm, the ENC becomes

$$\left(\frac{\mathrm{ENC}}{e^-}\right)^2 = 115\,\frac{\tau}{10\,\mathrm{ns}}\,\frac{I_{\mathrm{leak}}}{1\,\mathrm{nA}} \qquad (3.58\mathrm{a})$$

$$+\,388\,\frac{10\,\mathrm{ns}}{\tau}\,\frac{\mathrm{mS}}{g_{\mathrm{m}}}\left(\frac{C_{\mathrm{in}}}{100\,\mathrm{fF}}\right)^2 \qquad (3.58\mathrm{b})$$

$$+\,74\left(\frac{C_{\mathrm{in}}}{100\,\mathrm{fF}}\right)^2 \qquad (3.58\mathrm{c})$$

at room temperature. For a sensor with $C_{\mathrm{in}} = 200$ fF, a leakage current of $I_{\mathrm{leak}} = 1$ nA, a shaper peaking time of $\tau = 50$ ns, and a transconductance of 0.5 mS in the NMOS input transistor, the three contributions to ENC^2 are 575, 621, and 296, respectively, and so the total theoretical ENC becomes ENC = 40 e$^-$.

The sensor must have a small leakage current and a small capacitance in order to reduce the noise. Note that other noise contributions like the intrinsic input capacitance of the amplifier due to its gate capacitance, stray capacitances, etc., have been neglected in this treatment and so the above result presents only a lower limit. The dominant contribution in this example is from white channel noise which can be reduced with more current in the device. A reduction of L also leads to an increase of g_{m} but short channel effects may then worsen the $1/f$-contribution. An increase in the input transistor width W also helps g_m, but the effective input capacitance is also increased. This capacitance adds to C_{det} so that an optimum W can be found for a given detector capacitance [240]. Using a PMOS input device with a smaller K_f decreases the $1/f$-noise, but the transconductance drops and so this would not be a good choice in this example. The contribution of the leakage current is independent of the amplifier details and depends only on the shaping time. A shorter shaping accumulates less leakage current noise, but unfortunately, the channel noise contribution increases in that case.

3.3.8 Bipolar Amplifier

A bipolar transistor can be characterized by its current gain $\beta = I_c/I_b$, where I_c and I_b are collector and base current, respectively. The transconductance is simply given by $g_m = qI_c/kT$ [234]. It is assumed here for simplicity that the detector leakage current is small compared to the base current, that the contribution of a feedback network can be neglected, and that thermal noise from the base-spreading resistance r_b can be ignored. (This is the case when $r_b \ll 1/2g_m$, which is usually the case for small collector currents.) $1/f$-noise is small in bipolar devices and is therefore neglected here ($V_{-1} = 0$). A more detailed treatment of noise in a bipolar charge amplifier can be found for instance in [241, 242]. The most important noise contributions in (3.56) are

From base shot noise: $\quad I_0 = 2qI_b = \dfrac{2qI_c}{\beta}$ (3.59a)

From transistor channel noise: $\quad V_0 = \dfrac{2qI_c}{g_m^2} = \dfrac{2q}{I_c}\left(\dfrac{kT}{q}\right)^2$ (3.59b)

and so

$$\text{ENC}^2(I_c) = \frac{e^2}{4q}\left[\frac{I_c \tau}{\beta} + \frac{C_{\text{in}}^2}{I_c \tau}\left(\frac{kT}{q}\right)^2\right]. \qquad (3.60)$$

The ENC becomes a linear function of the detector capacitance for large C_{in} as before. For a given C_{in}, the noise increases for small collector currents because the current noise referred back to the input becomes dominant. The base current shot noise, on the other hand, contributes significantly at large collector currents. This behavior is particular for the bipolar input device where an optimum collector current

$$I_{c,\text{opt}}(\tau) = \frac{kT}{q} \times \sqrt{\beta} \times \frac{1}{\tau} \times C_{\text{in}} \approx 0.25\ \mu\text{A} \times \sqrt{\beta} \times \frac{10\ \text{ns}}{\tau} \times \frac{C_{\text{in}}}{100\ \text{fF}} \qquad (3.61)$$

exists. More current is needed for faster shaping and larger input capacitance. Both noise sources give the same contribution for this optimal collector current. The resulting optimal total noise is independent of τ:

$$\text{ENC}_{\text{opt}}^2 = \text{ENC}^2(I_{c,\text{opt}}) = \frac{e^2}{2}\frac{kT}{q}\frac{C_{\text{in}}}{q\sqrt{\beta}} \quad \text{for}\quad N = M = 1 \qquad (3.62)$$

or
$$\mathrm{ENC_{opt}} \approx \frac{115\ e^-}{\sqrt[4]{\beta}} \sqrt{\frac{C_{in}}{100\ \mathrm{fF}}}. \tag{3.63}$$

For example, for $C_{in} = 400$ fF, a peaking time of 100 ns, and $\beta = 100$, the best possible noise of $\approx 75\ e^-$ requires a collector current of $\approx 1\ \mu A$. Smaller noise values are not possible with this bipolar transistor and with CR–RC shaping.

3.3.8.1 Influence of the Collector Current

The noise varies only slowly with I_c and so a smaller collector current is often chosen [241] in order to reduce the power dissipation of the preamplifier. The resulting noise increase is shown in Fig. 3.33a where the excess noise factor $E = \mathrm{ENC}/\mathrm{ENC_{opt}}$ is plotted as a function of $x = I_c/I_{c,opt}$. Equation (3.60) becomes

$$E = \frac{\mathrm{ENC}}{\mathrm{ENC_{opt}}} = \sqrt{\frac{x + 1/x}{2}} \quad \text{with} \quad x = \frac{I_c}{I_{c,opt}}. \tag{3.64}$$

The current saving can be calculated by solving the above expression for x for a given excess noise factor $E > 1$. The resulting expression $x = E^2 - \sqrt{E^4 - 1}$ is plotted in Fig. 3.33b. A reduction of the power consumption by 50% leads to a noise increase of only 11%.

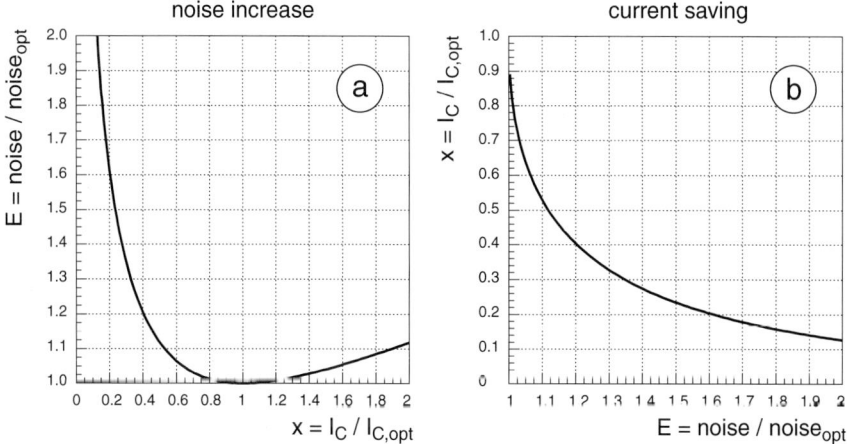

Fig. 3.33. Noise increase as a function of the normalized collector current in the bipolar input transistor (**a**) as given by (3.64), and current saving to achieve a given excess noise (**b**), calculated using the same expression

3.3.8.2 Effect of Higher Order Shapers for Bipolar Input

The output noise can be reduced by removing more high-frequency components with additional low-pass filters. Evaluation of (3.54) for $N = 1$ and general M with the bipolar noise coefficients from (3.52) and (3.59) yields

$$\langle v_{\text{sha}}^2 \rangle = \frac{A^2 \sqrt{\pi}}{2\pi C_{\text{f}}^2} \frac{\Gamma\left(M - \frac{1}{2}\right)}{\Gamma(M+1)} \left[\frac{qI_{\text{c}}}{\beta\omega_0}\left(M - \frac{1}{2}\right) + \frac{1}{2}\frac{(kT)^2}{qI_{\text{c}}} C_{\text{in}}^2 \omega_0\right]. \quad (3.65)$$

After normalization to the peak signal amplitude (3.44) for an input charge of one electron

$$V_{\text{max}}^{(1,M)} = A\frac{q}{C_{\text{f}}} \frac{1}{M!} \left(\frac{M}{e}\right)^M$$

the ENC can be compared to the optimum ENC value of the simple CR–RC-shaper (3.62):

$$E = \frac{\text{ENC}^{(1,M)}}{\text{ENC}_{\text{opt}}^{(1,1)}} = \frac{e^{M-1}\sqrt{(2M)!}}{(2M)^M}\sqrt{x + \frac{1}{x(2M-1)}}, \quad (3.66)$$

where $x = I_{\text{c}}^{(1,M)}/I_{\text{c,opt}}^{(1,1)}$ is the ratio between the collector current and the optimal current for the CR–RC-shaper (3.61). The noise is minimal for

$$x_{\text{min}} = \frac{1}{\sqrt{2M-1}}, \quad \text{where} \quad E_{\text{min}} = \frac{e^{M-1}}{(2M)^M}\sqrt{\frac{2(2M)!}{\sqrt{2M-1}}}. \quad (3.67)$$

E_{min} converges to $\frac{\sqrt[4]{8\pi}}{e} \approx 0.825$ for large M. This value is the ultimate noise improvement achievable with high order CR–RC^M-shapers. Figure 3.34a shows the excess noise factor E from (3.66) as a function of the normalized collector current for $M = 1$–4 for constant shaper corner frequency ω_0. It seems that low noise values can be achieved with very low currents if the order of the shaper is increased. This apparent improvement is, unfortunately, caused by the linear increase of the shaper peaking time with increasing M. For a *constant* peaking time the center frequency of the shaper must be increased correspondingly. The excess noise factor for the case of a constant peaking time is plotted in Fig. 3.34b. It becomes apparent that the collector current must be increased for all higher order shapers in order to improve on the noise.

3.3.9 Summary

The noise of an idealized preamplifier followed by a semi-Gaussian CR^N–RC^M-shaper has been studied. The total rms shaper noise voltage is given by (3.54) as a function of the input noise sources and of the shaper characteristic. Equation (3.56) gives the ENC for the special case of a CR–RC-shaper with

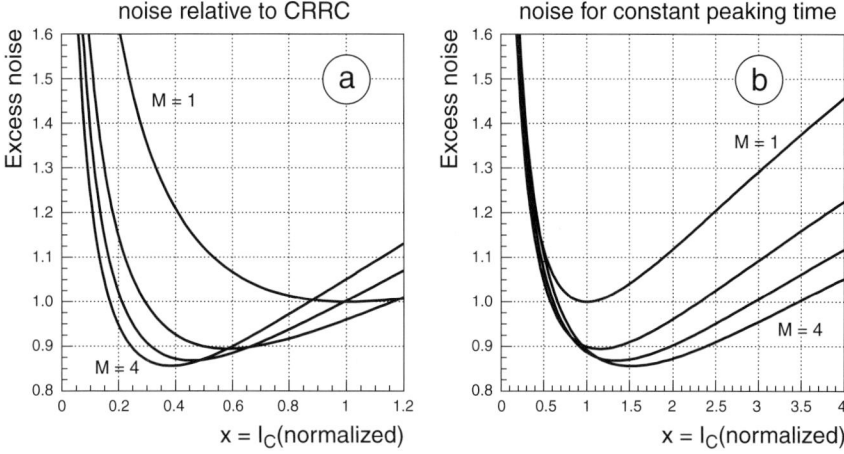

Fig. 3.34. Noise of CR–RC^M-shapers ($M = 1$–4) normalized to the noise of a CR–RC-shaper. The excess noise factor $E = \text{ENC}^{(1,M)}/\text{ENC}^{(1,1)}_{\text{opt}}$ is plotted as a function of the normalized collector current $x = I_c^{(1,M)}/I_{c,\text{opt}}^{(1,1)}$ for constant corner frequency ω_0 (**a**) and for constant peaking time (**b**)

$N = M = 1$. The consequences for a FET input device and for a bipolar input transistor have been discussed. The latter case allows a simple optimization of noise for a given detector capacitance by an appropriate choice of the collector current. The two dominating noise contributions in this case are the parallel noise current source due to the base current and a serial noise voltage source from the collector current referred back to the input. The former increases with increasing collector current while the latter decreases so that an optimum collector current $I_{c,\text{opt}}$ can be found, the exact value depending on the type of shaper. The simplest case of a CR–RC-shaper gives an ENC of 75 e⁻ for a detector capacitance of 400 fF and a bipolar current gain of $\beta = 100$. This value is achieved with a collector current of $I_c \approx 1$ µA for a peaking time of 100 ns. More current is needed for faster shapers with shorter peaking times but the minimal noise value remains the same. Higher order CR–RC^M-shapers can reduce the noise only by at most ≈15% with slightly increased collector current for a given peaking time.

3.4 Readout Architectures

The digital hit signals of the discriminators must be further processed by circuitry in the pixel and at the chip periphery. The architecture of this readout depends very much on the target application. The counting of the number of hits during a given time interval in every pixel can be sufficient in medical applications. This requires simple counters in the pixels and a mechanism to transfer the counter values to the periphery. More detailed information is required in applications in particle physics. The positions, often also the times and possibly the corresponding pulse amplitudes, of all hits belonging to an interaction must be provided. This requires a timing precision of 25 ns (the bunch crossing interval) for the detectors at LHC. Some experiments (like BTeV at FERMILAB) start the readout of every single event immediately after the interaction. Very often, however, a trigger system selects only a fraction of the events for readout in order to reduce the data volume sent to the DAQ. All hits must be identified and buffered for some time, in this case, because the trigger signal arrives with a significant delay. At the LHC experiments, the trigger latency is in the order of 2–3 μs which corresponds to $\approx$100 interactions. Almost all architectures perform an immediate zero suppression (i.e. process only pixels with amplitudes above a threshold) to reduce the size of the required buffers. The limited buffer space available can lead to a loss of hits.

The choice of a suited architecture mainly depends on the available chip technology and on the acceptable hit losses which can have very different characteristics for different readout concepts. Detailed simulations of the hit losses are therefore required before a choice can be made. An important decision to be made is whether the analog pulse height information of every hit is required. Some architectures are not suited for an analog readout.

3.4.1 Chips Without Data Buffering

The first pixel readout chips were designed for the relatively low interaction rates of the LEP accelerator so that every event could be read out. Figure 3.35 shows schematically the x–y scanning scheme [243] which has been implemented in the DELPHI experiment to find the hit pixels in the matrix. The discriminator signals set the hit flip-flop in the pixels. Horizontal "stop"-lines are pulled high if a hit is present in the row. An asynchronous vertical scan is initiated by injecting a "start row scan" token into a scan chain. The token propagates through the scan units until a stop signal is encountered in the first row with hits. The "stopped" output signal from the scan unit is used to generate the row address and to select the active row for the secondary column scan. This scan stops at the first hit column so that the x–y coordinate of the first hit is obtained. The hit flipflops are reset successively so that the scan skips from hit pixel to hit pixel.

3.4.2 Chips with Zero Suppression and Data Buffering

A large family of pixel readout chips is used for tracking in high energy physics where accelerated particles collide with a fixed target or with other particles from a beam running in the opposite direction. The collisions occur at regular, well-known intervals of 25 ns in the case of the LHC collider at

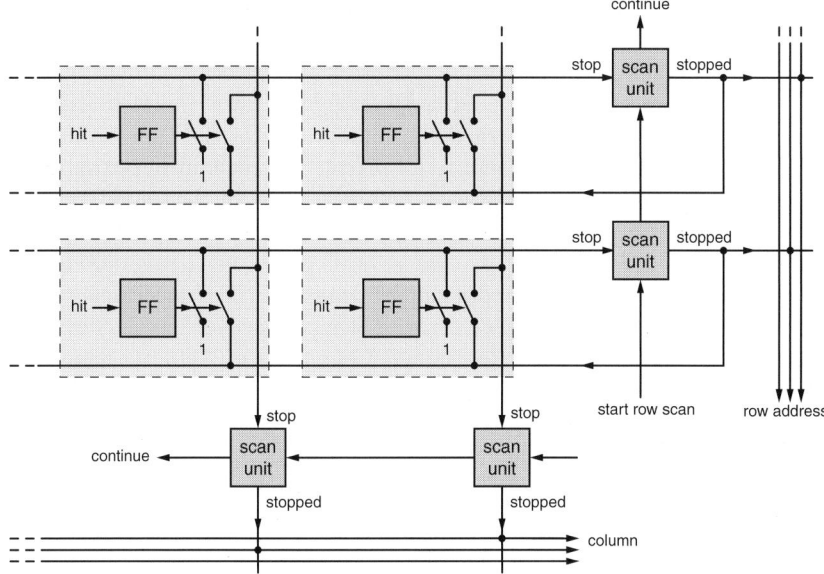

Fig. 3.35. The readout architecture of the DELPHI pixel chip

CERN, and so a bunch crossing clock (of 40 MHz at LHC) can be used to synchronize the data taking.

Most collision events do not contain relevant information, however, because the interesting physics processes are very rare. A selection of potentially interesting events is therefore made by other detectors of the experiment. This *trigger* decision, which is based on the analysis of many thousands of detector channels, requires several microseconds of processing time and so the trigger signal, which selects the interesting events for readout, arrives at the pixel chips with a delay of typically more than 100 bunch crossing clock cycles, i.e. some microseconds. The hit information of many events must therefore be stored *on the pixel chip* during this fixed *latency* interval. The event buffering is possible because only a small number of pixels ($\approx 10^{-4}$) are hit in one event. The storage of the hits only requires their real time detection, or in other words, a "zero suppression." Only triggered hit information is sent to the DAQ of the experiment; all other hit information is discarded after the latency. Although the trigger rate is low (<1% of the events), several nearly consecutive readout requests may occur. The chips must therefore be capable of accepting new triggers before the data of a previous request has been completely sent out.

Several different readout concept that have been implemented to solve the problems of hit buffering during the trigger latency are presented in the next sections. A brief comparison can also be found in [244].

3.4.2.1 Timer in the Pixel

A very simple and elegant approach is to store the hit information in the pixel by starting a timer which elapses after the trigger latency. A hit belongs to the bunch crossing of interest when the falling edge of the timer coincides with the trigger signal which is sent to all pixels. The readout of the valid hit pattern after a trigger is achieved with a shift register in the OMEGA [193] chip family. The timers in this chip are implemented as chains of current-deprived inverters (Fig. 3.36b) so that several pulses can travel along the delay line simultaneously, thus reducing the dead time. The timer delay must be precise enough to elapse with an error of at most one clock cycle, however. This requires a precision of better than 1% which is difficult to achieve in all pixels due to component mismatch. The practical implementations of the timer therefore offers a trimming possibility in every pixel to fine-tune the delays. A peak-to-peak variation of 30 ns has been achieved with 3 trim bits in the OMEGA3/LHC1 chip [201]. More recent implementations of the timer concept (for the ALICE and LHCB experiments [225]) use two timers per pixel activated alternately to further decrease the dead time in the pixel (Fig. 3.36a). Multiple hit flags can be buffered in the pixel to reduce dead time during the readout and to allow several closely spaced triggers. In order to solve the problem of delay time variations, a digital circuit (Fig. 3.36c) can be used for the timers at the expense of increased power consumption. It

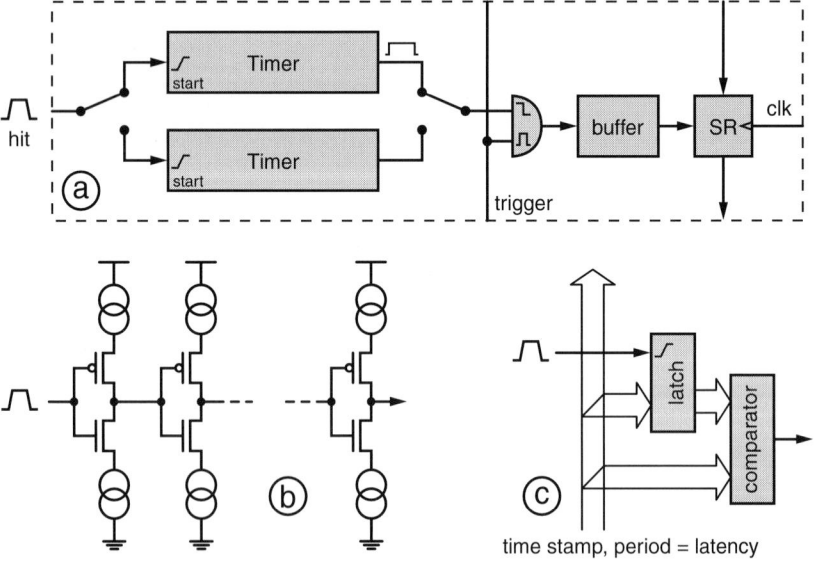

Fig. 3.36. Readout using timers in the pixel (**a**). The timers can be implemented as chains of current-deprived inverters (**b**) or by using an external time stamp in a purely digital solution (**c**)

uses a Gray-coded time stamp with a period of the latency distributed to all timers. The time stamp is latched when a hit occurs. The timer has elapsed when the distributed time stamp coincides again with the stored value.

3.4.2.2 Conveyor Belt Architecture

Another simple readout concept uses a vertically running shift register in every column to clock the row number of a pixel down the column (therefore the name "conveyor belt" proposed in [244]) as soon as a rising edge of the pixel discriminator occurs [245]. The shift register must be 6 bit wide, for instance, if the column is 63 pixels high. A shift register value ID arriving at the bottom signifies that the pixel in row ID has been hit. Furthermore, it is known that the hit has occurred ID clock cycles in the past (provided that the row numbering starts at the bottom), i.e. that the "age" of the hit is ID. The hit position is stored in a latch in one of several buffers at the bottom of the column as illustrated in Fig. 3.37. The age of the hit is reduced by the trigger latency and the result is written to a counter in the same buffer unit. The counter is incremented with the system clock. The total time since the hit has occurred equals the latency when the counter value reaches zero. The trigger coincidence is therefore made at that moment. The hits are flagged as valid for later serial readout. Several modifications (not shown in Fig. 3.37) have been made to this scheme in the ATLAS FEA chip [245] in order to improve the performance, to squeeze the logic in the available space, and to reduce power consumption. The presence of a valid ID in the shift register is flagged with a separate "full" bit which is also used to enable the clock of the shift register. The cells are therefore inactive when no hits are present, thus reducing the dynamic power consumption. A hit would be lost in this architecture in the event that the shift register cell is full at the moment where a new ID is to be written. Several tries to write the ID are therefore allowed, hoping that an empty shift register cell will pass by. The number of tries required is sent down the column in additional "late" bits for a correction of the hit time. The implementation in FEA shares the shift register between a pair of columns and between two consecutive rows in the column, i.e. one shift register cell serves four pixels with additional "left/right" and "up/down" flags to identify the hit cells.

The falling edges of the pixel discriminators can be used to determine the width of the discriminator output pulse (ToT) as a measure of the primary charge. The readout of the falling edges can be implemented as before, with an additional flag to distinguish the type of information. The buffers become more complicated in this case because the correct buffer unit for storing a falling edge information must be found.

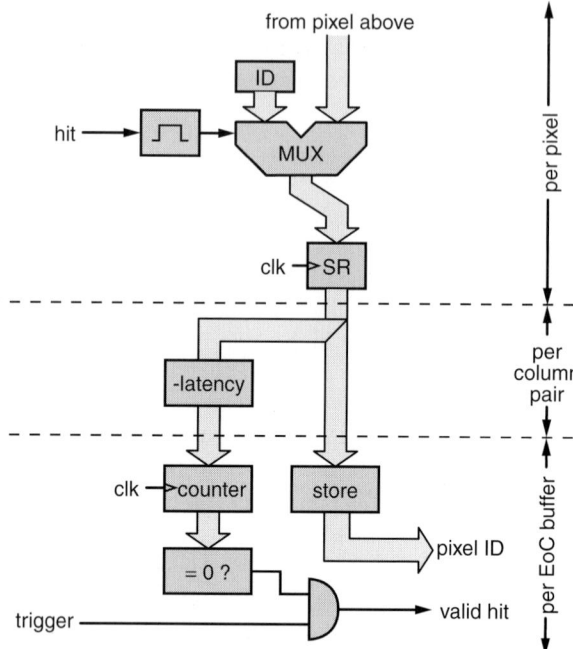

Fig. 3.37. The "conveyor belt" architecture uses a digital shift register to transport the ID of a hit pixel to the bottom of the column where it arrives exactly after ID clocks pulses. The trigger coincidence is performed after (latency-ID) further clock pulses in buffers at the bottom of the chip

3.4.2.3 Time Stamp Readout

The basic idea of this approach is to record the time at which a hit has occurred in digital form, the so-called *time stamp* [246]. When the trigger signal selects a certain bunch crossing for readout, the time information of all accumulated hits is compared to the interesting time (or time interval) referred to by the trigger signal. Hits with the correct time stamp are read out, all older hits are rejected. The time of the trailing edge of the discriminator output can be memorized as well so that the pulse width (the ToT) can be determined digitally by calculating the difference of the two values. The time stamp for a hit could be stored, in principle, in the pixel and the trigger coincidence made there. This would block a hit pixel, however, during the full latency so that a significant inefficiency would be introduced. The time stamp values are therefore transferred to buffers at the bottom of the pixel columns as fast as possible in the architecture of the chip used in the ATLAS experiment [202]. The most important elements of the digital readout are sketched in Fig. 3.38, where four elementary tasks are running in parallel:

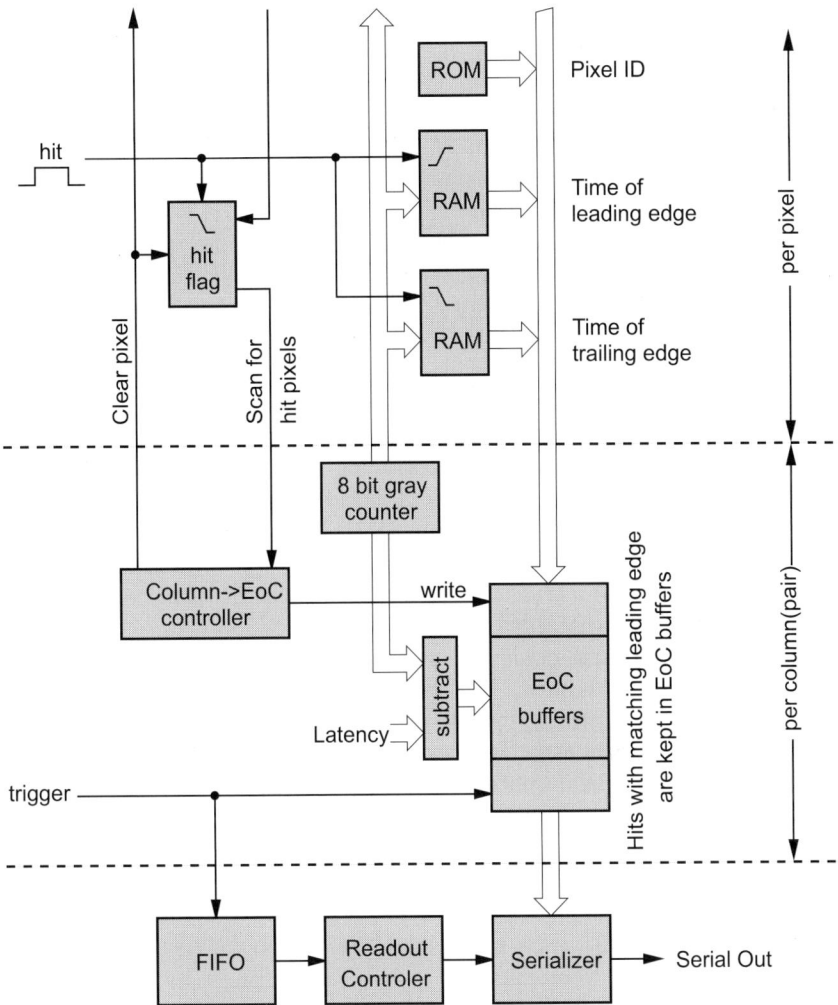

Fig. 3.38. Simplified diagram of the time stamp readout used in the FEI chip family for the ATLAS experiment

1. An 8-bit-wide time stamp counter operated at the 40 MHz bunch crossing rate generates the time reference which is distributed to all pixels in the chip. The counter uses Gray encoding to avoid out-of-sequence values during transitions and to minimize the number of bit transitions and thus the power consumption. When a pixel is hit, the time stamps of the rising and of the falling edge of the discriminator output signal are stored in memory cells in the pixel. The pixel data are ready to be processed when the falling edge has occurred and a "hit" flag in the pixel is set.

2. All hit flags in a column are connected by a fast asynchronous priority scan which indicates to a control logic at the bottom of the column that at least one hit is ready for readout. The control logic requests the uppermost hit pixel in the column to send its rising and falling edge time stamp and its row ID (hard coded in a ROM in the pixel) down the column on a 24-bit-wide bus. This information is stored in a free location of the end-of-column (EoC) buffer pool. The processed pixel is cleared and the scan continues the search for other hit pixels. Hits are thus transferred from the pixels to the EoC buffers at a programmable rate of 5–20 MHz. The left and the right halves of a column pair share the time stamp and readout busses and the EoC buffer pool consisting of 64 locations in the ATLAS FEI chip [210] to save layout area and power. The two sides are served alternately by the column control logic.
3. The hits must stay in the EoC buffers until the trigger latency, i.e. the time difference between the hit occurring in the sensor and the arrival of the trigger signal, has elapsed. The leading edge time stamp value in every EoC buffer is therefore permanently compared to the value of the time stamp counter minus the (programmable) latency value. When the values coincide and a trigger signal is present, the hit is flagged as "valid for readout." It is discarded (the EoC buffer location is freed) otherwise. A list of pending triggers is kept in a FIFO where they are distinguished by a 4-bit trigger number. The trigger number of a particular trigger is stored together with the "valid for readout" flag in the EoC buffer so that the hits can be associated with selected triggers later.
4. A readout controller initiates the serial readout of the hit data as long as pending triggers are present in the trigger FIFO. The EoC buffers are searched for valid data with the corresponding trigger number. The column and row address and the ToT of these hits are serialized and sent to the MCC. The readout controller adds a "start of event" and an "end of event" word (with error and status bits) to the data stream.

3.4.2.4 Column Drain Architecture

The "column drain architecture" [195] developed for the CMS experiment transfers all hits occurring within one clock cycle to a buffer pool in the periphery. The concept has been first implemented in a 0.8 μm technology with two metal routing layers only so that the number of devices and bus signals had to be minimized. The final readout chip was realized in a 0.25 μm technology [204] which allowed the implementation of additional features to ensure efficient operation of the innermost pixel layer at full LHC luminosity.

The main parts of the readout are illustrated in Fig. 3.39. One or several hits produced by the pixel discriminators in a column pair are flagged to the EoC by a fast OR. A single time stamp for *all* these hits is stored in 1 of 12 available digital buffer locations. The hit information of this group of pixels is now transferred immediately to a buffer pool, and hence the name "column

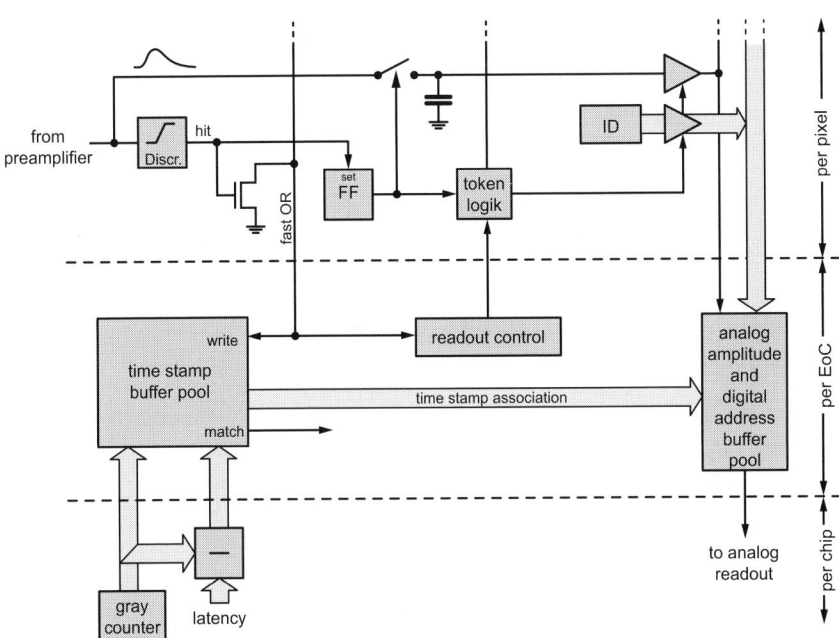

Fig. 3.39. The column drain readout transfers the amplitude and address of hits buffers in the EoC which are associated with a single digital buffer holding the time stamp of the event

drain." The hit pixels in the column pair are found by a fast scan mechanism which basically passes a token from cell to cell through closed switches until a hit cell is found. This mechanism has been designed to operate at above 1.5 GHz in order to reduce the dead time introduced by the scan. Pixels without a hit are still sensitive during the scan. The CMS readout provides the *analog* information stored by a sampling circuit on capacitors in the pixels. This analog value is output by the pixel which has been identified through the token scan. All hits are transferred sequentially to a pool of 32 buffers per column pair each containing an analog cell for the amplitude information and 9 digital cells for the pixel address. They are associated with the corresponding time stamp in the digital buffer by a pointer mechanism. Many buffers are therefore available for a single event so that high local multiplicities (as they can occur in jets of particles) can be coped with.

The time stamps in the digital buffers are permanently compared with a second time stamp value offset by the latency. When a match occurs and no trigger signal is present, the event is discarded and all buffers are freed. A serial analog readout is started otherwise. For the readout the pixel address is translated into five analog signals (two for the double column and three for the pixel) with six possible levels. The data for one hit can be transferred within six clock cycles. The readout of several chips is controlled by a token which is passed from chip to chip. It is generated by a separate "token manager" chip.

3.4.2.5 BTeV Readout

The BTeV readout implemented in the FPIX chip family [194] is shown in Fig. 3.40. Four identical EoC readout controllers at the bottom of the columns communicate with the pixels in the column through 4×2 command lines. The controllers can issue one of the four states "look for data," "idle," "output," and "reset." These states are sent to a command interpreter in every pixel. The readout can be divided into several elementary steps:

- One of the EoC controllers is requested by a priority encoder to send the "look for data" command to the pixels on its pair of command lines. This controller will be responsible for processing the next event in the column.
- When a pixel is hit, the command interpreter in that pixel is linked to the command set which carries the "look for data" pattern. The hit fast-OR line is activated. This informs the EoC that one or more hits have occurred in a column. The active EoC controller stores the time stamp (bunch crossing number) in a register and switches from the "look for data" state to the "idle" state. Another free EoC controller is selected immediately by the priority logic to issue "look for data."
- All "idle" EoC controllers and the associated pixels wait until their readout is requested. In the triggered operation mode, a desired event time stamp is presented to a comparator in the EoC controller (not shown in Fig. 3.40) which compares it to the stored time stamp value. Readout is started if the values are equal. If no match is found after a programmable timeout, the event is discarded. An operation mode without any trigger has also been implemented. All events are read out in this case.
- When a readout is requested, at most one controller issues an "output" command to its associated pixel set. All hit pixels in the set pull the read fast-OR line low. The EoC bus controller starts a token scan to find the first pixel in the selected set. The first pixel found outputs its address and ADC data onto the column bus, resets itself, and withdraws its assertion of the read fast-OR.
- All pixels have been found when the read fast-OR goes high. The EoC controller informs the priority logic that it is ready again to issue a "look for data."
- The EoC controllers can also sent a "reset" signal to reset all associated pixels if an event has been discarded, for instance, due to a timeout.

The FPIX readout uses a low-resolution flash ADC (3 bit) to digitize the hit amplitude immediately in the pixel. This information is sent to the readout bus together with the pixel address. The architecture is implemented in such a way that all hits can be read out immediately as long as the rate

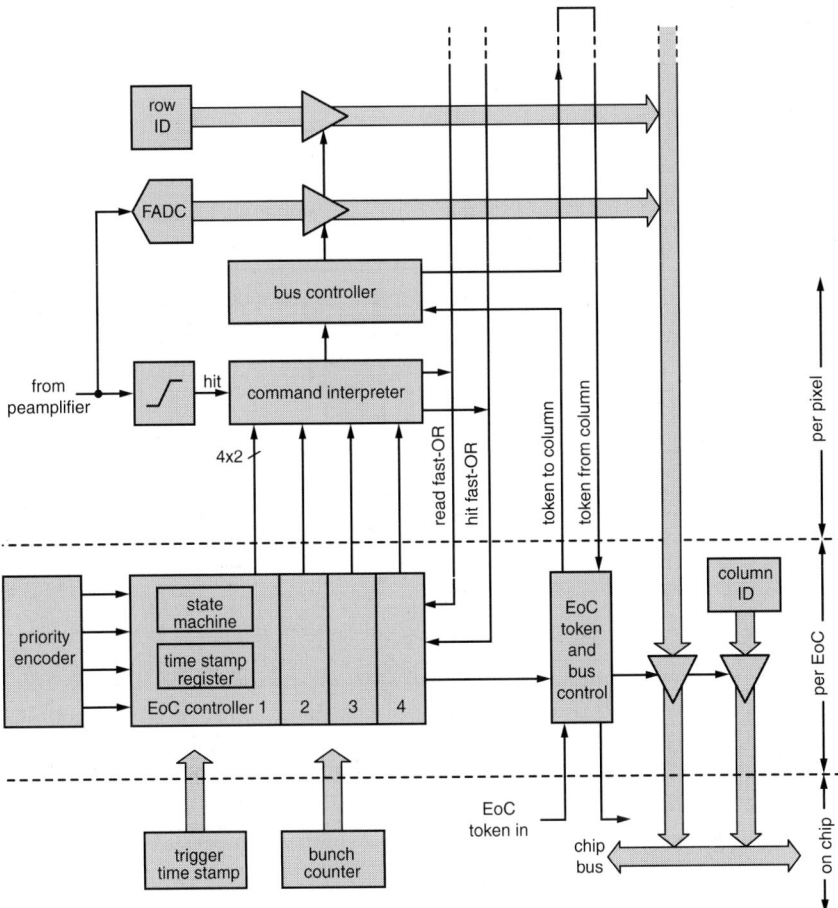

Fig. 3.40. The FPIX chip family uses several simple readout controllers in the end of the column part. All hits within a column occurring at the same bunch crossing are associated with one of the four available controllers where the time stamp is stored

is not too high. This allows the pixel information to be used for early trigger decisions. A throttling mechanism is available to discard hits for some time when the hit rate exceeds the readout bandwidth.

3.4.3 Counting Chips

For applications in biomedical (X-ray) imaging, synchrotron radiation experiments, autoradiography, and others, the number of particles absorbed in every pixel during a given time interval must be determined. The hit signals are therefore counted in every pixel and read out after the measurement interval as indicated in Fig. 3.41a. The practical implementation of this simple

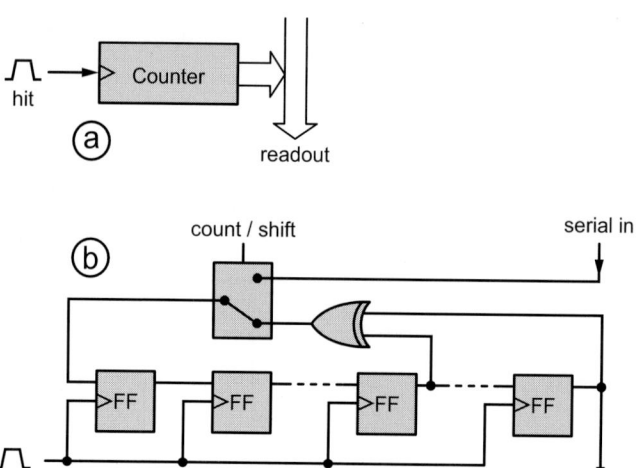

Fig. 3.41. (a) Pixel with a counter. (b) The counter can be implemented as a simple linear feedback shift register. This solution has the additional benefit of a simple serial readout

goal requires very compact counters of ≥ 16 bit to cope with the hit rates in brightly illuminated pixels. Classical binary counters are space-consuming and require a dedicated readout. Any state machine which generates as many states as possible out of N available bits can be used, in principle, to implement the counter. A particular simple design is a linear shift register fed back with an exclusive OR gate from two or more taps [247]. If the taps are chosen correctly, this leads to $2^N - 1$ states ("maximum length shift register counter"), and so the bits are used very efficiently. The bit patterns generated in this arrangement are pseudorandom numbers, unfortunately, which makes it difficult to determine the number of clock cycles from a given pattern. Lookup tables are therefore used best to determine the corresponding number of counts. Different taps can lead to slightly shorter sequences which can be decoded more easily [248]. This might therefore be a better choice for very long counters. The shift counter has the additional advantage of a simple serial readout by means of a multiplexer as shown in Fig. 3.41b. The flipflops used can be very simple dynamic cells [206] if all counters are clocked at a minimum rate with a regular refresh clock during the acquisition interval. The number of refresh clock pulses is subtracted from the count value after the serial readout of the counters. Most counting pixel chips are using this technique [162, 192, 196, 197].

The limitation of the dynamic range given by the maximum number of counts can be overcome by letting the counter "wrap around." An overflow bit is set by the 16-bit wide counters in the XPAD chip [207, 249] when this happens. External circuitry regularly scans the overflow bits of the chips and increments external counters. A virtually infinite dynamic range can be achieved with this method if it is guaranteed that every overflow bit can be recorded and cleared before the counter wraps a second time. At the end of the measurement, all counters are read out (this requires 1 ms for the XPAD chips). The total pixel count is calculated from the number of overflows and the remaining counter value.

Some counting chips use two discriminators per pixel with different thresholds so that incoming hits can be sorted into low and high energies. The analysis of the two images obtained should lead to a significant contrast enhancement for monoenergetic illumination (synchrotron) and is also expected to improve images taken with normal X-ray sources with a broader energy spectrum [198]. The latest Medipix chip [192] has two discriminators but only one counter while the MPEC2 [198] has two truly independent channels albeit with larger pixels. Systems using counting pixel chips are presented in Sect. 5.4.1.

4 Integration and System Aspects

Abstract. This chapter describes how the sensors (described in Chap. 2) and the electronics (described in Chap. 3) are put together to form a module. In particular, the various bumping methods are presented. Each module is a multichannel detector that can be fully characterized and operated independently. In practice, many modules are operated together to cover the area of interest with the necessary superposition and redundance. This chapter also deals with the problem of integrating many modules into a larger system with long-term operation stability and without losses of performance.

4.1 Introduction

Construction of large pixel systems is possible using small detector assemblies, called modules, which are arranged in a mosaic to fit the shape required by the experiment. In this section we describe the concepts which are at the base of the design of hybrid pixel modules and we present an overview of the mechanical structures used to support and cool the modules. These concepts and examples are presented having in mind applications for the next generation of high-energy hadron colliders. Related information can be found in some conference overviews [250, 251].

As already mentioned in Chap. 1, a hybrid pixel module consists of integrated circuits and sensors and allows independent optimization of sensor and electronics characteristics.

Building a module means to establish a huge number of closely packed electrical connections. These connections should not appreciably alter the sensor parameters or the front-end electronics operation; they should have low resistance ($\lesssim 100\ \Omega$) and very low capacitance ($\lesssim 0.1$ pF). A large fraction of this chapter will therefore deal with connectivity issues. The techniques most commonly used to join together electronics and sensors will be presented. The integration of power and signal connections to the remote electronics will also be addressed.

Because of their intrinsic accuracy and of the high power density required for their operation, pixel detectors require a high level of integration of the various electrical and mechanical elements ("system integration"). Some examples of this integrated approach are presented. The pixel modules must be supported by a stable mechanical structure but one in which the radiation length is minimized so as to prevent unwanted interactions of particles. The modules themselves and the power and signal cables must be as thin as practical in order to minimize the passive mass added to the system.

The heat generated by the integrated circuits (ICs), by the power and signal connections to these circuits, and by the leakage current of the sensors under a bias voltage must be controlled. This is necessary not only to avoid catastrophic thermal runaway,[1] but also to guarantee mechanical stability and long lifetime under irradiation. The local cooling needed to provide temperature control is most often integrated into the mechanical structure.

Finally, we briefly mention issues associated with radiation resistance and radio-activation of materials in pixel detector systems at high-luminosity hadron colliders.

4.2 Modules

The module is an identical (or nearly identical) unit replicated many times to create the active part of the pixel system. Only hybrid pixel modules are planned for use in the next generation of pixel detectors for the Large Hadron Collider (LHC) at CERN [12–14] and the Tevatron collider at FERMILAB [252]. This approach, already successfully used in fixed target [10] and collider [11] experiments, allows independent optimization of sensor and electronics characteristics and fabrication processes and it is the only design which is today validated by a detailed R&D program [5]. High-speed operation after hundreds of kilogray accumulated dose can only be obtained with specially designed sensors and electronics.

The basic concept of a hybrid pixel module is illustrated in Fig. 4.1. A module is made of a sensor of $\approx$10 cm^2 connected to up to 16 front-end chips. The dimension of each chip depends on the production and connection yield and is currently around 1 cm^2.

The dimension of each sensing element (pixel) should be equal to the dimension of its front-end circuit, as individual connections must be established between them over the whole active area.

The signals, once discriminated and coded, are sent out via a thin printed circuit board ("flex hybrid") to an event building chip (module control chip) that organizes the information coming from all front-end chips before sending it out through a thin cable (sometimes referred to as "pigtail"). A pixel module is therefore characterized by the huge number of closely spaced electrical bonds: many individual sensing elements need to be connected to their

---

[1] A temperature increase of 8°C doubles the bulk dark current of a silicon detector. After large fluences (e.g. several 10^{14} n$_{eq}$/cm^2) this current becomes so large as to significantly contribute to the heat load. A positive feedback, known as *thermal runaway*, is therefore possible in the case of poor heat transfer out of the pixel module

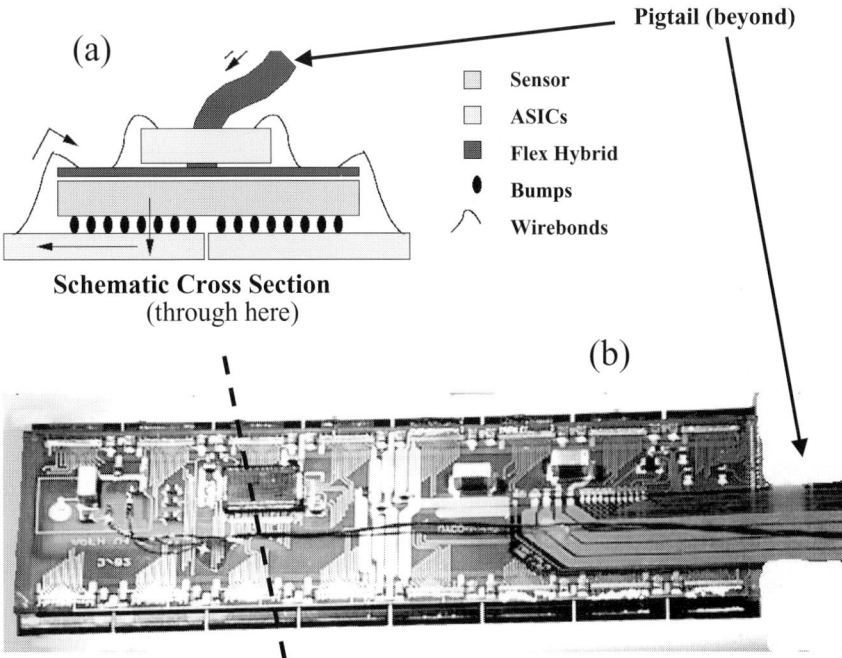

Fig. 4.1. Schematics of a hybrid module cross section (**a**) is shown together with a photograph of a real module (**b**). The module consists of (*bottom* to *top*) readout chips, bump- and wire-bond connections, sensor, flex-hybrid circuit, module control chip, and pigtail

mating readout circuit; one of such connections is shown in Fig. 4.2. Given the high connection density ($\approx$5,000/cm^2) and the minimum pitch (50 μm), the only available technique which meets the specifications is bump bonding. This technique consists of the following steps:

(a) Deposition of droplets or layers of metal on the input (output) pads of the electronics (sensors).
(b) Then flip of the electronics chip onto the sensor in such a way to precisely face the pad of the pixel (i, j) to the pad of the electronics channel (i, j).
(c) Finally, application of the proper pressure and heat cycle to guarantee that the electric contact is permanently established.

4.3 Bump Bonding

Bump bonding is extensively used in the electronics industry for the attachment of IC dies to printed circuit boards or other substrates [253]. The use of this technique is steadily increasing as a consequence of the need for compact

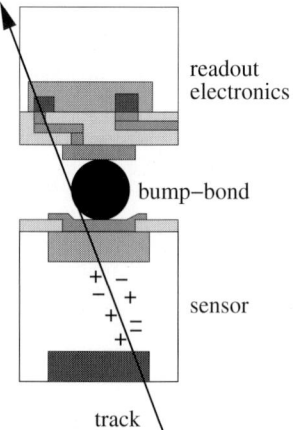

Fig. 4.2. Blow-up sketch (not to scale) of the cross section of a hybrid pixel detector, showing one of the many thousands of connections between sensor and electronics together with a particle track releasing ionization in the sensor volume

packaging of large dies with many input and output pads. The bump-bonding techniques which have proven to work at the contact density required for particle physics use either electroplated solder bumps or indium bumps deposited by evaporation. The bump deposition process is done at the wafer level; i.e., all the bumps are deposited at once on all the dies (or sensors) present on a wafer. Single die bump deposition can be done using gold studs. This technique is useful for prototype work, especially for exotic substrates.

4.3.1 Solder Bumping and Bonding Process

It is useful to start with the description of the Controlled Collapse Chip Connection (C4) [36] process introduced by IBM more than 30 years ago to overcome the limitations inherent in the wire bonding connection technique. This bumping process is illustrated in Fig. 4.3. The C4 bump deposition process is made of the following steps:

(i) Under bump metal (UBM) layers are grown on the contact pads by evaporation through a molybdenum mask with openings larger than 100 μm. These layers are made of different metals that serve different purposes. Moving out of the contact aluminum pad we find a layer of Cr that favors a good adhesion with the pad itself, a layer of Cr–Cu that acts as a barrier preventing the solder, to reach the metals underneath, a layer of Cu (or Ni) wettable by the solder, and a sacrificial thin Au layer to prevent oxidation.

(ii) A lead-rich solder (typically 95Pb/Sn) is evaporated through the mask. Typical solder "islands" of 100-μm diameter and 100-μm thickness are deposited on the UBM layers.

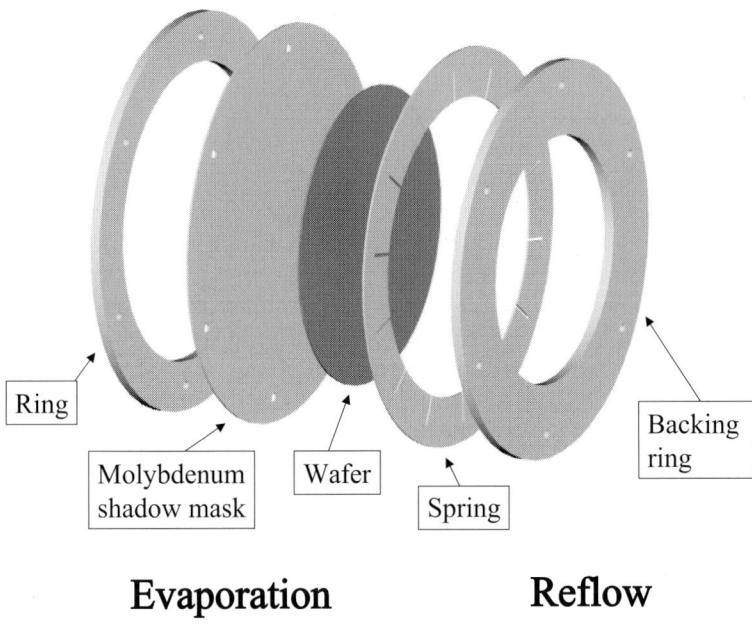

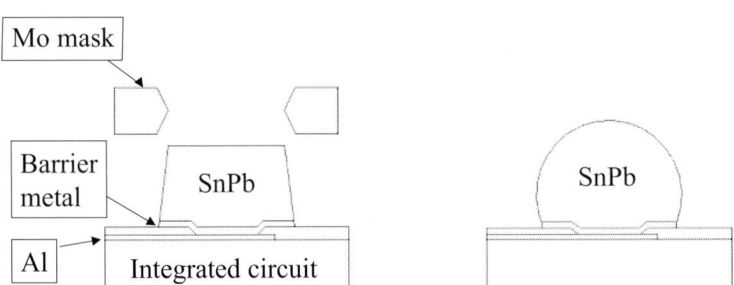

Fig. 4.3. *Top*: The mask assembly for the wafer level deposition of bumps in the C4; *bottom*: details of the deposited metal layers: a bump is shown before and after the reflow process (steps (ii) and (iii) in the text)

(iii) The wafer is heated up to the phase transition temperature ($\approx 350°C$) where, due to the surface tension, the bump will undergo a collapse to a truncated sphere shape. This operation, called "reflow," is performed in a flux or organic compound environment.

The pitch limitation (250 μm or higher), which makes C4 not useful in most particle physics applications, is due to the molybdenum mask thickness, which must exceed 100 μm to avoid warpage during the evaporation process, and to its thermal mismatch with the silicon substrate. This limitation can be overcome using UV patterned photoresist masks [254].

More recent industrial trends [255, 256, 258] favor eutectic solder (e.g. 37Pb/Sn), which is largely used for surface-mounting devices. This compound is more brittle than the Pb-rich solder, but it allows to easily mount components other than the bumped chips on the same substrate. The eutectic solder can hardly be evaporated because of the low Sn vapor pressure; the most recent bump deposition processes therefore deposit 37Pb/Sn through electroplating. This process requires good control of stoichiometry and an adequate plating system with bath analysis capability. From the Al contact pad on the IC wafer, first a thin Ti/W adhesion layer (200 nm) and diffusion barrier is sputtered before the Cu plating base ($\approx$300 nm). A wettable 1–5-μm-thick Cu metallization follows on which the PbSn cylinder is galvanically grown. The set of layers under the PbSn bump is known as *under bump metallization* (UBM) and is also built up on the pads of the sensor with an addition of Ni and Au layers for better wettability. The control of the thickness uniformity of the deposited layer and the reflow of very small droplets which result from this process are the challenges to be faced when the pitch is reduced to 50 μm. The process flow, from sputter etching to reflow, is shown in Fig. 4.4.

The PbSn cylinders are turned into 25-μm-diameter spheres by the reflow process. After processing the bump on the chip and a UBM on the sensor, both parts are merged in the flip-chip process. The chips are tacked to the UBMs of the sensors by solder flux. The subsequent reflow process requires no additional weight. The required temperature cycle reaches a peak temperature of 250°C. The distance between chip and sensor is larger than $\approx$20 μm, thus minimizing the cross talk between electronics and sensor. The reflow process which mates chip and sensor self-aligns the bumps by the surface tension once the mating parts are brought sufficiently close. This is one of the main advantages of the reflow step. The connection resistance is smaller than about 1 Ω and the ultimate shear stress is $\approx$50 MPa. Pictures of a small dimension PbSn bump after reflow and of a sequence of bumps on an ATLAS front-end chip are shown in Fig. 4.5.

A disadvantage of the solder bumping is the complexity of the technological steps. The UBM steps must be well controlled to guarantee good adhesion and long connection lifetime, especially if mechanical or thermal stresses are expected to play a significant role. The process requires comparatively high temperature cycles with temperature peaks between 250 and 350°C (vendor-dependent), which may not be acceptable for all possible mating parts.

4.3.2 Indium Bump-Bonding Process

The indium bump deposition has been developed for infrared sensor arrays, which require good mechanical and electrical properties of the connection down to liquid nitrogen temperature. While electroplating of indium is feasible, typically the bumps are grown by evaporation through polyimide masks spun on top of the wafers. Indium thickness uniformity better than 0.2 μm

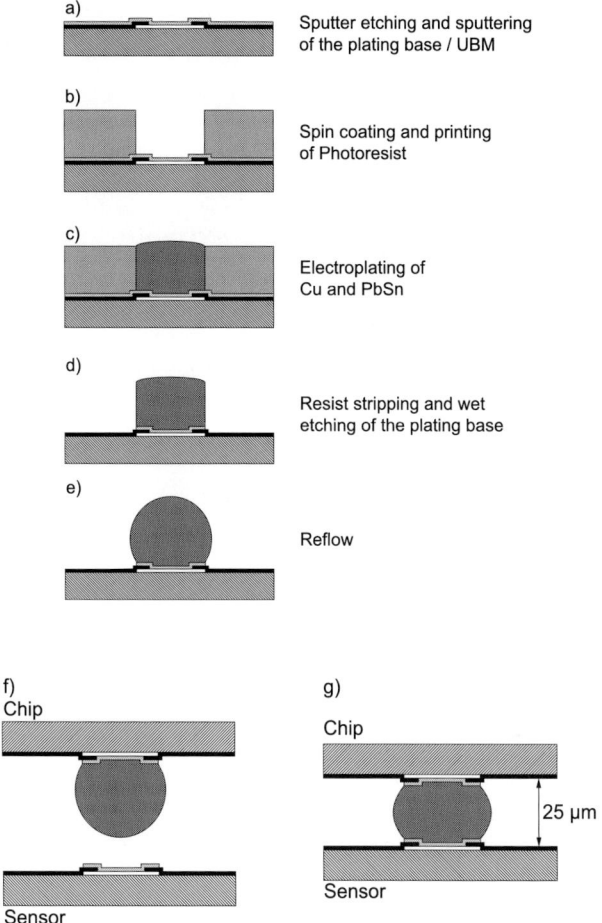

Fig. 4.4. Process flow in the PbSn (solder) bump deposition process (**a**)–(**e**) and flip-chip of the electronics die (with UBM and bump) to the sensor (with UBM only) (**f**)–(**g**) (Redrawn according to [257])

over an 8-in. wafer has been obtained routinely [259]. The openings of the masks face the pads of the dies, which must be covered by a Cr adhesion layer.

The indium bump deposition process is sketched in Fig. 4.6. Bump pitches can be as low as 30 µm, but the bump height is limited to 10 µm. This limit is due to the use of a lift-off process for the removal of the polyimide evaporation mask. Pulling off the mask becomes difficult when the thickness of the indium exceeds a certain limit. In this approach bumps must be deposited both on the sensor and on the electronics because mating is obtained by In–In thermocompression. Having bumps on both sides is also beneficial to decrease

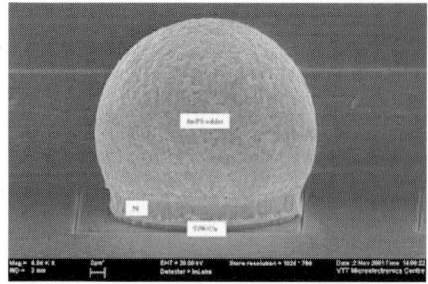

Fig. 4.5. Scanning electron micrograph of a single solder bump, showing the metal composition described in the text (courtesy of VTT Electronics, Helsinki), and of a row of solder bumps at 50-μm pitch (courtesy of FhG IZM, Berlin)

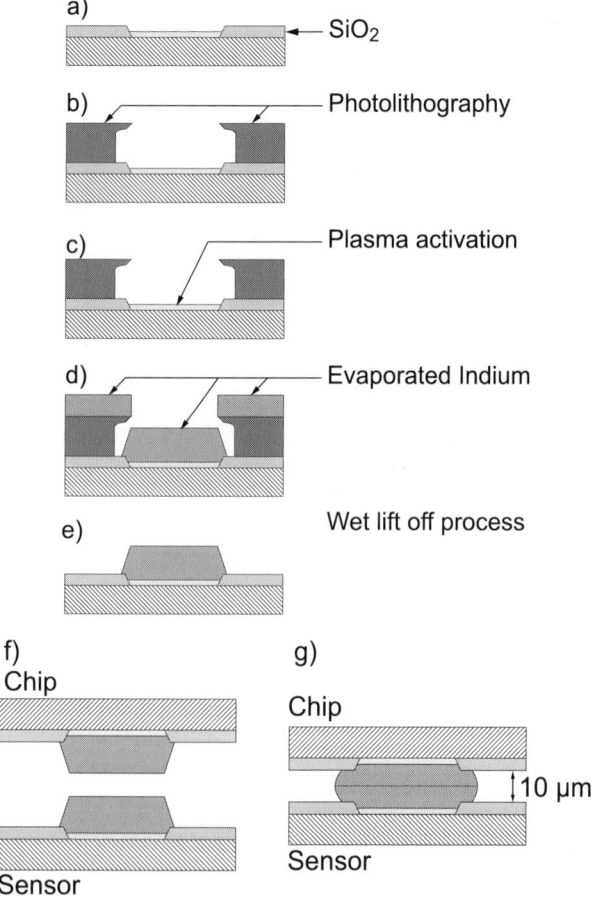

Fig. 4.6. Sketch of the standard indium bump deposition process [259] ((**a**)–(**e**)) and flip chip ((**f**)–(**g**))

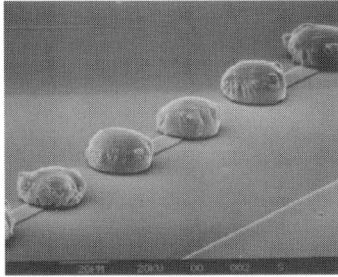
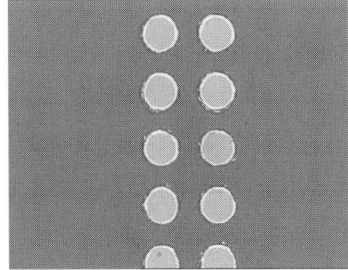

Fig. 4.7. Micrograph of an indium bump deposition on silicon at 50-μm pitch (*left*) and of a flip-chip assembly of two 50-μm pitch bump arrays (*right*) on glass substrates. (Courtesy of AMS, Rome) [259]

cross coupling between the sensor and the electronics. Figure 4.7 shows a micrograph of 50-μm pitch indium bumps deposited on two glass samples and then flip-chipped together [259].

Belonging to the third group of the periodic table of elements indium oxidizes, very easily forming an insulating In_2O_3 crust which is difficult to break. This is why the standard In process does not include reflow and also causes the connection resistance to be as high as hundreds of kiloohms before the oxide layer is broken. A voltage (≈ 500 mV) across each bump[2] is however sufficient to permanently break the oxide layer and restore a low ($\approx 10 \ \Omega$) contact resistance. In most cases the polarization voltage at the input of the charge amplifier is sufficient to break the oxide.

The flip-chip operation can be performed at room temperature, even if the best results are obtained at around 100°C. The bonding is obtained applying a pressure of about 2000 N/cm^2, i.e. slightly less than 1 g per 20-μm diameter bump. The bumps are then squashed together; their cross section increases by a factor of ≈ 2 and therefore the standoff of the two mating parts, each having 10-μm-tall deposited bumps, is of ≈ 10 μm. It is important that the pressure is applied uniformly on all the bumps of the flip-chipped parts. The parallelism between dies, the planarity, and the bump uniformity must all together give "out of plane" effects below 1 μm to avoid excessive bump squashing and consequent risk of short circuits or high cross talk. The most critical parameter is the parallelism of the flip-chip machine that must always be kept below 0.1 mrad.

The plastic properties of indium help in coping with thermal mismatches, but the poor mechanical properties of indium (e.g. ultimate shear strength of ≈ 6 MPa and ultimate tensile strength of ≈ 2 MPa) make this kind of

---

[2] This value depends strongly on the oxide thickness and therefore on the details (impurities, vacuum level, and temperature during evaporation) of the deposition process

connection quite fragile and so careful handling is required. The advantage of the indium bump bonding is the relative simplicity of the process, which implies minimal wafer handling and therefore intrinsically high yield. Moreover, the process will never expose parts to temperatures higher than 100°C.

A variant of the indium bumping includes a reflow process step after deposition. This has been especially developed [260] to increase the distance between dies after flip-chip still using a limited deposition thickness. The flip-chip operation happens, as in the case of the PbSn, in an oven where the parts, previously mated precisely, are connected through a high-temperature cycle. The bumps melt to establish connectivity and this allows for final self-alignment driven by surface tension forces.

The In deposition itself is similar to the standard indium process as illustrated in Fig. 4.8. However, it is a two-mask process: one for the sputtering of the base and one for the indium evaporation, which has much larger openings in the photoresist coating. The thin indium layers will form into spheres during reflow, as is shown in Fig. 4.9. Indium is deposited on both the sensor and the electronics chips. The layer on the sensor pads ($\approx$5-µm thickness and $\approx$50-µm diameter) undergoes reflow at $\approx$180°C, while the layer on the electronics pads ($\approx$2-µm thickness and $\approx$20-µm diameter) serves to establish a better connection. The readout chip is pressed down with about 1 g per bump for sufficient adhesion in the subsequent handling. The electromechanical connection is finally established by the reflow which also provides self-alignment.

The bump shape is roughly spherical. The size of the ball is defined by the diameter of the base metallization and by the volume of evaporated indium.

The indium bumping has the disadvantage, as any other deposition process, of being wasteful. Only a tiny fraction of the evaporated indium is used to grow bumps; the rest is deposited on the surrounding surfaces. The cost of cleaning the surfaces adds to the cost of the raw material and makes this process unsuited for large volume bumping. Only special applications justify and sometimes require the use of such a technique.

4.3.3 Gold-Stud Bump-Bonding Process

Among the materials that can be used to build bumps, gold has two interesting properties: it does not oxidize and it deforms easily allowing good electrical contact on all bumps even with some degree of bump height variation. Gold bumps can be electroplated on Ti/W/Au seed layers or, more interesting for pixel applications, can be placed through stud bumping. This last technique is a variation of the traditional ultrasonic wire bonding and is illustrated in Fig. 4.10.

The gold wire is intentionally broken after thermocompression attachment to the die pad and a coining process is used to make the bump more planar. The bumps are then attached, one after the other, to the die. Au-stud bumping allows for easier experimentation when the mating parts are not available

4.3 Bump Bonding   211

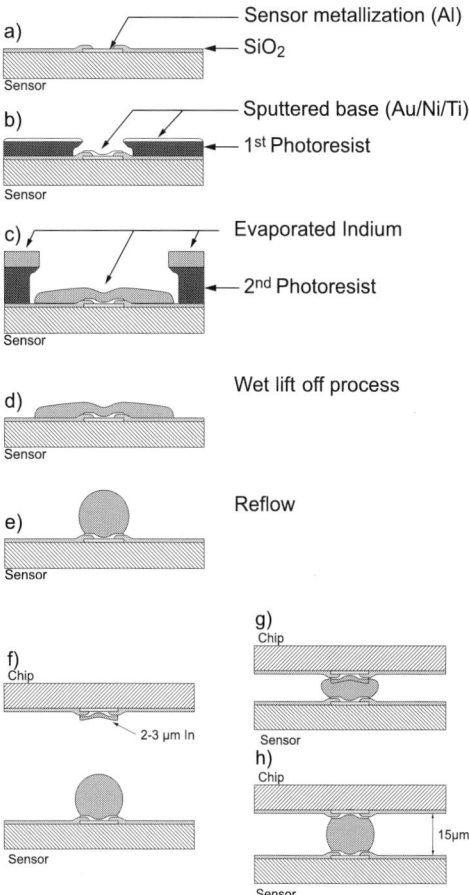

Fig. 4.8. Sketch of the indium bump deposition processes followed by reflow (**a**)–(**e**) and flip-chip operation (**f**)–(**h**) [260]

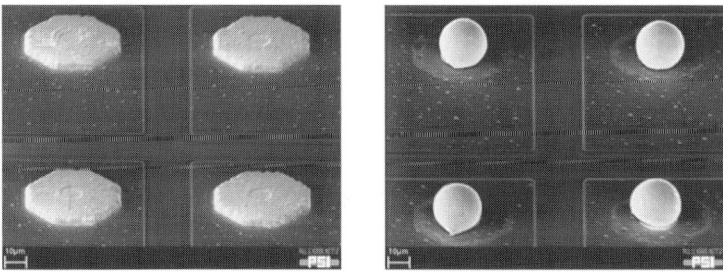

Fig. 4.9. A scanning electron micrograph of indium bumps before (*left*) and after (*right*) reflow [260] (courtesy by S. Ritter, PSI). The distance between the bumps is 100 µm

212   4 Integration and System Aspects

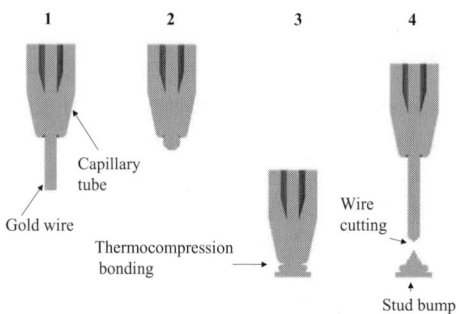

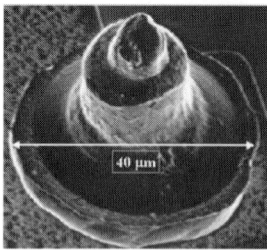

Fig. 4.10. The *top* figure illustrates the Au-stud bump formation process: a gold wire passes through a capillary tube (*1*), is heated (*2*), and pressed against the bonding pad (*3*). The wire is then pulled away and is cut close to the stud bump (*4*). The *bottom* figure shows two scanning electron microphotographs: on the *left*, one Au-stud bump; on the *right*, the bump once squashed in the flip-chip operation

on wafer scale or when materials softer and/or less homogeneous than Si are used. This process with double Au studs plus an indium topping has been used, for instance, for CdTe X-ray pixel detectors [261].

4.3.4 Quality Control of Bump-Bonded Assemblies

Inspections before and after flip-chip assembly are crucial to obtain the highest yield for functional pixel modules. Automated inspection of bumped wafers with the combined use of a television camera and laser interferometry [262] allows one to find missing bumps, merged bumps, deformed bumps, or other defects as well as to measure bump height on wafers. Inspection with high-resolution (2 µm) X-ray machines allows one to detect misalignment or merged/bridged bumps previously not detected or caused by the flip-chip process. An example of an X-ray image is shown in Fig. 4.11.

Both solder and indium bump bonding have been used to produce hundreds of pixel modules by different experimental groups and different firms,

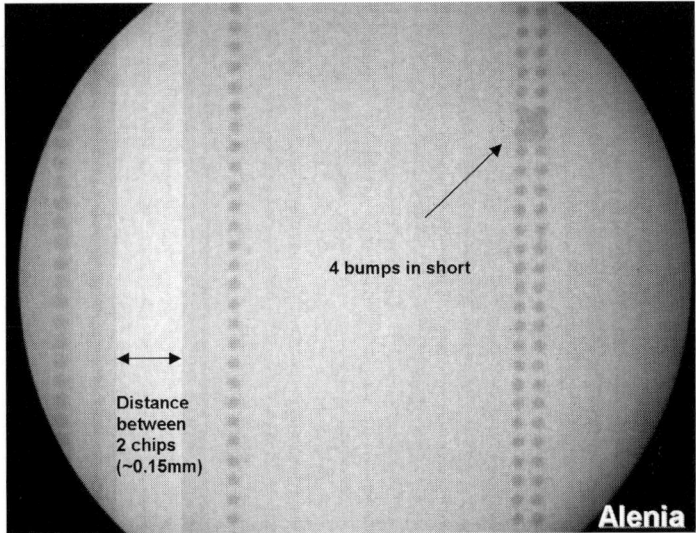

Fig. 4.11. Microradiography of an ATLAS module built with In bumps [259]. The region between two front-end chips is shown. The columns of bumps are clearly visible and some of the most common defects (missing bumps, misalignments, shorts, etc.) can be detected

with bump defect rate of $\approx 10^{-4} - 10^{-5}$ at the wafer level and $\approx 10^{-3} - 10^{-4}$ after flip-chip [263, 264].

4.3.5 Rework of Bump-Bonded Assemblies

A module is built with known good dies; i.e., all dies are tested prior to flip-chip and only the good ones are used. This is a crucial requirement as the module yield goes with the nth power of the electronics chip yield, n being the number of chips per module. It is nevertheless possible that some chips are damaged during the flip-chip operation. Reworking an assembly to change a defective chip must be considered as an option one can use in the module production process to keep the yield high.

Both solder and indium [265] bump-bonded modules have been reworked successfully. In both cases the operation requires heating and the application of a force to pull off the die. If the metal left on the sensor pads have uniform thickness, a new die can be flipped with high likelihood to properly connect all pixels. The reworking yield can be above 99% if special care is taken to avoid chips from the silicon cutting, which can destroy the surfaces of the mating parts when they stick out above the bump heights.

4.3.6 Thinning of Electronics Wafers

The silicon thickness occupied by the CMOS structures of the electronics wafer is only a few micrometers. Ninety-nine percent of the wafer thickness is needed to supply the mechanical strength during the wafer processing, i.e. to handle large size wafers with minimal breakage in the highly automatized production chain of the chip foundries. After the wafer has been processed its thickness can be reduced considerably. Thinning of IC wafers is a standard low-cost process widely used for miniaturized consumer electronics and is mostly done through backside grinding with the wafer held in place by means of UV-releasing tape glued on its active side. No special care is necessary down to 250-µm thickness. Chemical etching stress release techniques can become necessary if thicknesses much below this limit are required.

Thinning of electronics wafers to the lowest practical thickness is of obvious interest for pixel detectors builders in order to minimize the material in the path of particles. A complication for hybrid pixel detectors arises from the fact that the bump deposition process requires much wafer handling which necessitates a thickness of 700 µm for 8-in. wafers, especially if for some steps the handling is manual or semiautomatic. Electronics wafers for pixel module fabrication must therefore be thinned with bumps already deposited. This is possible by protecting the bumps by a photoresist layer which must assure that the forces applied during thinning are not transmitted to the fragile bumps. The UV-releasing tape is then applied on top of this resist layer. The actual thinning is done by backside grinding of bumped wafers as is sketched in Fig. 4.12, while the wafer side with protected bumps is turned over to the bottom. Dicing of the wafer and singularizing the individual dies should preferentially be done immediately after the thinning and, if possible, in the same production line. Nowadays, hybrid pixel modules are routinely built using IC dies thinned down to 150–180 µm [266].

4.4 "Dressing" the Modules

After the flip-chip operation is done, one has an assembly, called *bare module*, that still needs an extra layer of connectivity before becoming usable in a detector system. Adding this layer of connectivity and, eventually, of data handling is often referred to as "dressing" the module. Before starting to dress, the module should first be tested under needles in a probe station, one chip after the other, to screen defective chips and eventually send the module back for reworking. The reworking operation can, in fact, be done only at this point.

At present there are two ways of dressing a module: using a thin printed circuit (see Sect. 4.4.1) or embedding most of the connectivity on the sensor itself (see Sect. 4.4.2).

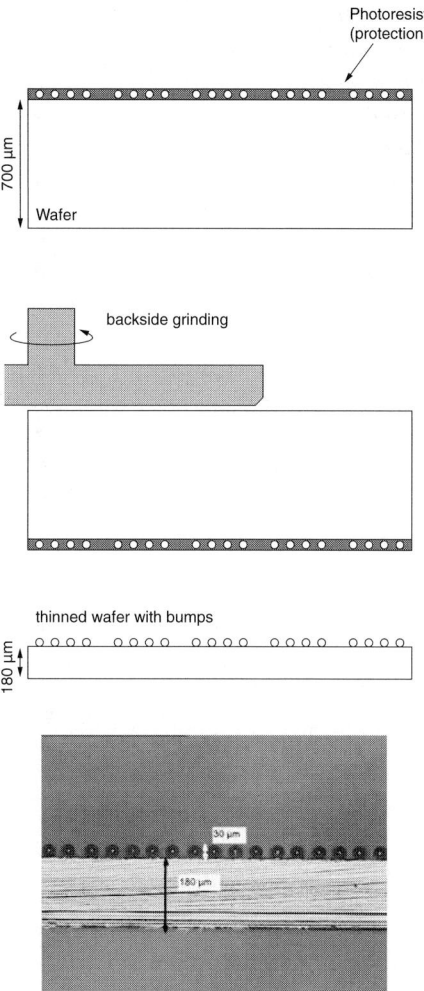

Fig. 4.12. The thinning procedure of bumped wafers by backside grinding. *Top*: bump protection by a photoresist layer; *Second from top*: backside grinding; *Third from top*: thinned wafer with bumps; *Bottom*: microphotograph (courtesy IZM) of a cut through a thinned wafer with bumps

4.4.1 Flex Hybrid

Signal and power pads of the front-end chips must be connected to remote electronics and power sources in order to operate a pixel module. The simplest solution is to use a printed circuit for routing. This can be a rigid printed circuit, a ceramic hybrid, or a flexible circuit. The latter has been adopted in all LHC pixel detectors under construction, while ceramics has been used, for instance, in WA97.

Flexible circuits for pixel modules are produced making use of conventional techniques if copper lines are acceptable. If, on the contrary, material limitation considerations impose the use of aluminum traces, only processes done in specialized laboratories [267] can be considered. Even if copper lines are adopted, the geometrical constraints (line width, spacing, via diameter, and substrate and conductor thicknesses) required by the typical pixel module design are aggressive and place this application at the edge of what is currently achievable with commercial processes. The substrate material must have a low pin-hole probability, as it must also house high-voltage ($\approx$600 V) traces for the sensor bias, and must not degrade after large (500 kGy) doses. Kapton[3] is the most commonly used substrate material. Attention must also be payed to the coverlay material; its linear CTE (coefficient of thermal expansion) and CME (coefficient of moisture expansion) must be as close as possible to those of the bare flexible circuit to avoid warpage and therefore difficulties when gluing and connecting to the bare module. Also this gluing process must avoid stresses due to the unavoidable thermal mismatch with the silicon substrate. The use of "flexible" glues (i.e. with shear modulus around 0.1 N/mm^2) is mandatory to minimize stresses on bumps.

The design of the flexible circuit [268] should take the following into account:

(a) The power distribution. Each pixel front-end chip needs $\approx$0.1 A current on each supply voltage (typically one for the analog part and one for the digital part of the chip). The power lines must therefore be dimensioned such as the voltage drop dispersion on the flex be limited to $\approx$50 mV in order to have all chips in the same operating conditions.
(b) The decoupling of the front-end chips. Optimal filtering of noise requires capacitors to be placed as close as possible to each front-end chip.
(c) The data transmission and signal integrity. Attenuation on opposite ends of an active trace and cross talk between neighboring traces should be simulated in detail taking into account all the structures and discontinuities. Correct termination of transmission lines must eventually be provided.

The flexible circuit may also house a chip that organizes the readout of the individual front-end chips and a temperature sensor to trigger rapid corrective actions if the module temperature exceeds some safety limits.

4.4.2 Multichip Module Deposited

A possible alternative to the flex-hybrid approach to make intramodule connections is to use the multichip module deposited (MCM-D) technology to create these interconnections directly on the sensor. Details of this new technology are given in Sect. 6.2.1. The concept is illustrated in Fig. 4.13.

---

[3]Kapton is a special polyimide film trademark of DuPont

4.4 "Dressing" the Modules

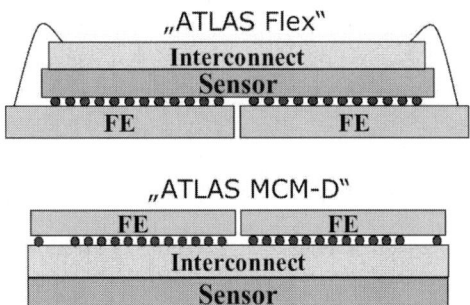

Fig. 4.13. The MCM-D concept (*bottom*) compared with the flex-hybrid approach (*top*) for module interconnections

The interconnect volume contains multiple layers of dielectric and metal (copper) which are deposited and patterned on the sensor. Bump bonding is applied afterward to connect the front-end chip to the sensor. In the case of an MCM-D module the sensor is wider than for the flex-hybrid case and extends up to the edge of the front-end ICs to allow the bump-bond connection of the chip's I/O pads to the bus system of the MCM-D. This extra 2–3 mm per side is also necessary to contain the MCM-D bus, which cannot enter the sensor active zone where the feed-through system (see Fig. 4.14) covers all the active surface.

The sensor must also be longer than in the flex-hybrid case as the module control chip and a minimum of discrete components (decoupling capacitors, termination resistors, temperature sensors) must be bump-bonded to the MCM-D bus and therefore some free sensor surface must be allocated. As the module control chip is bump-bonded to the MCM-D module together with the front-end chips, the final assembly only requires adding the discrete components and the cables.

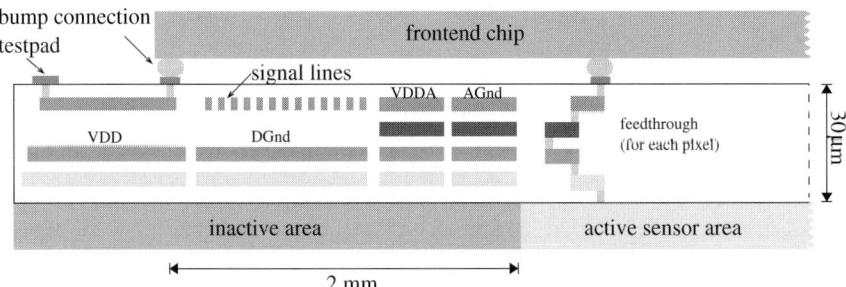

Fig. 4.14. Sketch of a cross section of a generic MCM-D assembly. The via structure connecting each pixel to its electronics channel is shown together with the signal and power busses in the sensor inactive area

The MCM-D pixel module has no wire-bond connections, which makes it easier to handle and possibly faster to assemble after the bare module has been produced. Since the process is applied on valuable sensor wafers, the yield requirements are quite severe. Preliminary results on modules built using the ATLAS readout electronics [199] indicate that noise and threshold dispersion of an MCM-D module are comparable to those of a flex-hybrid module.

The MCM-D approach is currently in the R&D stage and is not planned for use in the next generation of hadron-collider pixel detectors, but it is regarded as a promising development for future applications.

4.5 Support Mechanics and Cooling

A colliding beam experiment pixel detector is required to have large hermetic coverage. The hermeticity is obtained superimposing the modules for a small fraction (few percent) of their surface, while the volume coverage is obtained merging a barrel geometry (for polar angles larger than ≈500 mrad) and a disk geometry for smaller polar angles. This is shown, for instance, in Figs. 5.8 and 5.10.

In spite of large efforts to minimize material only about 0.1% of the mass is "active" in a pixel system; the rest is necessary to support the pixel modules precisely, and to a larger extent for services. Services are both electrical (cables to bring in power together with control and trigger signals and cables or fibers to send out data) and hydraulic (tubes and pipes to circulate a coolant).

High-speed pixel detectors have a high power consumption. The main contribution to the power budget is due to the front-end electronics that requires ≈0.5–1.0 W/cm^2. The voltage to polarize the sensor and the resulting reverse current give a power contribution that is negligible before type inversion (i.e. fluxes of ≈10^{13} m.i.p./cm^2) but may reach up to ≈0.1 W/cm^2 after hundreds of kilograys. The power dissipated in the cables is also not negligible as the operational voltage of the electronics is typically around 2.0 V and currents of ≈300 mA per chip might be necessary, especially after irradiation.

The cross section of the traces of the low voltage lines are minimized to avoid adding heavy materials in the vicinity of the interaction region; this unavoidably implies larger heat dissipations. The contribution of power cables to the module power budget is in the order of ≈0.1 W/cm^2.

As pointed out in Chap. 2, silicon sensors must be kept below temperatures of about 0°C to increase their lifetime under significant radiation.

These two facts, considerable heat load and the requirement to keep silicon sensors cold, lead to the development of powerful systems with cooling capacities of tens of kilowatts at temperatures well below 0°C. The mechanical and cooling systems must be tightly integrated to meet these requirements with minimal material.

In the sections below we describe some examples of elements of the mechanical and cooling systems for collider pixel detectors; more details can be found in [269].

4.5.1 Mechanical Supports

Requirements of the mechanical supports are numerous and often conflicting. The supports must be:

(a) Lightweight, to minimize unwanted particle interactions.
(b) Stable to few micrometers, to avoid the deterioration of the detector space resolution. This requirement must also include very small coefficients of thermal and moisture expansion.
(c) Radiation hard, to survive close to the interaction region of a high-luminosity collider.
(d) Stiff, to have small deflections and high fundamental vibration frequencies.
(e) Good conductors of heat. Transverse heat conduction coefficient of at least 1 W/mK is necessary to avoid too large temperature drops through the mechanics support when the heat load approaches 1 W/cm^2.

For practical reasons the supports are subdivided into local supports, which house few modules, and global supports, which constitute a skeleton to which the various local supports are then attached. The local supports can be more easily fabricated and handled; they must contain cooling elements and must guarantee precise placement of modules on their surface. The global support joins the local supports into a unique detector system and also houses all the pipes and cables necessary for its operation.

The schematics of a local support is presented in Fig. 4.15. The heat, produced almost uniformly over the module surface, must be funneled through the cooling channel where a low-temperature fluid is permanently circulating. This requires an in-plane thermal conductivity of the support material

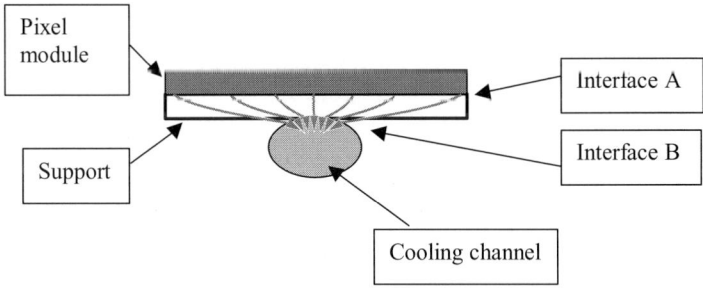

Fig. 4.15. Scheme of a local support for a pixel module with indications of the heat flow lines. Interface B is the most critical for heat transfer as the heat load is more concentrated

exceeding 3–4 W/mK. A good transversal thermal conductivity of the interfaces A and B and of the support material itself are necessary to efficiently extract heat out of the pixel module. The interfaces A and B must also be compliant to survive the thermal stresses due to the fabrication and operational conditions; typical temperature boundaries of $+50°C$ to $-25°C$ must be considered. The degree of compliance (i.e. the Young's modulus of the glue joints used for these interfaces) depends on the CTE, the CME, and the size of the parts.

Not very many materials can be considered for building pixel mechanical supports. Low-Z metals (like Be or Al) can be considered only for small structural parts as the bimetallic distortions increase linearly with the part size.

Use of carbon-based materials represents a valid alternative to low-Z metals. Carbon fibers reinforced polymers (CFRP) meet all requirements but the low transverse thermal conductivity (≈ 0.05 W/mK); they can therefore be used only if the thermal constraints are not severe. A long-term and complex process of "burning" in CO_2-filled furnaces enriches the CFRP glue matrix in carbon. The resulting material, known as carbon–carbon, keeps the excellent mechanical properties of the CFRP but with a transverse thermal conductivity (≈ 4 W/mK), which is adequate[4] for large thermal loads. Carbon–carbon is very stable with an almost negligible, but slightly negative, CTE ($\approx -1.0 \times 10^{-6} K^{-1}$), but it is porous and therefore surface treatment (for instance with bencocyclobutene) is recommended once the supports are machined to avoid the risks due to conductive dust.

4.5.2 Cooling

The large heat load of a typical pixel system (≈ 1 W/cm^2) imposes to use a liquid cooling system, either monophase or biphase. The operating temperature of a silicon pixel detector designed to survive large doses is a few degrees below zero. If one takes all temperature drops in the support system, including the glue joints, into account, the above requirement means that the coolant must circulate at about $-25°C$.

Fluorocarbon fluids, like perfluoro-n-propane (C_3F_8) or perfluoro-n-butane (C_4F_{10}), have several remarkable properties:

(a) Low viscosity at low temperature
(b) Large latent heat
(c) Excellent chemical stability
(d) Nonflammable, noncorrosive, nontoxic insulators with zero ozone depletion potential.

---

[4]The temperature drop through a 1-mm-thick support made out of such a carbon–carbon under a thermal load of 1 W/cm^2 is of 2.5°C

These properties make them an excellent choice as a cooling medium for a pixel system. Biphase (or evaporative) systems [270] profit of the large fluorocarbon fluid latent heat (≈ 100 J/g) to operate with minimal amount of coolant. It is in fact sufficient to have a liquid flow rate of 20 g/s (a factor of ≈ 10 less than for a monophase system) to extract 1 kW. The complexity of the evaporative approach, which requires a sophisticated control system to operate, is then justified by the reduction of the passive material close to the interaction region. Evaporative cooling systems tend to favor C_3F_8 as the amount of gas produced per cubic centimeter of liquid is small (71 cm^3 versus 242 cm^3 for C_4F_{10}) and therefore long cooling tubes can be operated without large pressure drops [271]. The principle of operation of such a system is illustrated in Fig. 4.16 and is not much different from the principle of operation of the refrigerator one has at home. The evaporation happens after expansion through a 0.6–0.8-mm-diameter capillary; then the gas is pumped out of the system and compressed to the liquid phase and injected into the closed loop system again.

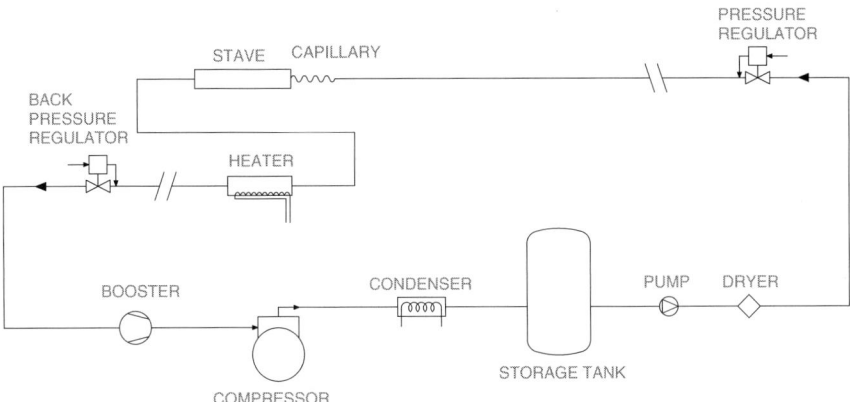

Fig. 4.16. Scheme of an evaporative cooling system. This is a closed loop circuit where a fluorocarbon (e.g. C_3F_8) is stored as liquid in a tank under pressure. The liquid is pumped into the system until expansion at the output of a capillary injects a spray into the coolant tube (of a stave in this case) directly in contact with the pixel detectors. Evaporation of the liquid droplets extracts heat efficiently from the detectors. After heating to evaporate residual liquid, the gas is then recuperated and compressed again into the cooling tank

The monophase liquid cooling system, adopted, for instance, by ALICE and CMS, is simpler to operate. This system is intrinsically more stable than the evaporative one and can more easily cope with large thermal load changes that happen at the start-up and may happen in some faulty conditions (e.g. a group of modules suffer a power cut).

Moisture condensation on the coldest pixel detector parts could create low resistance paths (especially on the sensors or on unprotected metallic lines) and then stop proper operation. The detector must therefore be surrounded by a nitrogen atmosphere with low enough water contamination to have a dew point lower than $\approx -25°C$. The dry gas must be contained in an environmental box that is quite complex as it must allow for many hermetic feed-throughs and can also operate as a thermal screen. A thermal screen is a thermal insulation surrounded by a set of resistances (e.g. thin copper lines printed on a kapton foil) heated by the Joule effect that guarantees thermal neutrality between the pixel detector and the environment.

4.6 Power and Signal Interconnect in High-Luminosity Colliders

The critical system aspects for a pixel detector are primarily related to the high power density necessary to operate it. The previous section dealt with the problem: how can one efficiently extract the power (i.e. the heat) out of the detectors? This section, on the contrary, addresses the questions: how is the power supplied? And how are the signals extracted?

A pixel system in a particle physics experiment is made of hundreds or thousands of "modules," each fed with 5–10 W at ≈ 2 V (plus up to 1 W at $\lesssim 600$ V) requiring to transmit megabits of data per second. To complicate the issue this system is relegated in a high-radiation environment, where working conditions are severe. Large voltage drops in the supply cables are unavoidable as the power supply system can hardly be located close to the high-radiation region where the detectors operate. This requires all power supplies to have individual floating channels to allow for different voltage drops along the cables. One module may in fact use more or less current depending on the accumulated dose or the specific operating conditions required. Nevertheless, the safe operation of a pixel detector requires no floating parts in order to avoid uncontrolled conditions to happen. All low-impedance grounding must preferentially be done where the front-end electronics operates, i.e. at the module level, while high-impedance safety grounding should be done at the supply end. This star topology should avoid ground loops and therefore large currents flowing through unwanted and uncontrolled paths.

While complex power supplies can hardly be made such as to survive large radiation doses, voltage regulators can be built using technologies able to survive tens of kilograys [272]. These regulators can then be located a few meters apart from the detector and can safely stabilize the electrical operating conditions of each module.

Due to their very low capacitance and good shielding, individual pixels can hardly pick up noise directly. The long power lines, on the contrary, have a much larger electromagnetic coupling to the environment. To minimize the common mode coupling (and therefore the noise increase on the pixel electronics) power and sense lines must be connected, whenever possible, using (eventually shielded) twisted pair lines.

One unavoidable source of noise is due to the beam image currents accompanying the beam crossings inside the beam pipe of the accelerator. An aluminum Faraday cage with skin depth of $\gtrsim 50$ μm surrounding the beam pipe itself should be sufficient to dump the beam image current at the LHC crossing regions.

The data transmission is preferentially done through optical fibers to improve the immunity to noise and to increase the data throughput. Some distance (up to a few meters) can be covered with differential electrical signals without performance penalties, but electro-optical conversion of signals must still happen in proximity of the pixel modules. This means that the design of chips [273] transforming the light into electrical signals (and vice versa) and of optical transmitters and receivers [274] must be implemented using radiation-hard technologies. The solutions adopted differ if only digital signals have to be transmitted or if analog information is necessary too, the linearity requirements being more severe in the second case. The optical transmitters are semiconductor laser diodes, either VCSEL (vertical cavity surface emitting lasers) or MQW (multiquantum well), while the optical receivers are p-i-n diodes. All these devices have long lifetime and can be built to survive doses of hundreds of kilograys with minor deterioration of performance [274, 275].

Optopackages have two interfaces: one electrical and one optical, the latter being, by far, the most difficult to realize with good repeatability and good yield as the light source (or target) and the fiber must be aligned within a few micrometers. Having a good scheme to couple the optopackage with the optical fiber is therefore a crucial element in the whole optoelectronic readout system.

Many readout channels (tens of thousands) go through one optical fiber. Therefore, this optoelectrical interface constitutes a single point failure issue with potential large effects on the detector performance. Optical fibers must be protected by sheaths as they are quite fragile and incorrect handling, especially excessive bending, may damage them.

4.7 Operation in High-Radiation Environment

Only some components and some materials can survive 500 kGy, the dose expected after 10 years of LHC operation at 10-cm radial distance from the interaction point. A detailed qualification program must be carried out before a part is allowed to stay in this environment. This is a complex and lengthy operation and consequently parts or functions that must not necessarily stay on the module should be moved as far as practical.

Radiation resistance qualification is done using pion and/or proton beams. Pions are the dominating component of the particle flux in the vicinity of the

LHC interaction region, while the neutron flux only becomes important at larger radii (up to a few meters) due to hadronic interactions, mainly in the calorimeters, resulting in a large secondary neutron flux. The neutron-induced damage can, in the first approximation, be neglected for pixels. After irradiation to 500 kGy or more, the samples are stored for the time necessary for the radioactive isotopes to decay. Once the samples can be safely handled they are characterized by measurements to verify if they still fulfill the design specifications. The measurements are done in great detail for the key detector components (e.g. sensor and electronics) that are also continuously monitored during the radiation exposure. Proper checks must also be performed on other parts (e.g. passive electronics components, glue interfaces, mechanical supports, coolant and cooling circuits) that enter into the detector construction.

The dose rate used for these kind of tests is typically 2 orders of magnitude larger than the one expected during normal operation. Reverse annealing effects, if any, must then be taken into account when evaluating the results.

A pixel detector will, very likely, require maintenance after some accumulated dose. Choice of materials should also take this into account [276], minimizing the content of elements with large atomic number and, in particular, of those elements (e.g. Ag) that may produce, under a hadronic flux, long-lifetime radioactive isotopes (e.g. ^{110m}Ag, a γ- and β-emitter with a half-life of 250 days). Isotopes with lifetime shorter than a few days are less of a concern as the access to the pixel detector is a quite complex operation that requires halting the accelerator for weeks.

It is interesting to observe that the requirements on the radiation hardness for the long-term operation of a pixel detector at the LHC also enhances the likelihood to survive catastrophic beam losses. This kind of accidents did happen in some collider experiments in the past, causing permanent deterioration of the performance of the vertex detector. To mimic a beam loss accident at LHC far upstream of the detector, bunches of 10^{11} protons of 24 GeV, lasting 42 ns, have been sent at grazing angle toward an ATLAS pixel module. The instantaneous dose provided in this configuration is 30 times the expected maximum instantaneous dose in the ATLAS pixel detectors in case of a beam accident at LHC. The only noticeable effect [277] was a drop of all supply voltages due to the large current peak in the pixel diodes. Once the voltages were restored the module did work as before. The results remained unchanged using up to 200 proton bunches, demonstrating the relative immunity of this kind of detector to beam accidents.

5 Pixel Detector Applications

Abstract. Pixel detectors have been originally developed for particle physics applications. The next generation of experiments at hadron colliders can greatly profit from the unique characteristics of the pixel detectors, provided these can be tailored to the demanding environment expected at these machines. These requirements boosted the activities around pixel detectors and also favored their development for imaging applications in biomedicine and in astronomy.

5.1 Pixel Detectors for High-Luminosity Collider Experiments

The driving projects for the next generation of pixel detectors are the experiments at the Large Hadron Collider (LHC). Three of the four large area pixel detectors currently in construction will constitute the innermost layers of three LHC experiments: ATLAS [12], CMS [13], and ALICE [14]. The fourth pixel detector is also designed for a hadron collider, FERMILAB'S Tevatron, in the BTeV [278] experiment. This is no surprise as pixel detectors can efficiently cope with the high particle densities and the large integral particle fluxes, both generated close to the interaction region of high-luminosity hadron accelerators.

In the LHC, protons are accelerated from 350 GeV up to 7 TeV and are then maintained at this energy while rotating in the 27-km-long accelerator vacuum pipe 100 m underground. One beam of protons rotates clockwise, and the other beam counterclockwise in separate but close orbits and they can be forced to collide almost "head-on" in specific regions around which the experiments are located. In order to accelerate the protons and control their trajectories, they are grouped into "bunches," each containing about 10^{11} particles. Each bunch is confined in cylindrical volume of ≈30-μm diameter and a few centimeter length and is 7.5 m apart from the next bunch. Since their speed is very close to the speed of light, they cross each other every 25 ns. Most of the protons pass unaffected at each crossing, only about 20 of them collide, and each interaction generates ≈7 charged particles per unit of rapidity.[1]

LHC detectors cover 5 units of rapidity and therefore should measure ≈700 charged particles every 25 ns, which translates, for a pixel detector of 5-cm radius, to a hit rate of ≈50 MHz/cm^2 or ≈10 kHz/pixel . This is just the beginning of a difficult hunt for the needle in the hay stack as the number of events of interest are few per second, while, as we have just seen, the frequency of "any" event is ≈1 GHz. A multistep selection criteria must therefore be envisaged to filter out the less interesting events. Since about 100 events per second (corresponding to 250 Mbyte/s) can be saved for off-line analysis, most of the selection process should happen online and therefore requires proper design of the readout architecture to minimize data losses. It also requires high speed information exchange as the definition of "potentially interesting" is obtained combining data extracted from several independent detectors, each looking at different characteristics of the same event (i.e. track parameters, energy flow, particle masses).

The first step of the selection process, often termed "level-1 trigger," is the most critical since a decision, which must reduce the number of events by a factor close to 1,000, should be taken in a couple of microseconds. Events must be continuously stored while waiting for the level-1 trigger decision and, when this is positive, the related hits must be retrieved. Storage and retrieval capabilities must be, in the case of pixels, embedded in the readout architecture of the front-end chip as there is no time to move the several terabit of raw data per second anywhere else.

The signal processing steps in a pixel readout circuit can therefore be summarized as follows [244](see Fig. 5.1):

1. Charge amplification
2. Signal discrimination (is there a hit?)
3. Temporary hit storage
4. Hit retrieval and readout according to trigger selection

The main constraints coming from the operation at LHC are related to uniformity of response, speed, hit density, and raw data storage. In detail:

1. Several millions of channels should operate together after large, and possibly nonuniform, accumulated radiation doses. Good efficiency throughout the lifetime of the detector can only be obtained if the discriminating threshold can be set low enough to collect the small charge collected after irradiation. In practice this means operating at a threshold of ≈3,000 e$^-$. The typical noise of one pixel (initially ≈200 e$^-$ rms increasing to ≈300 e$^-$ rms after 500 kGy of radiation dose) is low enough that any statistical fluctuation can hardly overcome the threshold. The threshold

---

[1]Rapidity is a kinematical variable that can be approximated, for highly relativistic particles, to pseudorapidity $\eta = -\ln\tan(\theta/2)$, where θ is the particle zenith angle referenced to the direction of the crossing beams

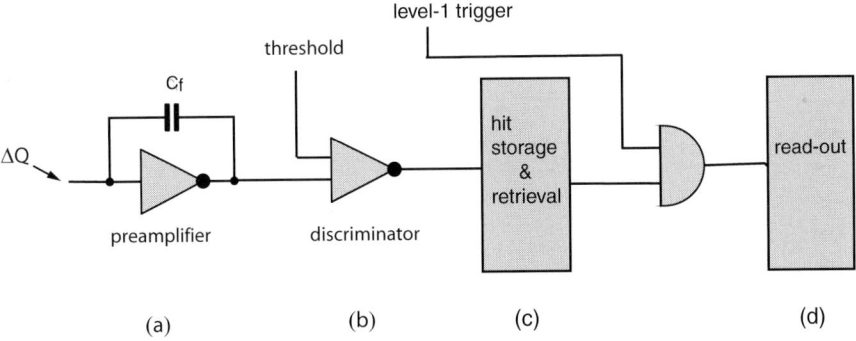

Fig. 5.1. The signal processing steps in a pixel readout circuit. One can distinguish (**a**) charge amplification, (**b**) signal discrimination, (**c**) hit storage and retrieval, and (**d**) readout for the events selected by the trigger logic

dispersion r.m.s. over all channels in a chip may, on the contrary, be as large as ≈500 e⁻ and therefore becomes the dominant cause of spurious hits.² This dispersion is due to the parameter variation in the integrated circuit fabrication and increases with the accumulated radiation dose. A mechanism for threshold equalization, such as to render this contribution negligible, is implemented in the electronics design (see Fig. 5.2) and allows for low threshold operation throughout the lifetime of the experiment.

2. The signal should have a fast rise time (<50 ns), while it may have a relatively long fall time (<2 μs). The fast rising edge is necessary to associate pulses of different heights to the correct beam crossing. Pulses with the same peaking time, but different heights do, in fact, pass the discriminator threshold at different times. The longer falling edge can be tolerated as the mean hit per pixel is $\approx 5 \times 10^{-4}$, i.e. less than half a percent integrated over 2 μs. Signals should also have minimal over- and undershoots to avoid uncontrollable pulse height biases.

3. The discriminator should be fast, i.e. should not add more than a few nanoseconds of time jitter. It is not only important that the threshold be set as low as 3,000 e⁻, but also that the overdrive required to have an in-time efficiency should be minimal. Hits corresponding to low pulse heights can be associated to wrong (i.e. late) beam crossings unless a sophisticated logic circuitry is implemented. The time behavior of a pixel cell can be described by its time walk curve, as illustrated in Fig. 5.3, where the response delay of the discriminator is plotted against the charge injected at the input of the pixel preamplifier.

4. Hits have to be stored for up to 128 crossings, waiting for the level-1 trigger decision. The low hit probability per pixel requires zero suppression

---

²The "effective" average noise in a chip or in a module is $\sqrt{N^2 + T^2}$, where N is the mean value of the noise and T is the root mean square of the threshold distribution

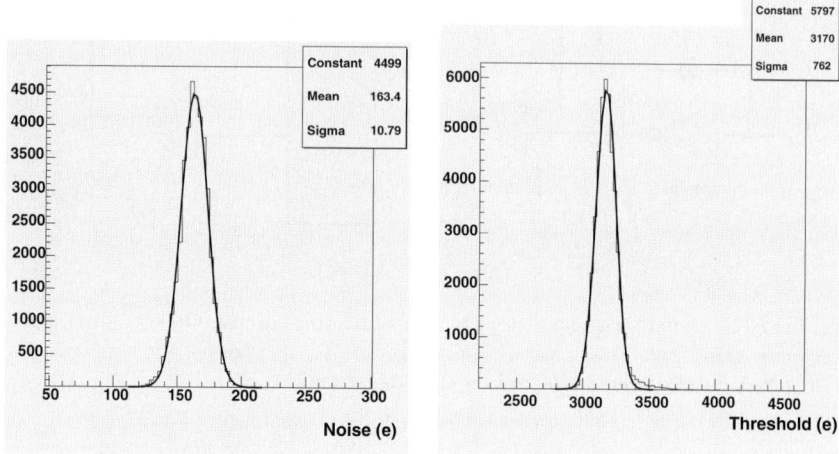

Fig. 5.2. Noise (*left*) and threshold (*right*) distributions for a module built with ATLAS FEI3 pixel chips. The Gaussian noise distribution has a mean value of $\approx$160 e$^-$ while the threshold distribution, measured after channel equalization, has a dispersion (rms) of $\approx$80 e$^-$. The module has 46,080 channels and the threshold and noise could be automatically determined for 46,057 channels. This figure proves safe operation of a pixel module at $\approx$3,000 e$^-$ threshold

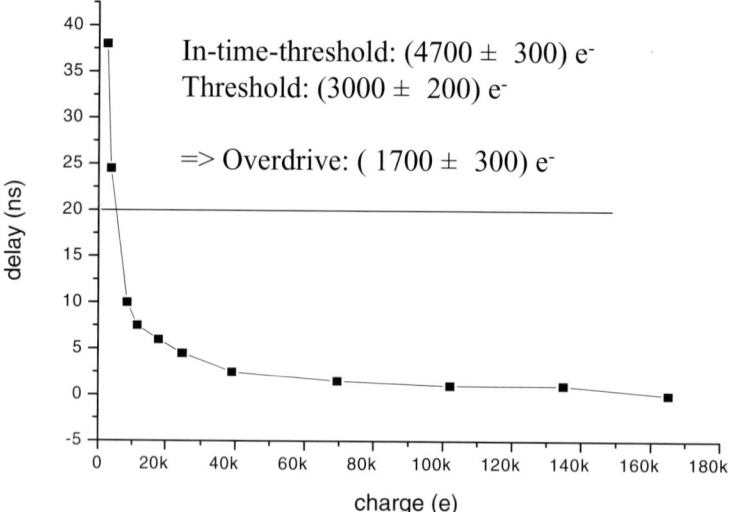

Fig. 5.3. Response time of the ATLAS FEI2 pixel preamplifier–discriminator as a function of the injected charge above a fixed 3 ke$^-$ threshold. If the time delay relative to very large pulses exceeds 20 ns, the hit is likely to be associated with the next beam crossing. The "in-time" efficiency will, in this case, be reached at a charge Q of $\approx$4.7 ke$^-$. An overdrive of $\approx$1.7 ke$^-$ is needed to move from the efficiency to the "in-time" efficiency

circuitry and storage of coded nonnull data. This storage is more efficiently implemented in the chip periphery, also because only "necessary circuitry" should be implemented in the pixel cell in order to reduce its size. In this approach the pixel cells have to be continuously emptied. This is done through high-speed column busses. The peripheral circuitry is thereafter often named "end of column (EoC) logic." If hits, at the output of the column busses, find all buffers full, they are lost contributing to the pixel inefficiency. Dimensioning of the buffers should then be done taking into account fluctuations of hit densities, which may happen in high momentum particle jets, i.e. those kind of events which may bring physics information of special interest.

5. The association of each event with a given beam crossing has to be stored until the level-1 trigger signal is available. Then only the information selected by the trigger is transmitted outside the detector to the remote processors dedicated to the next triggering step, and the rest of the information is discarded. The logic must therefore be organized to respond to the trigger commands with minimal time losses.

So far only the effect of particle fluxes on the data transfer rate has been considered. High radiation doses also change sensor and electronics properties. These modifications, described in detail in Chaps. 2 and 3, have to be taken into account to optimize the lifetime of the detector and its safe operation. Most of the irradiation effects are cumulative; for instance, the sensor reverse bias current increases with the dose. Current densities as high as $0.5\,\mathrm{mA/cm^2}$ are expected after a flux of 10^{15} minimum ionizing particles per cm^2. This is one of the reasons to favor high segmentation in silicon sensors when survivability to high doses is required. Tolerance to large leakage currents is therefore an additional requirement the pixel amplifier must meet to operate at the LHC.

Another effect that becomes more important after large radiation doses is the reduction of the charge collected in a silicon sensor. This happens because the radiation-induced damages in the crystal act as trapping centers capturing about one third of the charge after a flux of 10^{15} particles/cm^2. The charge loss can be significantly higher if one is forced to operate the detector partially depleted. Since the voltage necessary for full depletion of the sensor increases with dose, after an initial decrease before type inversion (see Chap. 2), there exists an absorbed dose value above which the detector system cannot be safely operated at full depletion. Low threshold operation is therefore a system requirement that becomes increasingly important with the dose. Uniformity of all the critical front-end parameters (noise, time walk, discriminator threshold, etc.) over the whole detector system is of paramount importance and this is obtained by individual digital adjustments to be periodically done.

One irradiation effect is not cumulative: the single event upset (SEU). This happens when a large ionization deposit, often related to a nucleus breakup following a hadronic interaction, "flips" a storage node from one logic state to the other. The transition from logical "0" to logical "1" does not necessarily have the same probability as the opposite transition. The flipping asymmetry depends very much on the particular design. Figure 5.4 shows measured cross sections for two different designs and also the strong dependence of the SEU effect from the supply voltage.

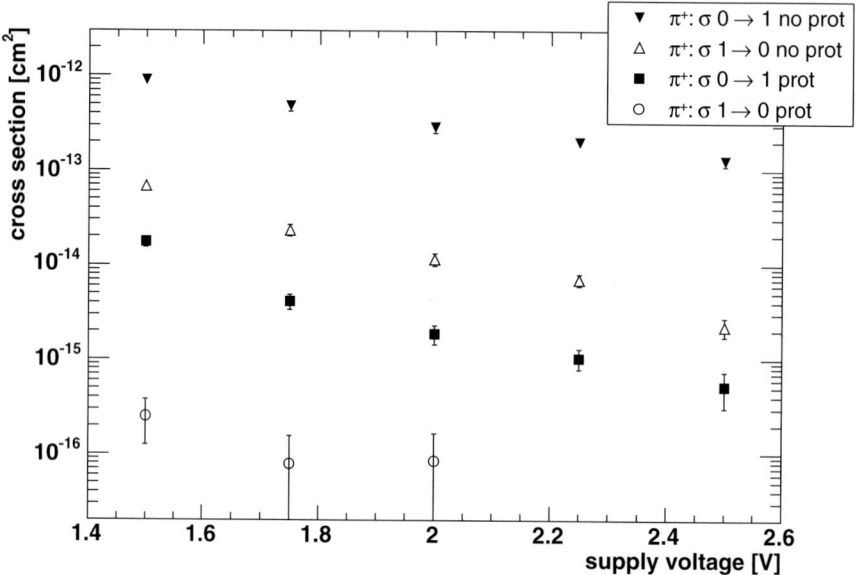

Fig. 5.4. The SEU cross section measured by CMS [221] for test structures made of 500 flip-flops realized in 0.25 μm CMOS technology. Some test structures are protected by 75-fF capacitors; this improves their immunity to SEU by 2 orders of magnitude

A system with 100 million channels easily has 200 Mbyte of configuration data and 500 Mbyte of temporary storage. The effect of changing one bit might be largely different: it depends what this bit is used for. If, for instance, it belongs to a command stream, the operation of the detector may be deeply affected. The logic circuits must therefore be designed in such a way to be highly insensitive to SEUs (see Chap. 3). For some critical bits that may affect the operation of a sizeable part of the detector, this is normally obtained using redundancy, e.g. storing them in three places and using a majority logic to define their true value. SEU cross sections of $\lesssim 10^{-14}$ cm^2 are necessary to avoid too frequent data refresh in the innermost layers of an LHC pixel detector.

All the above considerations are valid for ATLAS, CMS, and, to a large extent, for BTeV too. As will become evident in Sect. 5.2, this last experiment is characterized by a smaller dose, a higher particle density, and the requirement that the pixel detector should take part in the level-1 trigger decision.

The ALICE pixel detector lives, on the contrary, in a quite different environment. The expected dose is smaller by 2 orders of magnitude, and the number of interactions per second is only a few thousands, but 8,000 charged particles per unit of rapidity are typically produced in each interaction (which here is between lead ions, rather than between protons as in all other cases). The pixel system must therefore cope with particle densities as high as $100/cm^2$, but it has quite some time to digest all these data as only few thousand interactions per second are expected. Here the challenge is more on the reconstruction of those very complicated events and a premium is on minimizing the amount of material since any kind of secondary interaction will complicate further the event topology and will hamper the precise reconstruction of very low momentum tracks.

5.2 Large Systems in Construction

In this section the four large pixel detectors presently under construction are described in some detail. They will be identified by the name of the experiment in which they are used: ATLAS, CMS, ALICE, and BTeV. Only prototype parts of these pixel detectors exist at the time of writing this book. The construction of the final setup is either underway or should begin soon. It will then not be surprising if the detectors in operation in few years from now may look somewhat different from the description that follows, as their unprecedented complexity may require some modification during the production phase.

The role of the pixel detectors in the next generation of particle physics experiments will be crucial to precisely measure the track parameters in the vicinity of the interaction vertex. This primarily means accurate (i.e. within $50\,\mu m$) determination of the position of the primary and, eventually, of the secondary vertices. The discussion of the physics measurements that become accessible thanks to the pixel detectors is beyond the scope of this book; the interested reader is referred, for instance, to [279].

The description that follows stressed the system aspects and is not intended to go much into details as more specific information can be found elsewhere in this book. ATLAS will be used to illustrate the general problems of a large area pixel detector. In the description of the other three detectors, emphasis will be placed on the approaches and the solutions different from those adopted in ATLAS.

5.2.1 ATLAS

As its name suggests ATLAS[3] is the largest particle physics experiment in construction for the LHC and it also contains the largest pixel detector [280]. The pixel sensitive area of $1.7\,\text{m}^2$ is made out of 1,744 modules of $6.04 \times 1.64\,\text{cm}^2$ active area. All modules are equal and are built using $6.34 \times 2.44 \times 0.025\,\text{cm}^3$ n-doped silicon sensor "tiles" containing 47,532 n^+-pixel implants. The active area is surrounded by 16 guard rings to allow for high-voltage operation. The use of n^+-implants allows partially depleted operation after type inversion, should this be needed at the end of the detector lifetime. The large majority ($\approx 98\%$) of the pixels covers a surface of $50 \times 400\,\mu\text{m}^2$, some are bigger to allow for a continuous sensitive area also in-between[4] the front-end chips as illustrated in Fig. 5.5. Having rectangular pixels warrants better resolution in the plane where tracks are bent by a 2-T uniform magnetic field and where therefore the track momentum can be measured.

The ATLAS pixel module has been shown in Fig. 1.9. Sixteen front-end chips are connected through bump bonding (i.e. individual connections between each sensor and its mating readout cell) to the sensor. Each chip reads out 2,880 pixels, covers an area of $0.74 \times 1.09\,\text{cm}^2$, and is thinned down to $\approx 200\,\mu\text{m}$ to minimize dead material. Every pixel is read out through an amplifier followed by a discriminator which detects when the pulse exceeds an adjustable threshold. The time resolution is below 25 ns as requested to unambiguously associate the pixel hits with a given LHC beam–beam interaction. Crossing the threshold level is detected, in 25-ns units, both for the rising and for the falling edge of the pulse; this allows one to measure how long a given pulse has been "over the threshold" (see Chap. 3 for details). The larger the pulse, the longer it stays above threshold: this relationship allows one to translate the "time over threshold" into a 7-bit pulse height information as shown in Fig. 5.6.

A sizeable ($\approx 25\%$) fraction of the front-end chip is dedicated to the EoC logic. This circuit drains continuously the information from the pixel front-end circuits. It can store up to 64 hits per column while waiting for the

---

[3]The ATLAS acronym stands for A Toroidal Lhc ApparatuS, but it is also a reminder of the unprecedented size of this detector

[4]The outer edge of an electronics chip cannot be closer to the active part of the circuit by less than 100 µm. Otherwise damage is likely to occur when the die is singled out of the wafer. It is therefore necessary to keep a safety distance of 400 µm between the active parts of two contiguous chips. The matching between readout electronics and pixels requires some solution to have the detector continuously sensitive. The solution is simple in between two columns: it is enough to elongate the pixel of these columns to cover $50\,\mu\text{m} \times 600\,\mu\text{m}$. When the gap is at the top of the columns, 4+4 pixel per column cannot match any electronics cell directly and must therefore be connected, through metal lines laid on the sensor, to 4+4 neighboring pixel at the end of the column and directly connected to the electronics. The ambiguity (which of the two pixels connected by the metal line was hit?) can be resolved only after the track is reconstructed by the entire detector

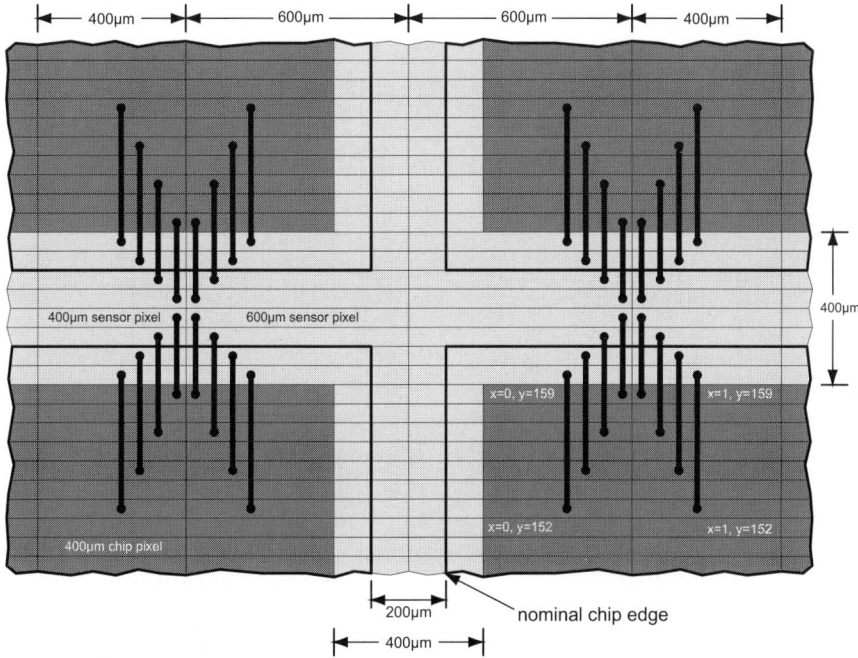

Fig. 5.5. The solution adopted by ATLAS to have a continuously sensitive pixel detector. Figure shows the region of the pixel detector at the edge of four front-end chips; the area of the sensor covered by the active part of the four chip edges is marked in gray. The pixels inside the interchip regions *(white rectangles)* are connected, through metal lines, with mating pixels directly readout from the front-end chip. Ambiguity is later resolved using the full track reconstruction.

level-1 trigger to be delivered. Once bonded, most of the EoC logic extends out of the sensor area. Wire bonding pads at the output of the EoC logic are then accessible and allow one to connect each front-end chip to the readout system (see Sect. 4.2). This connection happens through another radiation-hard chip, the module control chip (MCC) [211], placed on each module. The connection between the front-end chips and the MCC is shown in Fig. 5.7.

The MCC is a state machine with 45-kbit storage memory dedicated to the event building at the module level. The MCC must therefore be designed to be SEU-tolerant to avoid significant data corruption. All front-end chips are connected with the MCC in a star-like topology; this allows operation of a module even if one chip stops working and therefore reduces the effects of some of the possible failure modes.

The MCC receives and transmits digital data via optical fibers. The transformation of the electrical signals into modulated light pulses is done through ASICs [281] which are located up to 1 m from the MCCs in the first of the

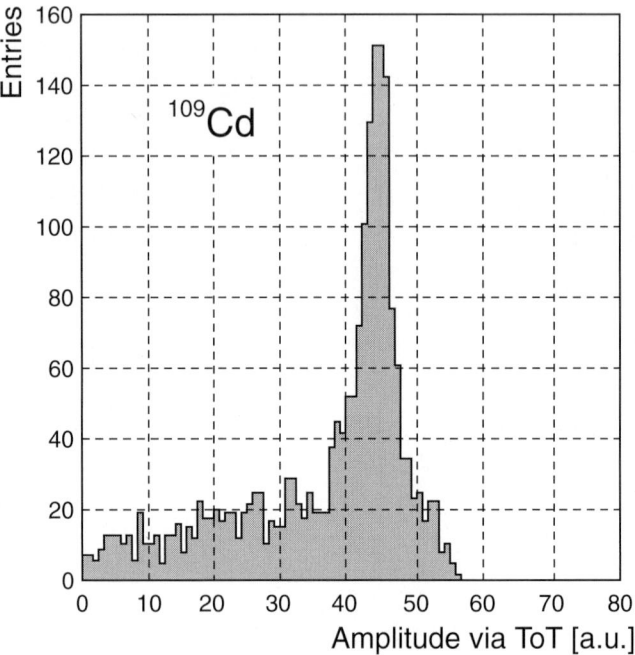

Fig. 5.6. Pulse height distribution resulting from ^{109}Cd photon conversions inside an ATLAS pixel assembly. The measurement is performed using the correlation between the charge collected by one pixel and the time over threshold (ToT) of the output of its amplifier. The 22.3-keV ^{109}Cd photopeak is measured with an energy resolution of $\approx 5\%$

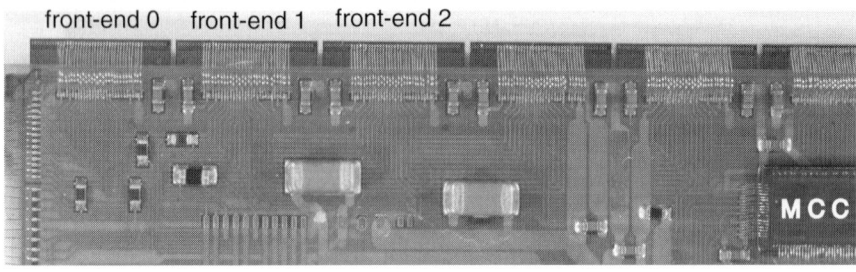

Fig. 5.7. Detail of an ATLAS pixel module showing the front-end circuits extending out of the sensor area (*upper edge* of the photograph) and connected to the MCC (the chip on the *right-hand side* of the photograph) via wire bonds and copper lines on a kapton circuit

patch panels which are necessary to break the services for the detector installation.

A thin, double-sided kapton circuit glued on the backside of the sensor supports the MCC and brings to it all the front-end signals through copper lines and aluminum wire bonds. The kapton circuit is also used to distribute properly decoupled low voltage to all chips (a total power of 3.5 W is required prior to irradiation, possibly increasing to 5 W after an accumulated dose of 500 kGy). The module temperature is remotely monitored via a negative temperature coefficient resistor placed on the kapton circuit and a fast interlock will power off a module if overheating begins. This is necessary as the large power density and the low mass of the support and cooling structure imply temperature rises of up to 10 degrees per second in the case of a cooling fault.

To obtain more than a square meter of continuously sensitive area one must build up a mosaic of modules precisely placed on a light and stable support structure. This support must also satisfy other, often conflicting, criteria like being a good heat sink (10 kW/m^2 are produced by the modules which have to be operated below 0°C to minimize the effect of the irradiation [105]) and an electric insulator to allow for a properly designed grounding scheme. The ATLAS support structure is largely made of carbon–carbon, a composite material obtained exposing a carbon fiber layup to high temperature in CO_2 atmosphere. This process results in a carbon-rich material with excellent mechanical and thermal characteristics.

The support layout, shown in Fig. 5.8, is made of a barrel part covering up to pseudorapidity of $\eta \approx 1.5$ and a forward part covering up to pseudorapidity of $\eta = 2.5$. The barrel part is built using $\approx$80-cm-long and $\approx$2-cm-wide local supports called staves and arranged in a turbine geometry to allow for hermetic track coverage. The tilt angle has been chosen to counterbalance (at least in part) the effect of the Lorentz force and therefore to minimize

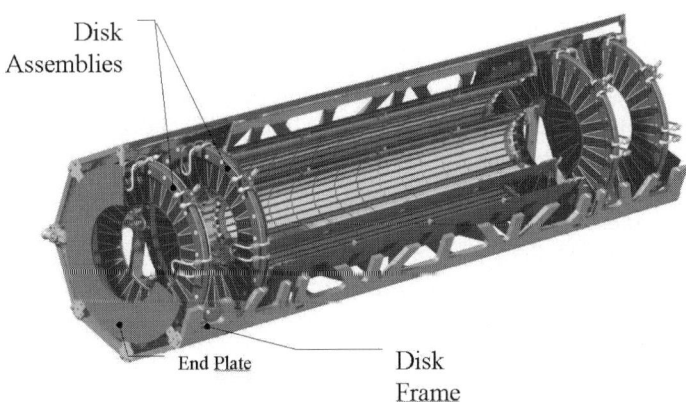

Fig. 5.8. Three-dimensional view of the ATLAS pixel layout in its initial configuration foreseen for the low luminosity operation of LHC. Two cylindrical barrel layers cover the central rapidity region and two disk layers cover the forward rapidity region. The transition is at $\eta \approx 1.5$. In the final layout a third pixel layer is added in between the two layers shown here

Fig. 5.9. An ATLAS stave loaded with modules. The stave is on a temporary support necessary for the module loading operation. A stave holds 13 modules and covers an area of $\approx$125 cm^2 with $\approx 6 \times 10^5$ channels

charge sharing between pixels. Robustness of operation has been privileged over space resolution obtainable with charge interpolation. Thirteen modules are placed on each stave as shown in Fig. 5.9 and face the interaction point.

The central pixel detector is made of three barrel layers of radii 5.0, 9.8, and 12.2 cm, built with, respectively, 22, 38, and 52 staves. The innermost layer is of special importance as it determines the impact parameter resolution (i.e. the accuracy of the tracks extrapolated to the interaction vertex). The forward part is built using sectors of 8.9-cm inner radius and 15.0-cm outer radius. Each sector has six modules (three in front and three on the back) and the hermeticity is obtained superimposing the front and back modules. Eight sectors make a disk and six disks make the entire pixel forward system. All particles with a momentum above 2 GeV/c in the central 5 rapidity units should cross the active regions of at least three modules. The modules are superimposed by few millimeters to allow for detector alignment using high-momentum tracks during normal data taking. Cooling is provided by evaporation of a fluorocarbon (C_3F_8) which expands in the aluminum tube embedded in each local support after having passed through a capillary.

As it is apparent from the description above, the heart of the pixel system is quite complex with a huge number of channels to be powered, tuned, controlled, and readout. Services, control systems, and remote readout are a real technical challenge especially taking into consideration the space and material limitations which should be met in the interest of the whole experiment.

The ATLAS pixel project is characterized by a high degree of standardization meant to ease the production process: all modules are equal, and all stave and all sectors are equal.

The pixel detector can be installed independently once the rest of the ATLAS tracker (made of silicon microstrips and straw tubes [282, 283]) is in place. This facilitates repairs and upgrades, which have been frequently needed for vertex detectors of previous experiments.

5.2.2 CMS

CMS (Compact Muon Solenoid) is the experiment that will compete with ATLAS for the study of the high p_T processes at LHC. The reader will then not be surprised by the similarity of the two detector designs. It is also not surprising that not only the whole apparatus, but also the pixel system [284] is more "compact" in CMS than in ATLAS. In the final configuration three 57-cm-long barrels of radii 4.4, 7.3, and 10.2 cm and three disks at ±34.5, ±46.5, and ±58.5 cm contribute to cover the necessary acceptance. One of the three barrel layers and one disk pair may be missing at the experiment start-up when the accelerator luminosity will be much below its nominal value. Figure 5.10 shows one intermediate layout with three barrel and two disk layers.

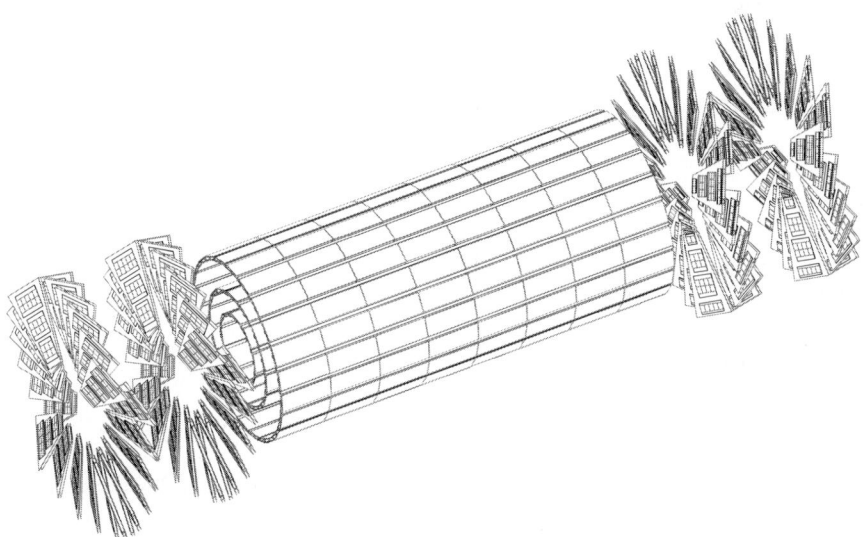

Fig. 5.10. Three-dimensional view of one of the possible configurations of the CMS pixel layout. Three cylindrical barrel layers cover the central rapidity region and two disk layers cover the forward rapidity region. The transition happens at $\eta \approx 1.5$

The whole pixel system consists of 1,776 modules of seven different types (five for the disks and two for the barrels). The smallest module has 2 read-out chips; the largest module is read out by 16 chips. The variety of modules allows one to better fit the geometrical acceptance at the cost of some complication in the production process. The total number of channels is about 7.5×10^7 and 18,000 front-end chips are needed to be bump-bonded on the n^+-on-n sensors. The pixels of CMS are $100\,\mu\mathrm{m}\;(r\varphi) \times 150\,\mu\mathrm{m}\;(z)$; this choice (almost square shape rather than elongated rectangle) is motivated by the requirement to have similar resolution on both projections and implies the use of charge interpolation to find the charge centroid on both coordinates. This, in turn, requires:

- That the charge be always shared between two or more pixels and
- That the charge collected by each pixel be measured with the necessary resolution

The first requirement can be met orienting the detectors in the 4-T magnetic field of CMS such as to profit from the large drift due to the Lorentz force that will then spread the charge carriers over several pixels as shown in Fig. 5.11. The detectors are deliberately not tilted in the barrel layers, but are tilted in the end disks, resulting in a turbine-like geometry (see Fig. 5.10).

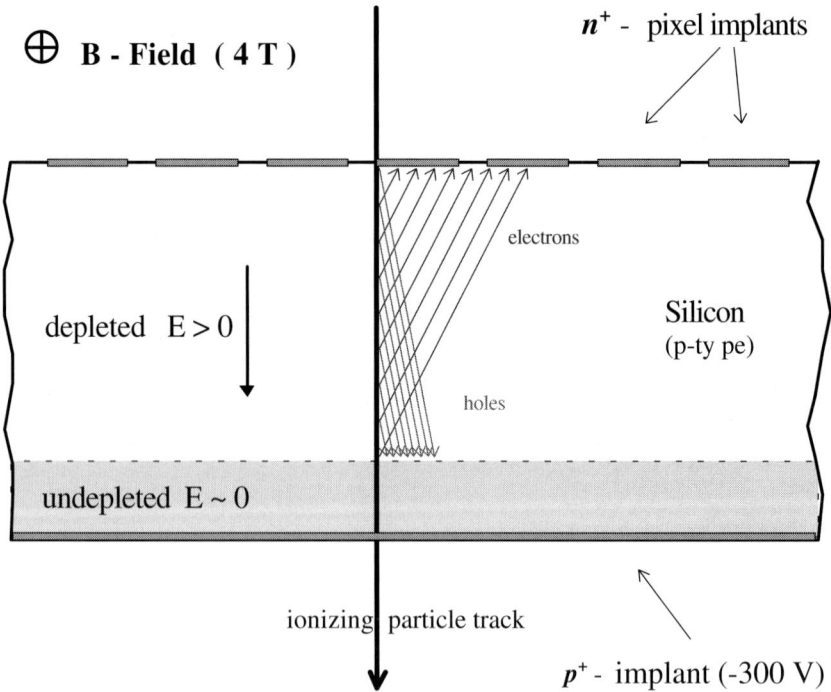

Fig. 5.11. Charge sharing induced by the Lorentz drift in the 4-T CMS solenoidal magnetic field. After a dose largely exceeding the type inversion point the Lorentz angle will be reduced [66] due to the combined effect of the smaller drift volume (undepleted detector) and of the larger electric field

The second requirement can be met with analog readout of the charge information of each pixel. Charge is stored on switched capacitors in each pixel cell and, later in the data stream, in the chip periphery until the trigger decision is taken (i.e. for less than 3.2 μs) and then transmitted out through a 40-MHz analog optical link that connects a group of 8 or 16 (barrel) and 22 or 23 (disks) pixel chips to the remote readout electronics.

To get the maximum information out of the analog signal interpolation the stability and calibration of the whole electronics chain must be well under control and the operation of the complete pixel detector at low ($\lesssim 2,500$ e$^-$) threshold must be possible.

The modules in the barrel region are quite similar to those of ATLAS: 16 chips are bump-bonded to one sensor and glued onto a silicon nitride baseplate, while a thin three-layer Cu/polyimide circuit connecting the chips to a control chip is placed on the sensor's backside (see Fig. 5.12). Some of the barrel modules are half-size with eight chips.

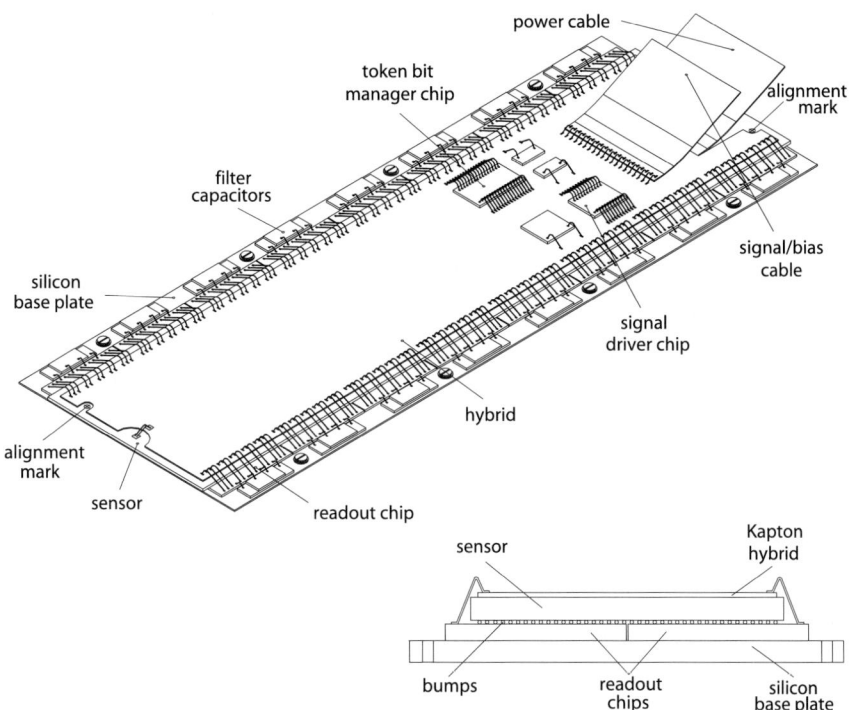

Fig. 5.12. Sketch of a CMS pixel barrel module

On the disks, tiles consisting of various sized sensors with two, five, six, eight, or ten chips are placed on both sides of the disk blades, ensuring a hermetic coverage. The tiles are glued onto a Cu/polyimide circuit having the shape of the blade and interconnecting the tiles.

The serial readout of triggered data is controlled by the token bit manager (TBM) chip located on the modules of the pixel barrel and on each side of the disk blades. A readout token is passed from chip to chip, connecting chip after chip to the analog readout bus. An arbitrary number of chips can be read out sequentially in one readout token scan. Contrary to the ATLAS MCC, the TBM chip has no event-building capabilities.

To connect readout chips and sensor, CMS is investigating in-house, rather than industrial, bumping and flipping operation. Of particular interest is the development going on at PSI for the deposition and reflow of indium described in some more detail in Chap. 4.

The support frames for barrels and disks are largely made of trapezoidal (barrels) or rectangular (disks) thin-walled Al cooling pipes interconnected by high modulus carbon fiber sheets. Cooling is obtained by circulating liquid fluorocarbon (C_6F_{14}). The construction is modular and both barrels and disks are built in halves (see Fig. 5.13) to facilitate insertion and extraction around the beam pipe independently from the rest of the apparatus. This is necessary whenever a beam pipe bakeout is required to restore the proper vacuum conditions in the interaction region. Pixel detector removal and insertion are expected to happen few times during the lifetime of the experiment. Repairs

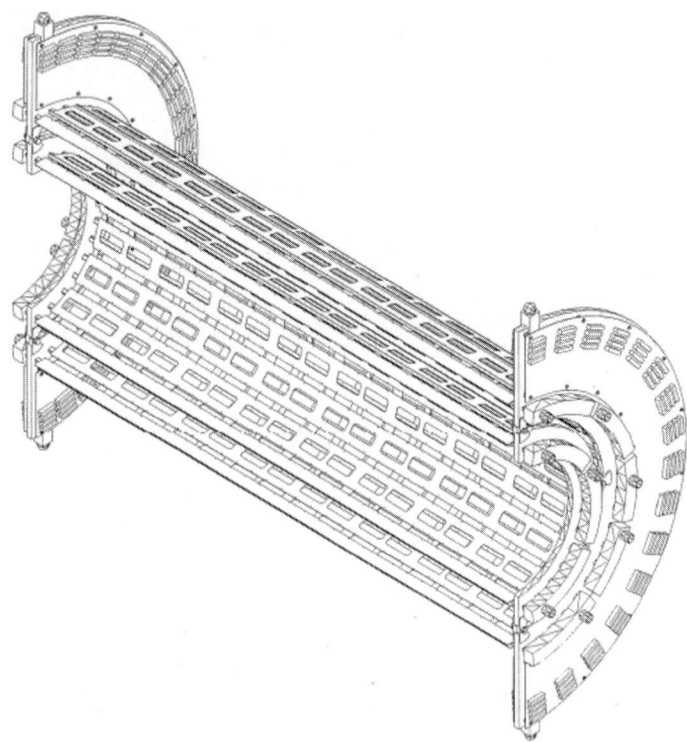

Fig. 5.13. Perspective view of a pixel half-barrel mechanical support for the CMS experiment. The cooling tubes, embedded in the structure, are not visible

or upgrades are likely to be needed for this novel type of detector intended to operate in a harsh environment.

Both ATLAS and CMS face the same experimental conditions, have the same physics objectives, and therefore aim at radiation-hard, high-speed pixel systems. It is natural that the solutions adopted are similar. It is, nevertheless, worth noticing that some strategic design choices are quite different: CMS privileges spatial accuracy while ATLAS favors standardization and ease of construction.

Only long-term operation will tell us which approach has been more effective.

5.2.3 ALICE

ALICE (a large ion collider experiment) is an LHC experiment designed for the study of nucleus–nucleus interactions at the center-of-mass energy of 5.5 TeV per nucleon.

While for proton–proton collisions (i.e. ATLAS and CMS) the interaction rate is close to 1 GHz, for nucleus–nucleus collisions it is a few kilohertz, but each interaction should be quite spectacular with $\approx$8,000 charged particles per unit of rapidity (to be compared with $\approx$7 in the case of proton–proton collisions).

The very large number of particles, each of relatively low momentum, demands sensors with high segmentation and low mass to minimize the effects of multiple scattering on the impact parameter and the momentum resolution. Pixel detectors are therefore ideal in this environment and the design effort here is specially dedicated to make them as thin as practical.

Contrary to the case of ATLAS and CMS, the radiation hardness in ALICE is less of an issue thanks to the low interaction rate. The innermost pixel layer should be designed to survive *only* 5 kGy. At this dose level, however, the electronics must be designed to be radiation-hard and special care must be taken to avoid SEUs, which are expected to be more frequent in the ion–ion collision environment than in the proton–proton collision environment. The ALICE silicon pixel detector (SPD) [285] is located inside a 0.4 T solenoidal magnetic field and consists of just two barrel layers of radii 3.9 and 7.6 cm and length 28.3 cm. The SPD covers about 2 units of rapidity (3 for the innermost layer) and has 0.24-m^2 silicon active surface. These are the first detectors seen by the particles emerging from the nucleus–nucleus interactions and are primarily dedicated to the reconstruction of short-lived charm and beauty particles. To minimize material the same CFRP structure holds both layers and hosts the services and cooling pipes. A set of ten identical CFRP structures builds up two hermetic barrels as shown schematically in Fig. 5.14.

Each pixel measures 50×425 μm^2, a size that leads to a typical occupancy of 2% in the case of central Pb–Pb collisions. Each module measures $\approx 70.7 \times 16.9$ mm^2 and is made of one sensor tile to which the 40,960 channels

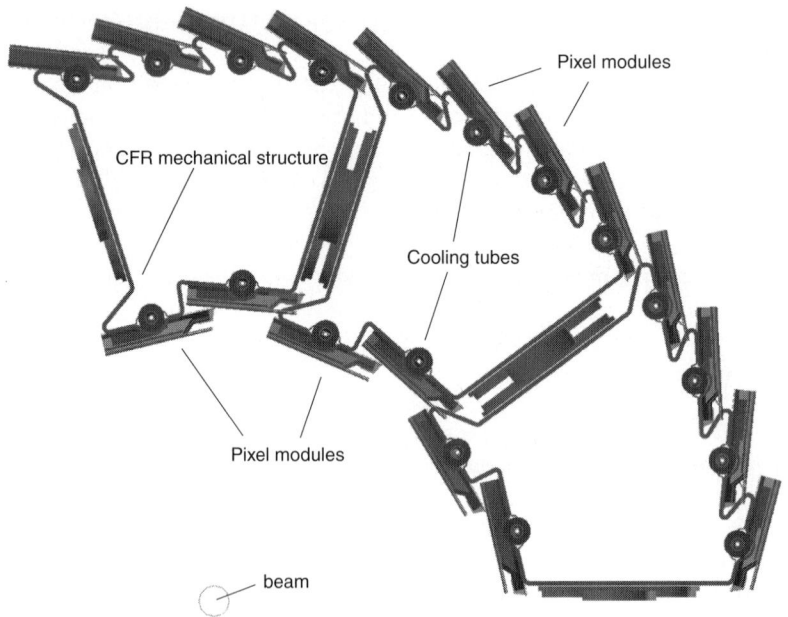

Fig. 5.14. Sketch of a part of the cross section of the ALICE pixel detector (3 CFRP structures out of 10). The region where the beams cross is also shown

of five front-end chips are connected through bump bonding. Two modules are placed one after the other and controlled by the same pilot and link driver chip as shown in Fig. 5.15.

The design of the modules is quite complex. The power and signals distribution must be done through a six-layer high-resolution aluminum-on-kapton circuit glued on top of the pixel sensor as shown in Fig. 5.16. This printed circuit technology, which is mastered in some research laboratories but is not an industrial standard, is crucial to keep the radiation length of each pixel

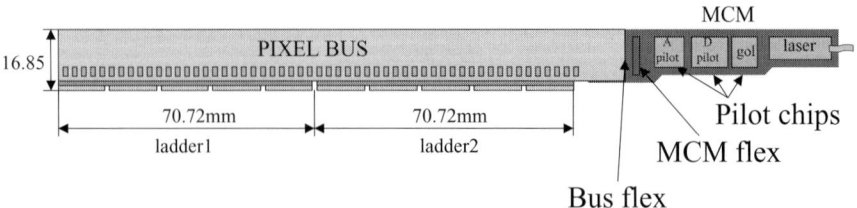

Fig. 5.15. Sketch of a pair of ALICE modules (also called a ladder). This is the building block of the barrel pixel system of ALICE

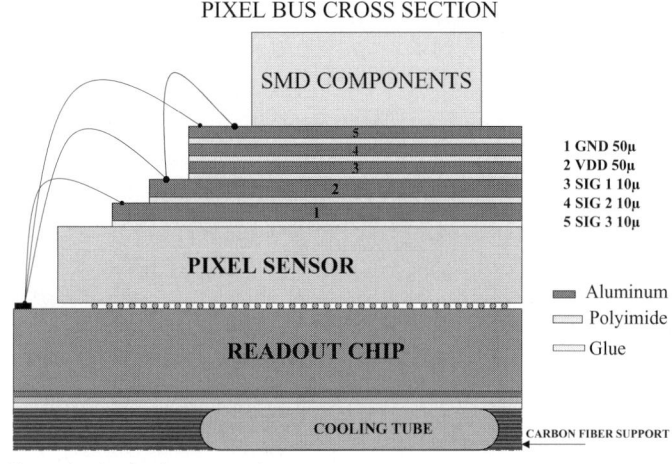

Fig. 5.16. Sketch of the ALICE module on the carbon fiber support

layer below 1%. This goal is quite challenging. ATLAS and CMS set, in fact, upper limits of the radiation length per pixel layer to about 2%. Services are coming out at each end of the barrel layout.

Sensors are of p^+-on-n design, a simpler approach justified by the expected radiation levels which should not cause type inversion. The sensor thickness is limited to 200 μm by the production yield of 5-in. wafers, which decreases rapidly with the thickness of the wafer (the standard being 300 μm). The electronics will be thinned down to 150 μm after bumping, a technique demonstrated by ATLAS [266] on 6-in. and 8-in. wafers.

The power load of the electronics chips and the power distribution lines is about $0.5\,\mathrm{W/cm^2}$ (i.e. similar to ATLAS and CMS). This requires an efficient cooling system even if, in this experiment, it is not required to operate the sensors at low temperature to minimize the radiation damage.

5.2.4 BTeV

The BTeV (Beauty at TeVatron) experiment[5] aims at the high statistics study of beauty particles produced in proton–antiproton collisions at 2-TeV center of mass energy. The strong forward–backward correlation of the beauty–antibeauty particles in this experimental conditions drives the apparatus design. This consists of a one arm spectrometer, shown in Fig. 5.17,

[5] At the time of printing this book, the construction of BTeV was stopped by the US Department of Energy (DOE). Nevertheless, the pixel detector developments done in the BTeV framework are considered relevant and it was then decided to keep this subsection

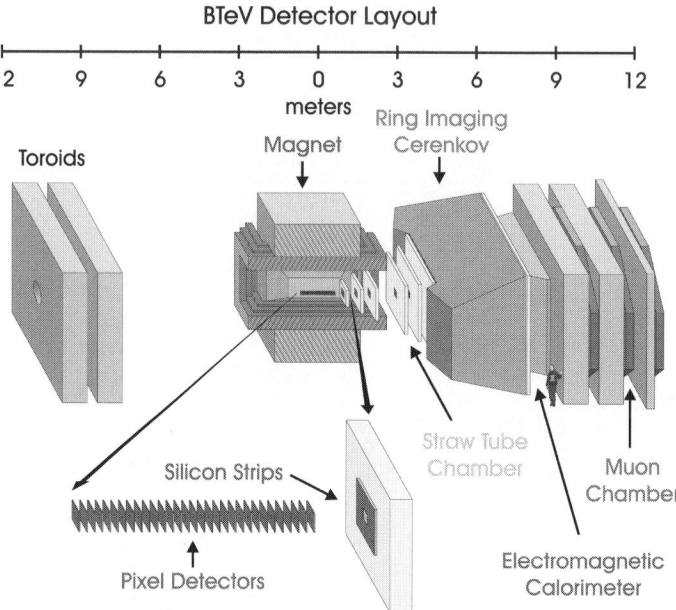

Fig. 5.17. Artist view of the BTEV spectrometer, with the blow-up of the vertex detector part made out of 31 pixel detector planes

covering ±300 mrad around the direction of the incoming proton (antiproton) and surrounding the C0 interaction region of the FERMILAB Tevatron.

The BTeV silicon pixel detector [286] will be placed close to C0 in the central dipolar magnetic field and inside the vacuum pipe of the accelerator. This is the first pixel detector that is designed to operate inside the accelerator beam pipe. This requirement adds further difficulties (e.g. high-density vacuum feed-throughs, choice of materials to minimize outgassing, electromagnetic interference with the beam) to the challenges a large scale pixel detector already has to face.

The BTeV pixel detector is made of 30 measuring stations uniformly distributed over 124 cm. A station, schematically shown in Fig. 5.18, is built with four detectors (or half-planes), two measuring precisely the horizontal coordinate and two the vertical one. The square inner hole is made to allow the circulating beams to cross without interacting in the detector itself; this hole can be enlarged, moving apart the plane halves, during the beam acceleration or tuning.

The pixel size is $50 \times 400\,\mu m^2$. The half-planes are made of modules of different sizes, with the largest one having up to $0.64 \times 7.32\,cm^2$ active area, placed close to each other on a thin mechanical support. Each module is made of a 250-μm-thick n^+-on-n sensor connected to four, five, six, or eight front-end chips through bump bonding. A thin flexible circuit brings power

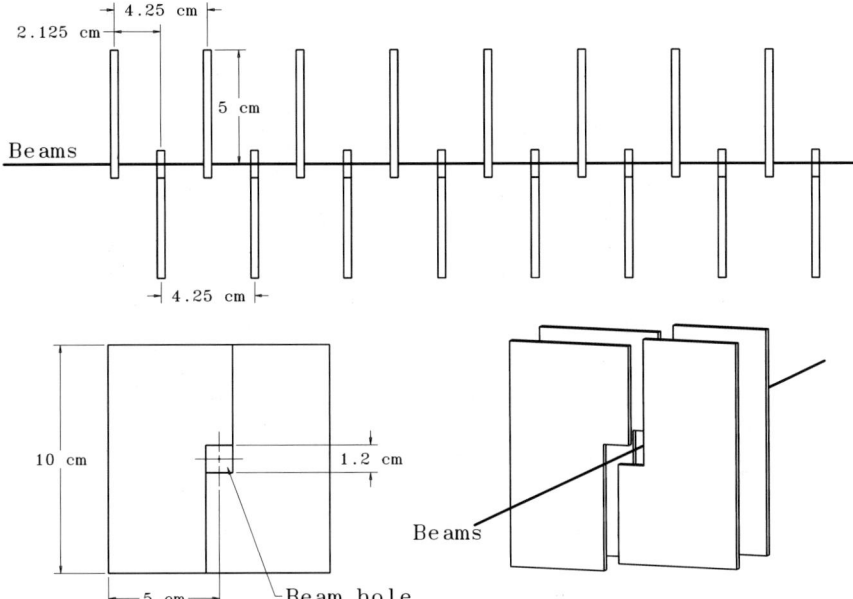

Fig. 5.18. Geometric layout of the BTeV pixel detector. In the *upper part* of the figure the side view of four pixel detector stations is shown; they do appear as a continuous set of half-planes uniformly distributed along the beam line. In the *lower part* of the figure one can see how these half-planes are organized in stations. A station is made of four half-planes, two measuring precisely the horizontal coordinate and two the vertical one

and control signals to the chips and sends the data signals to the remote electronics via vacuum feed-throughs. The choice of the sensor design is made to cope with the expected dose, which can be hundreds of kilograys close to the beam and some kilograys in the periphery of the sensor planes. Standing the electric field necessary to deplete a sensor with such a large accumulated dose variation is one of the challenges BTeV has to face. The front-end chips, fabricated using 0.25 µm technology, have 2,816 channels (128 rows × 22 columns) and will be thinned down to 200 µm. The BTeV pixel detector has a total of 23 million channels.

The thickness in radiation length of each detector plane must be minimized as typically 12 planes are crossed by each secondary particle before entering the rest of the spectrometer. This is possible by placing the relatively bulky cooling and power services outside the ±300-mrad spectrometer acceptance. Each pixel plane has a design thickness of ≈1.3% of a radiation length.

Even though BTeV is a colliding beam experiment, the pixel detector layout is similar to the one used in fixed target experiments. In both cases, in fact, pixels should cover a limited solid angle where high track density is expected. The role of the pixel detector is particularly important in BTeV as the primary aims of the experiment, i.e. the selection and the study of the beauty particle decays, are both heavily based on the capability of measuring secondary vertices and therefore on the operation of the pixel detector. The Tevatron beam crossing frequency of 7.5 MHz requires one to make available the hit information in 132 ns if this has to be used in a track trigger. This is the case for the BTeV pixel detector that should provide the information necessary to reconstruct track segments and verify if some of them are not emerging from the proton–antiproton interaction vertex, thus providing evidence for decays of short lifetime particles. The selection of events with offset tracks has to be done in real time as those events are relatively rare and only a small fraction of all the interactions can be recorded for further analysis. It must also be done very early in the selection chain in order to access, with minimum bias to the data, most of the decay modes of the beauty hadrons. This requires a data throughput of $\approx$2 terabit/s and a dead time-free readout system. Both these specifications are met in the prototype BTeV readout system.

5.3 Pixel Detectors for Imaging Applications

The development of hybrid pixel detectors for particle detection with high spatial resolution in high energy physics experiments has spun off a number of developments with applications in imaging, most notably biomedical imaging [287–290], and also imaging in X-ray astronomy [170, 291–295]. In the latter, the reconstruction of low-energy X-ray images originating from astronomical point sources with high spatial and high energy resolution at moderate data rates is the challenge. So far this goal has been met best by using fully depleted pn-CCD detectors [296]. For the next generation of X-ray satellites, however, monolithic or semimonolithic pixel devices with excellent spectroscopic performance are in the focus of development [293, 297] (see also Chap. 6).

The largest progress has recently been achieved in the development of pixel detectors for X-ray imaging for applications in radiology and in protein crystallography. Another field in which pixel detector developments have created new possibilities for real-time imaging is that of biomedical autoradiography, where most commonly low-energy β-radiation from radiolabeled tissue must be detected with high spatial resolution and high efficiency. Also γ-emitters are occasionally considered. The use and development of pixel detectors in medical or astrophysical imaging systems like Compton cameras [298–300] is also addressed by various research groups [301, 302].

Figure 5.19 sketches the principle of autoradiography of radiolabeled tissue. A thin tissue is placed as close as possible to the detector. The products from radioactive decays are emitted in all directions; emission to the bottom hemisphere can be detected. The attainable resolution is hampered by the

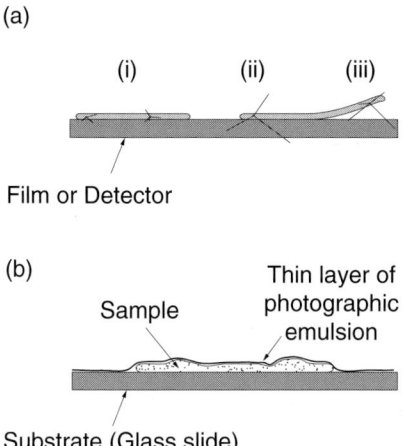

Fig. 5.19. (a) Autoradiography principle: the parallax effect is minimized for low-energy emitters and by a close contact to the detector (case i), the situation is worse for higher energy emitters (case ii), and/or less close contact of the tissue to the detector (case iii). (b) High spatial resolution autoradiography using thin layers of photographic emulsion by direct coating of the tissue

so-called parallax effect; i.e., particles which are emitted at shallow angles can be detected far away from the emission point (see Fig. 5.19a). The less deep the particle enters the detector and the closer to the detector the emission occurs, the better is the imaging resolution. Low penetration depth correlates, however, with low energy of the emitted radiation which is therefore most difficult to detect. While film-emulsion techniques [303, 304], using the principle as shown in Fig. 5.19b, excel in submicrometer spatial resolution, pixel detectors as electronics devices offer time-resolved imaging and, if the energy resolution is sufficient, the possibility of discriminating two different radiolabels in the same tissue. The detection of the ^{3}H radiolabel, which is attractive for biomedical autoradiography since it can be substituted in every biomolecule, is a real challenge for the detectors due to its low-energy β-spectrum, with an end point energy of 18.6 keV and a most probable emission energy around 3 keV, corresponding to less than 1,000 created electron–hole pairs in silicon.

The two mentioned application examples also show the main difference between particle tracking and imaging with pixel detectors. While in the former case individual charged particles, usually triggered by other detectors, have to be identified with high demands on spatial resolution and timing, in imaging applications an image is obtained by the usually untriggered accumulation (integrating or counting) of the quanta of the impinging radiation during some exposure time.

Also, the demands on the detector performance can be quite different. Silicon pixel detectors for high-energy charged particle detection can assume typical signal charges collected at an electrode in the order of 8,000–15,000 electrons even taking into account charge sharing between pixel cells and detector deterioration after irradiation to doses as high as 500 kGy. In tritium autoradiography, on the contrary, or in X-ray astronomy the amount of charge to be collected with high efficiency can be well below $1,000\,e^-$.

The spatial resolution is governed by the attainable pixel granularity from a few to about $10\,\mu m$ at best, obtained with pixel dimensions in the order of $50 \times 100\,\mu m^2$. The requirements from radiology are similar or even more relaxed than those in particle physics, while some applications in autoradiography require submicrometers resolutions, not attainable with present day pixel detectors. However, for applications with lower demands on the spatial resolution ($\mathcal{O}(10\,\mu m)$) but with demands on real-time and time-resolved data acquisition, semiconductor strip [305, 306] or pixel detectors are attractive. In addition, energy resolution capability may be provided at the same time.

While silicon is almost a perfect material for particle physics detectors, allowing the shaping of electric fields by spatially tailored impurity doping, the need of high photon absorption efficiency in radiological applications requires the study and use of semiconductor materials with high atomic charge, such as GaAs, Cd(Zn)Te, or HgI_2 (see also Sect. 2.6). Figure 5.20 demonstrates this, showing the absorption length of various high-Z materials in comparison to silicon. For such materials the charge collection properties are much less understood and mechanical issues, in particular those related to hybrid pixels, are abundant. The hybridization of detectors when they are not available in wafer scale sizes can be a real problem. Last but not least, the cost-to-performance ratio is an important factor to consider if an imaging application can be commercially interesting.

In imaging, the spatial resolution of a system is characterized differently than in particle tracking [307]. A common mathematical tool is the point spread function (PSF) which is defined as the image of a point source object $O(x,y) = \delta(x,y)$ with unit intensity. While for pixelated systems the treatment is more involved, for homogeneous and linear systems the image $I(x,y)$ of $O(x,y)$ is defined as

$$I(x,y) = \int_{-\infty}^{+\infty} \int_{-\infty}^{+\infty} O(x',y') P(x-x', y-y')\,dx'\,dy' \quad (5.1)$$

where $P(x,y)$ is the point spread function. For isotropic imaging systems the PSF is radially symmetric and depends on r only. The often used line spread function (LSF) is in this case defined as

$$L(x) = \int_{-\infty}^{+\infty} P(x,y)\,dy. \quad (5.2)$$

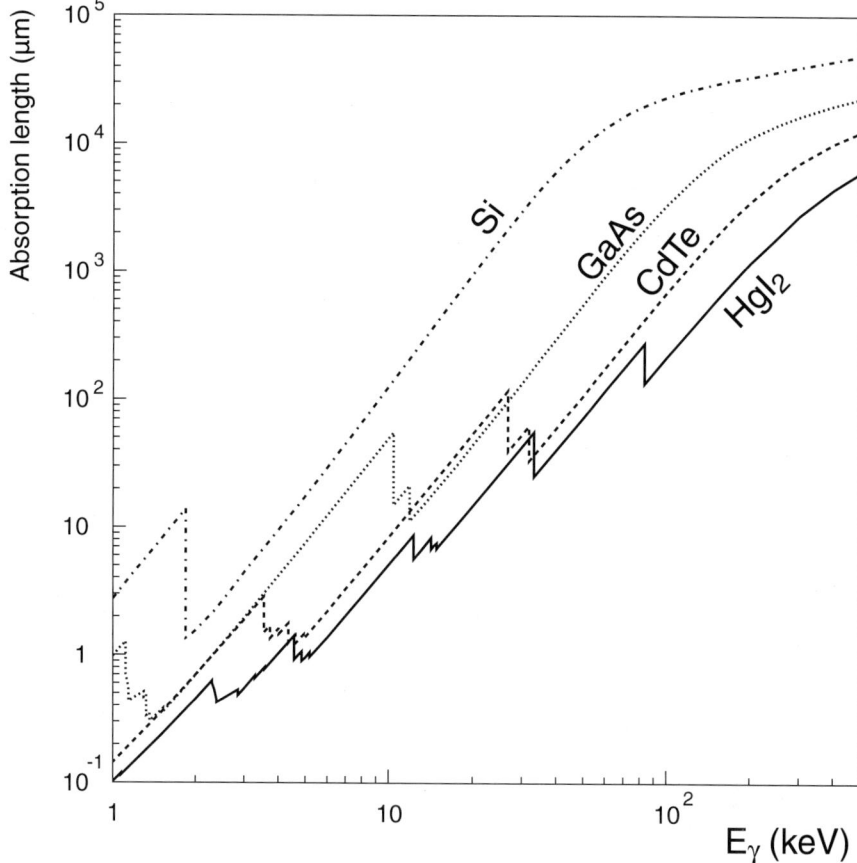

Fig. 5.20. Absorption lengths for various high-Z materials in comparison to silicon

The capability of transforming an object with a given object contrast or intensity modulation

$$C_{\text{object}}(f) = \frac{I_{\max} - I_{\min}}{I_{\max} + I_{\min}}, \tag{5.3}$$

where $I_{\max}$ and $I_{\min}$ are the maximal and minimal intensities of the object and f is its spatial frequency, into an image with the image contrast $C_{\text{image}}(f)$, is described by the *modulation transfer function*

$$\text{MTF}(f) = \frac{C_{\text{image}}(f)}{C_{\text{object}}(f)}. \tag{5.4}$$

The MTF is 1 for $f = 0$ and decreases for increasing spatial frequency toward 0. If the PSF is a Gaussian distribution

$$P(x,y) = \frac{1}{2\pi\sigma} \exp\left(-\frac{x^2+y^2}{2\sigma^2}\right), \tag{5.5}$$

it can be shown that the MTF takes the form

$$\mathrm{MTF}(f) = \exp\left(-\frac{f^2}{2\sigma_{\mathrm{MTF}}^2}\right) \quad \text{with} \quad \sigma_{\mathrm{MTF}} = \frac{1}{2\pi\sigma}. \tag{5.6}$$

Figure 5.21 shows the MTF for the case of a Gaussian PSF of different widths.

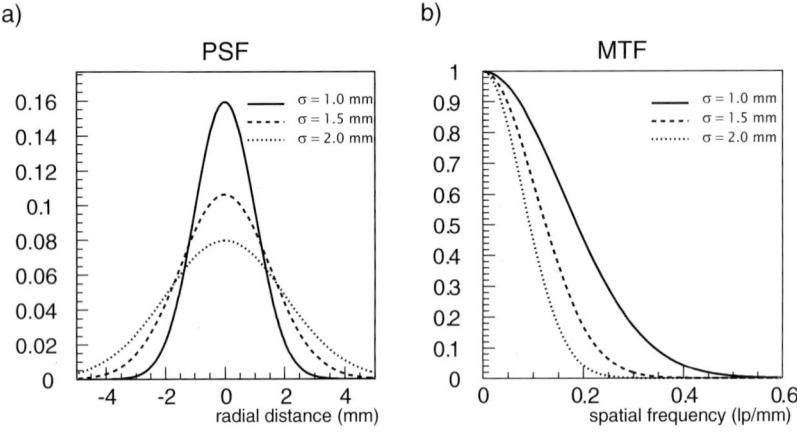

Fig. 5.21. (a) Gaussian PSFs with different widths σ and (b) their corresponding MTFs. The assumed widths are $\sigma = 1$, 1.5, and 2 mm, respectively

The resolution of a line pattern is often expressed as the number of resolved line pairs per mm (lp/mm), which is the spatial frequency f_x at which the MTF has decreased to a value $x < 1$. Most commonly, x is chosen to be 30%.

5.4 Pixel Imaging Systems in Operation

In this section the development of pixel detector systems for biomedical imaging is described. A restriction is again made to systems in or close to operation. As for particle physics the hybrid pixel technique is the most mature technology to date, opening up a new type of detector in radiology and a new class of experiments in protein crystallography [288]. New (nonhybrid) pixel developments with very interesting features for both imaging and tracking are deferred to Chap. 6.

5.4.1 Counting Pixels for Radiography

A vast amount of techniques for radio-detection and radiological imaging exist, among them also techniques using pixelated detectors. As the commercial state of the art, flat-panel imagers [308–311] can be considered. The principle of a typical flat-panel imager is shown in Fig. 5.22. The incoming X-ray photons are converted into optical photons in a scintillator, realized, for instance, by a columnar grown CsI crystal structure. Due to the columnar orientation the optical photons are kept directionally confined and are detected by a photodiode matrix. The generated charge is stored on a capacitance until the pixel cell is readout via address and readout lines. A typical pixel size is $150 \times 150\,\mu m^2$. Flat-panel imagers can be made in large area sizes ($\approx 50 \times 50\,cm^2$) suited for most human radiology applications. Other developments using amorphous selenium (a-Se) [310, 312] as a direct conversion material are also quite mature.

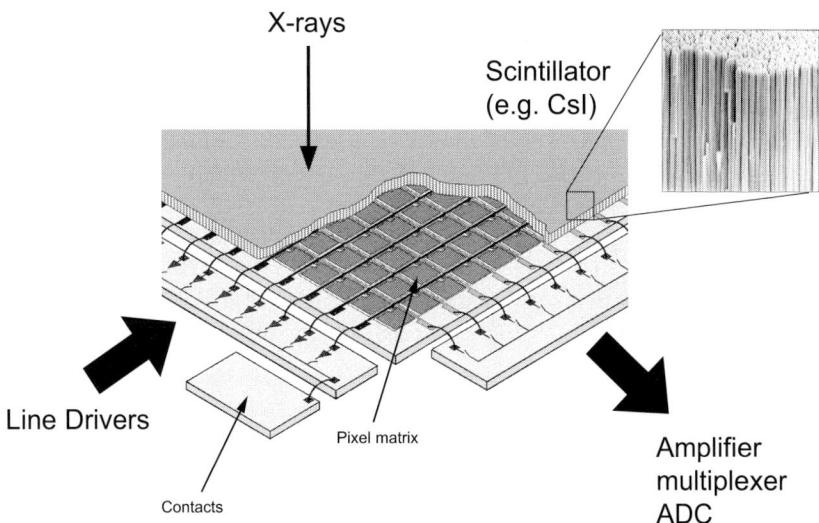

Fig. 5.22. A flat-panel imager based on indirect conversion of X-ray photons in a scintillator (e.g. CsI)

Pixel detectors as discussed in this book have been further developed for radiographical imaging mainly as counting pixel detectors. This is a direct spin-off from pixel detectors for particle physics. The analog part of the pixel electronics is similar to the one for LHC pixel detectors while the readout has been modified. Governed by the need to accumulate the hits of the radiation quanta rather than their individual readout as in tracking devices, the electronics circuitry of an imaging pixel cell needs to count the hits (counting pixels) (see Sect. 3.4.3) or integrate the charge of many X-ray photons of the impinging radiation. Several developments for individual counting of X-ray photons in every pixel cell are being carried out [16,192,313,314]. The principal advantages are the excellent linearity in the response function, the large dynamic range, optimal exposure times, and the good image contrast as compared to conventional film-foil-based radiography, thus avoiding over- and underexposed images.

The challenges to be addressed in order to be competitive with integrating radiography systems are high-speed counting ($\gtrsim 10$ MHz/pixel) necessary for high-resolution X-ray images and a counting range of typically 15–16 bits. Operation should be possible with little dead time due to readout. Low noise and particularly low threshold settings are required, with a small dispersion of the thresholds from pixel to pixel. The last requirement is important in order to allow homogeneous imaging of soft X-rays. It is also mandatory for a differential measurement of the energy of the X-ray photons, realized so far only as a double threshold with an energy windowing logic [93, 317, 318]. Figure 5.23a shows the threshold dispersions of the low and high threshold of a counting pixel chip (MPEC) with energy windowing before and after tuning. The narrow threshold spread is essential for X-ray counting applications. A differential measurement of the energy, exploiting the different shapes of X-ray spectra, for example behind tissue or bone, can enhance the contrast performance of an image as is shown in Fig. 5.23b. Finally, for radiography high photon absorption efficiency is mandatory, which requires the often difficult hybridization of high-Z materials.

Figure 5.24a shows an X-ray scan of a sardine taken in successive scanning steps with the MEDIPIX1 chip (64×64 pixels, $170 \times 170\,\mu m^2$) using a Si sensor. Figure. 5.24b displays the image of a ^{57}Co (122 keV γ) point source obtained with the MEDIPIX2 [192, 313] (256×256 pixels, $55 \times 55\,\mu m^2$) counting

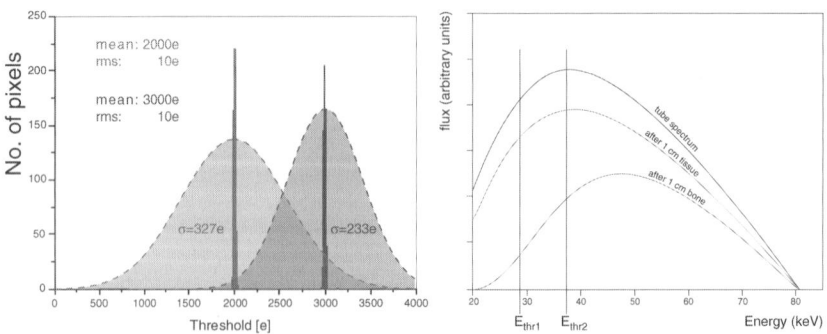

Fig. 5.23. *Left*: Threshold distributions of a counting pixel chip with two (low and high) thresholds [317] showing how tuning can narrow the threshold spread. *Right*: Schematic spectrum of an X-ray tube as observed without an absorber and behind soft tissue or bone, respectively. The difference of the observed intensities behind different absorbers depends on the energy window defined by the two thresholds

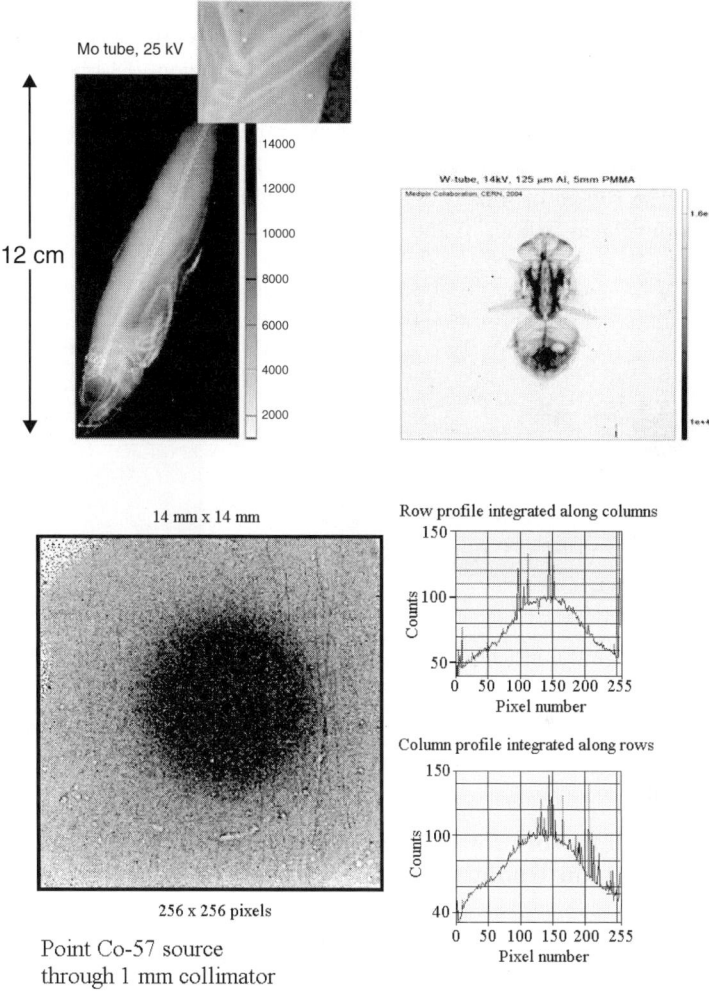

Fig. 5.24. *Top left*: Image of a sardine obtained with the MEDIPIX1 counting pixel chip with Si sensor obtained by successive scans. *Top right*: Image of an insect obtained with the MEDIPIX2 chip. *Bottom*: Image of a ^{57}Co (122 keV γ) point source taken with the MEDIPIX2 counting chip and a 1-mm-thick CdTe sensor [318]

chip bonded to a 14×14-mm^2, 1-mm-thick CdTe sensor [318]. This chip is fabricated in 0.25 μm technology, has energy windowing via two tunable discriminator thresholds, and a 13-bit counter. The maximum count rate per pixel achieved so far is about 1 MHz.

Imaging sensor systems, including modules with several readout chips [170], using the MPEC chip have been made with high-Z semiconductors for more efficient X-ray absorption. A technical issue is the bumping of individual die sensors of Cd(Zn)Te or GaAs. The MPEC chip [16, 317] features 32 × 32 pixels (200 × 200 μm^2), double threshold operation, 18-bit counting at ≈1 MHz/pixel, as well as low noise values (≈120 e$^-$ with CdTe sensor) and threshold dispersion (21 e$^-$ after tuning). A multichip module with 2 × 2 chips using high-Z CdTe sensors with the MPEC chip [317] is shown in Fig. 5.25a as well as the image of a cogwheel obtained with this module using 60 keV X-rays from ^{241}Am in Fig. 5.25b.

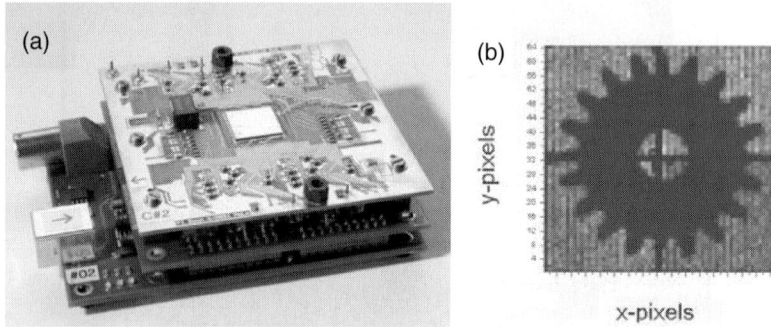

Fig. 5.25. (a) 2 × 2 multichip CdTe module with counting pixel electronics using the MPEC2.3 chip on a USB based readout board [317] (b) Image of a cogwheel taken with 60 keV X-rays

5.4.2 Pixel Detectors for Protein Crystallography with Synchrotron Radiation

Hybrid pixel detectors as counting imagers may lead the way to a new class of experiments in protein crystallography with synchrotron radiation [287, 289]. In this field of application the challenge is to image, with high rate (≈1–1.5 MHz/pixel) and high dynamic range, many thousands of Bragg spots from X-ray photons of ∼12 keV (corresponding to structure resolutions in the 1 Å range) or higher, scattered off, e.g., protein crystals. The typical size of a diffraction spot, for a target to detector distance in the order of 1 m, is 100×200 μm^2, calling for pixel sizes in the order of 100–300 μm^2. The high linearity of the hit counting method and the absence of so-called *blooming effects*, i.e. the response of nonhit pixels in the close neighborhood of a Bragg spot, makes counting pixel detectors very appealing for protein crystallography experiments.

Counting pixel developments are made for ESRF (European Synchrotron Radiation Facility, Grenoble, France) and SLS (Swiss Light Source at the Paul-Scherrer Institute, Switzerland) beam lines. The XPAD project [319,

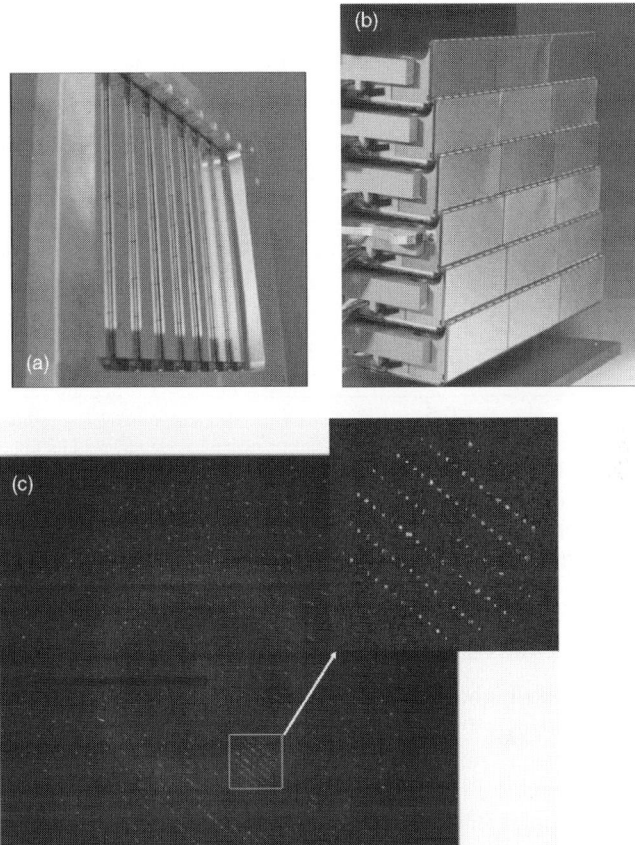

Fig. 5.26. (a) The eight-module XPAD2 detector in development for ESRF and (b) the $20 \times 24\,\text{cm}^2$ large PILATUS 1M detector for protein crystallography at the SLS, both using counting hybrid pixel detectors. (c) Image of lysozyme taken with PILATUS [321]. The *insert* shows details of the Bragg spots

320] is developed for ESRF synchrotron beam lines and for X-ray CT-scanning. The 6×6 cm^2 XPAD2 detector has modules with eight chips of 24×25 pixels with cell dimensions of $330 \times 330\,\mu\text{m}^2$, bonded to a 500-μm-thick silicon substrates (Fig. 5.26a). The PILATUS 1M detector [322] at the SLS ($217 \times 217\,\mu\text{m}^2$ pixels) is made of eighteen 16-chip- modules each covering $8 \times 3.5\,\text{cm}^2$. Figure 5.26b shows a photograph of this detector with $>10^6$ pixels covering a total area of $20 \times 24\,\text{cm}^2$ [322, 323].

A systematic limitation and difficulty of photon detection by counting over large areas is the following: A homogeneous hit/count response in all pixels, also for photons absorbed at the pixel boundaries or between pixels where charge sharing plays a role, must be maintained by delicate threshold tuning. Figure. 5.27 shows a study of this effect with the XPAD2 detector [320]. If

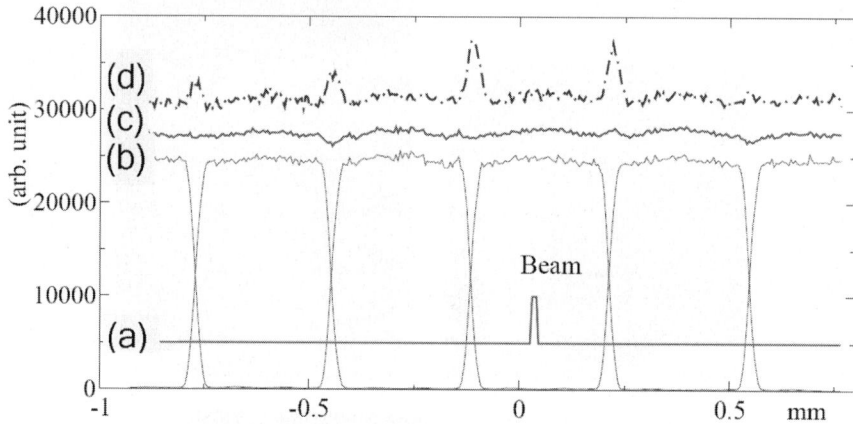

Fig. 5.27. In protein crystallography applications with synchrotron radiation the pixel thresholds need to be precisely tuned to provide homogeneous counting profiles. Shown is a measurement obtained for the XPAD detector [320] by scanning a 10-µm beam along the pixels indicated curve (**a**). (**b**) Count rate per pixel with proper threshold setting to 3,500 e⁻, (**c**) total count rate of both pixels, (**d**) total count rate with threshold set too low (3,200 e⁻)

the threshold is too high, photons entering between two pixels are lost due to the charge sharing; if the threshold is too low, double counting occurs in the regions between pixels (Fig. 5.27d). Only for proper threshold tuning (Fig. 5.27b and c) is the response homogeneous.

Figure 5.26c shows the diffraction pattern of a lysozyme crystal with 10-s exposure to 12 keV synchrotron X-rays [321]. Most of the spots are contained in one or two pixels only, showing that the chosen pixel area is well adapted to the spot size.

5.4.3 Autoradiography with Pixel Detectors

In biomedical autoradiography with tritium the amount of charge to be collected at the electrodes with high efficiency can be well below 1,000 e⁻. Tritium is attractive as a radiolabel because it can be widely applied replacing hydrogen atoms in molecules. With the counting hybrid pixel concept tritium autoradiography is very challenging since very low energy and very homogeneous threshold settings for every pixel are required. Using the MEDIPIX2 chip with a silicon detector [318] a demonstration is given in Fig. 5.28. The picture sequence shows the image of a ^{3}H-spot becoming fainter with increasing threshold settings from upper left (6 keV) to bottom right (18 keV). Note that the peak of the β-emission spectrum of tritium is around 3 keV, well below the lowest chosen threshold.

Spectroscopic pixel devices are necessary to obtain energy resolution which a counting detector system cannot provide. One such concept is the

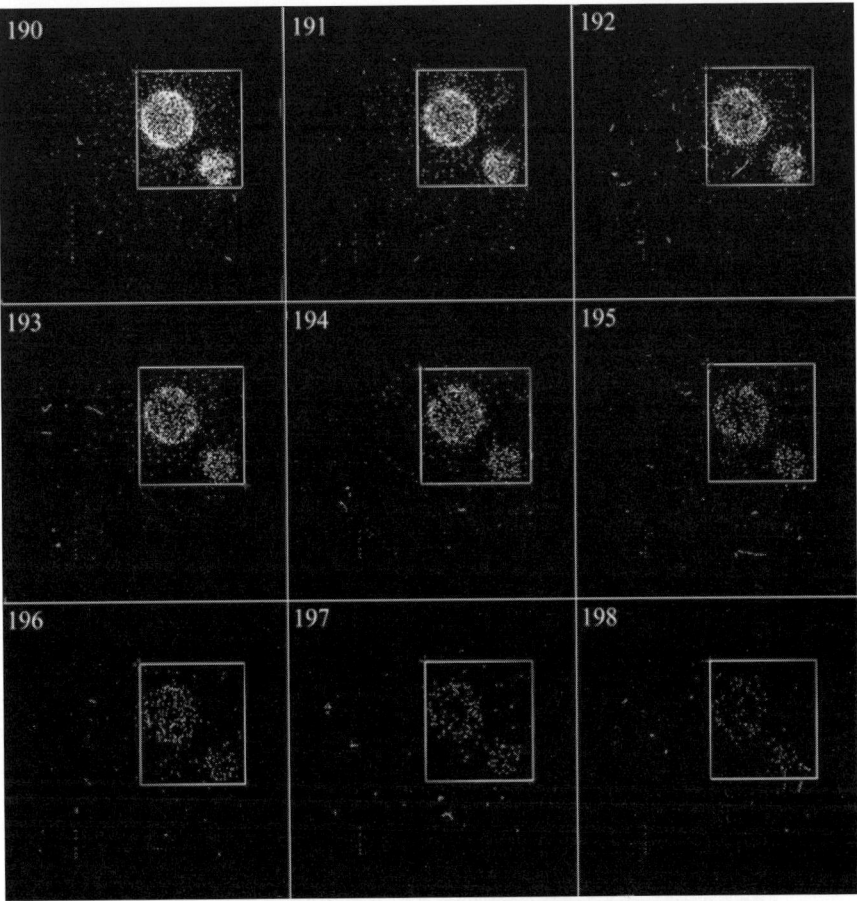

Fig. 5.28. Demonstration of tritium autoradiography with a counting hybrid pixel detector using the MEDIPIX2 chip [318]. The threshold setting is increased from 6 keV *(upper left corner)* to 18 keV *(bottom right corner)*

DEPFET pixel structure [4,324], which is described in Chap. 6. Every pixel cell contains an implanted amplifying transistor allowing very low noise operation due to the low input capacitance without external connections and the direct in-pixel amplification. The electronic noise of individual pixels (ENC $<5\,e^-$) is even below the Fano noise contribution of about 10 e^- (14 e^-) in silicon obtained at 300 K for 3 keV (6 keV) photons. With a ^{55}Fe (6 keV) X-ray source an energy resolution of 134 eV (FWHM) has been measured at room temperature [325]. A DEPFET pixel matrix of 64×64 pixels has been developed for autoradiography [17,326]. Detection and imaging of ^{3}H-labeled tissue as well as double-label imaging, exploiting the excellent energy resolution of DEPFET pixels, has been demonstrated. The spatial resolution of

Fig. 5.29. (a) Imaging of a tritium-labeled leaf with a DEPFET pixel matrix (BIOSCOPE) [17]; (b)–(e) distinction of ^{3}H (leaf) and ^{14}C (filament) radiolabels by energy discrimination [17]: (b) object, (c) image, (d) and (e) separates images of the objects demonstrating the separation capability. The pixel area has dimensions of $3.2 \times 3.2 \, \text{mm}^2$

37 lp/mm for 6 keV γ-rays obtained with 50×50 µm^2 size pixels is excellent [17]. Figure 5.29a shows a tritium-marked leaf, cut in pieces, and imaged with a DEPFET pixel matrix system [326]. A piece of the leaf together with a ^{14}C-marked filament is shown in Fig. 5.29b, demonstrating the capability of imaging doubly labeled tissue. The active area of the matrix is 3.2×3.2 mm^2.

6 Trends and New Developments for Pixel Detectors

Abstract Present trends in research and development on pixel detectors for future particle detectors and new imaging systems are described. The hybrid pixel technology can be viewed as the state of the art of pixel detectors today. Its further potential and its limitations are reviewed. Monolithic pixel detectors which integrate the detection sensor and the readout electronics are attractive but also challenging. Their development as seen today and compromises to be made are discussed.

6.1 Introduction

In particle physics pixel detectors have just begun to be built for large-scale applications, such as vertex detectors in collider experiments. The other, rather complementary field for pixel detectors is their use in imaging applications, most notably biomedical imaging with very different requirements and limitations [288]. In particle physics experiments individual charged particles, usually triggered by other subdetectors, must be identified with high demands on spatial resolution and timing. In imaging applications the image is obtained by the untriggered accumulation (integrating or counting) of the quanta of the impinging radiation, often also with high demands on rate (e.g. $\gtrsim 1$ MHz per pixel in certain radiography or CT applications) and on imaging resolution ($\gtrsim 10$ line pairs/mm) and contrast.

Pixel detectors in high energy physics typically have signal charges in the order of at least 8,000 electrons in worst case scenarios, including charge sharing and radiation damage effects. As described in Sect. 5.3, imaging applications often require efficient detection of signal charges almost an order of magnitude lower than this. The spatial resolution of pixel detectors is governed by the attainable pixel granularity. Resolutions of about 10 μm are obtained with pixel cell dimensions in the order of 50–100 μm. The requirements from radiology (mammography) are similar or even less demanding, while some applications in autoradiography require submicrometer resolutions, not attainable with the present day pixel detectors. For applications with lower demands on the spatial resolution (≈ 10 μm) but with demands on real-time and time-resolved data acquisition, semiconductor pixel detectors are however very attractive.

Thin detector assemblies are mandatory for the vertex detectors at collider experiments, in particular for the planned linear e^+e^- collider. While silicon is an almost perfect material for particle physics detectors, allowing the shaping of electric fields by tailored impurity doping, the need of high photon absorption efficiency in radiological applications requires the study and use of semiconductor materials with high atomic charge, such as GaAs or CdTe (see Chap. 2). For such materials the charge collection properties are much less understood and mechanical issues, in particular those related to hybrid pixels, exist abundantly, for the hybridization of detectors especially when they are not available in wafer scale sizes. Last but not least, the cost–performance ratio is an important factor to consider if an imaging application could shall be commercially interesting.

6.2 Limitations and Prospects of the Hybrid Pixel Technology

The hybrid pixel technology has been the pixel technology of choice for the next generation of detectors in high energy physics at the LHC and in high-rate fixed target experiments. This choice is governed by the fact that this technology is the only one which at present is sufficiently mature to build detectors with an area larger than a few square centimeters. The main advantages of this technology for large-scale detectors are the following:

- The hybrid assembly approach allows testing at several intermediate steps, thus offering a comparatively high yield in producing pixel modules with area sizes of tens of square centimeters.
- The chip, sensor, and interconnection technologies are industrial processes, matured during many years of experience. They are available from a variety of industrial vendors.
- Because the sensor and the amplifying/readout chip are separate items, materials other than silicon can be used for the sensor substrate.

The disadvantages of the hybrid technique become evident when one addresses the requirements of particle detectors at high-energy accelerators of the future, in particular the demands on high resolution in high-multiplicity environments (i.e. small cell size), on a low material budget, and on high speeds.

The technological limitations are mostly related to the bumping technology and the power density associated with the constraint that the electronics circuitry for amplification and logic is confined to the same area as the detecting electrode. In 0.1 μm CMOS technology the projected pixel size could be as small as $10 \times 80\,\mu m^2$ or $25 \times 25\,\mu m^2$ for a square geometry. This leads to an estimated power density of $30\,mW/mm^2$ or $30\,kW/m^2$ detector.

The pitch for the PbSn bumping technology is limited to about 10–15 μm due to mask alignment and galvanic process precision [327]. In the case of indium bumps the smallest pitch obtained so far is 15 µm [328]. The technological limit lies below 10 µm.

For LHC upgrade scenarios pixel detectors are also discussed for larger area coverage than today. In this case, the present hybrid pixel technology with one-to-one coverage of the area with sensor, readout chips, and a module interconnect layer (flex) not only constitutes a significant material issue but also drives the costs of pixel detectors at larger radii.

6.2.1 MCM-D Integration

With decreasing pixel cell sizes the high-density interconnection techniques become more complex, demanding more advanced technologies. A development for the next generation of hybrid pixel detectors regarding the power and signal distribution in a module is the so-called multichip module technology deposited on silicon substrate (MCM-D) [329]. This concept has been contrasted with the flex-hybrid approach for module hybridization in Fig. 4.13. Another illustration is shown in Fig. 6.1 which stresses the possibilities of the MCM-D technology with respect to the extra degrees of freedom for the routing of power lines [330].

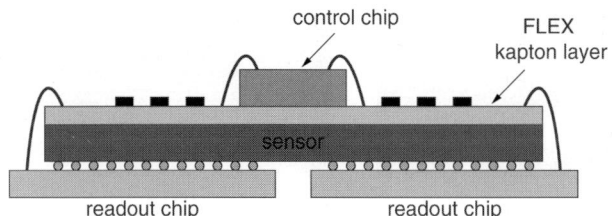

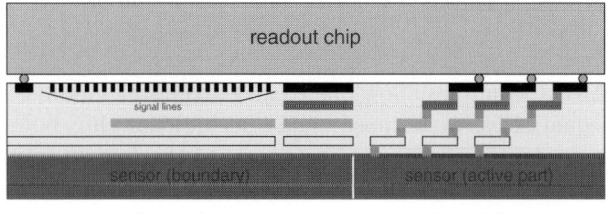

Fig. 6.1. (a) Cross-sectional view of a full hybrid pixel module with wire bonding from the readout chip to the flex kapton layer. (b) Cross section of the edge of a readout chip at the transition between the active detection area (*right*) and the outer periphery (*left*) in the MCM-D concept. In the periphery outside the active area the power bus lines are routed in four metal layers which are built up directly on the sensor substrate. In the active area the layers are simply crossed by vias to the chip side where bump bonding is done [329]

264   6 Trends and New Developments for Pixel Detectors

The MCM-D concept offers an elegant way to integrate ICs and sensors into a quasi-monolithic high-density-interconnect module. A multiconductor-layer structure is built up on the silicon sensor itself. Layers of a dielectric (e.g. photosensitive benzocyclobutene [BCB], $\epsilon_r = 2.7$) and metal (Cu) are deposited and patterned on the electrode (i.e. pixel) side of the sensor by using thin-film technologies. In the process presented in [329], BCB is spun on the silicon sensor with a thickness of 6 μm, while Cu is electroplated with 3-μm thickness. The processing temperature resides below 220°C at all times and has been proven not to compromise the sensor characteristics. The metal layers are connected by vias of 25-μm minimum diameter. Finally, the same bumping and flip-chip processing as described in Chap. 4 is applied to connect the front-end ICs to the sensor elements. Figure 6.2 shows scanning electron microphotographs of via structures made with MCM-D technology. In the displayed example four metal layers are built up from the silicon sensor substrate.

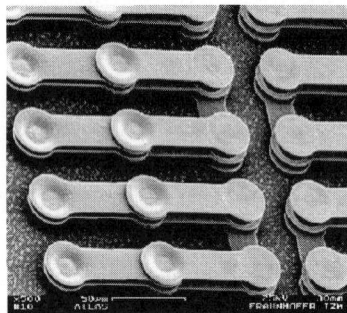
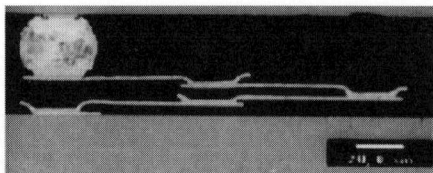

Fig. 6.2. (a) Scanning electron microphotograph of an MCM-D vias structure. (b) Cross section cut with a bump-bond connection to the readout chip. The dielectric is etched away in these samples for better visualization. (Courtesy of Fraunhofer Institut für Zuverlässigkeit und Mikrointegration, IZM, Berlin)

The MCM-D techniques allow one to make compact modules in which all pixels have equal sizes, even those which lie between chip boundaries, due to the increased routing freedom. Furthermore, the bus structures and the voltage supply lines can be buried in several layers under the inactive area at the edge of the module. The feed-through via system to the readout chip, which is necessary for every pixel cell, is shown in Fig. 6.1b. Since sensor and chips must cover the same area, the size of the sensor is wider for the same readout chip width than in the flex-hybrid concept and also longer in order to house, for instance, a module control chip and other necessary passive components for supply bypassing (see Fig. 6.3). Yield losses apply to the entire module. Therefore the yield requirements on MCM-D pixel modules are very high. In return, the MCM-D technology can provide a

6.2 Limitations and Prospects of the Hybrid Pixel Technology

Fig. 6.3. Photograph of a 16-chip MCM-D module developed as an alternative module concept for the ATLAS pixel detector. Note the absence of wire bonding and the position of the module control chip (*MCC*) on a balcony outside the actual module area. (Courtesy of University of Wuppertal)

compact multichip pixel module without external wire bonds [35] and flex circuitry. Using the ATLAS readout electronics several prototype MCM-D modules with 16 front-end chips have been built with good and encouraging preliminary results [35, 330], showing that complex MCM-D modules can be built with negligible increase in noise performance. At present, yield and perhaps cost issues are the limiting factors of the MCM-D technology for large pixel detectors.

6.2.1.1 Sparse CMOS Using MCM-D

The integrated multilayer possibilities provided by the MCM-D technology can, in the future, allow much wider applications than sketched above. The increased flexibility having become available in printed circuit board technology after the introduction of the multilayer techniques is in principle also expected here. One such application for imaging can be the use of the MCM-D multilayer structure to fan out chip I/O lines to larger sensor pixels as sketched in Fig. 6.4. In many X-ray applications, for example, the optimal pixel pitch and cell size of the image-taking sensor is much larger than the required pixel area for the chip circuitry, especially with the advancement of very small scale CMOS processing technologies. An $\approx$1-cm^2 chip can be used for readout of a much larger area. This might also be interesting for large area pixel detectors in high energy physics as the smaller number of readout chips per module offer a price reduction in places where a less precise position resolution is sufficient. However, special attention must be given to the additional capacitance introduced by the fan-out lines, which can be much larger than the actual pixel capacitance.

6.2.2 Interleaved Hybrid Pixels

An approach to cope with the pitch limitations of the hybridization process exploits capacitive coupling of the pixels to have a larger readout and

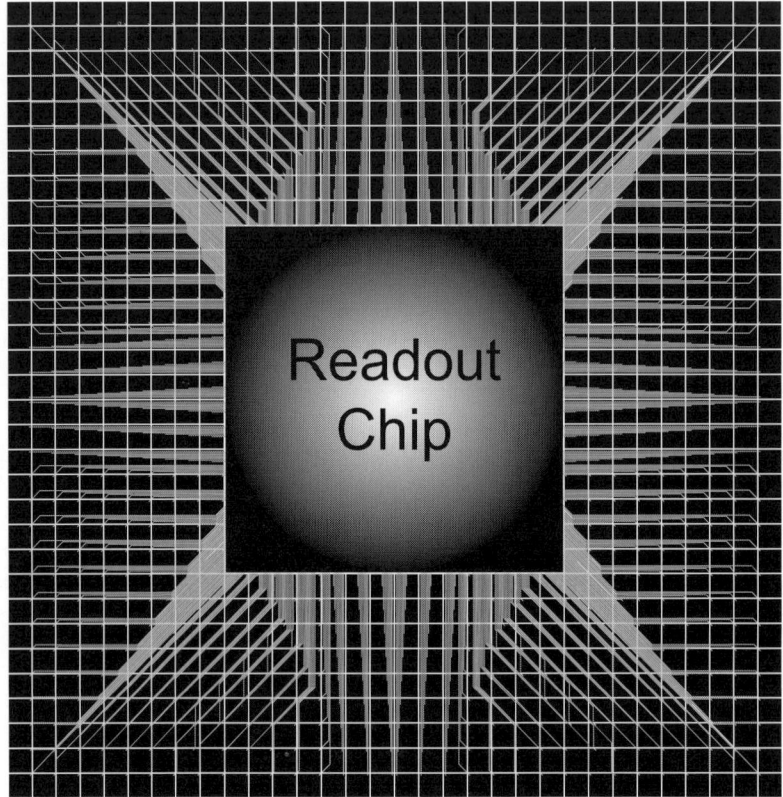

Fig. 6.4. Layout of a sparse CMOS fan-out routing

bump-bonding pitch than the actual sensor pixel pitch. This has been directly adopted from silicon microstrip detectors [331, 332] for which integrated capacitive charge division readout is very often employed to reduce the number of readout channels. The development is carried out in the context of pixel detector R&D for a linear collider detector under the label HAPS (hybrid active pixel sensors) [333, 334]. Figure 6.5a shows a layout of a pixel matrix with readout and sensor pixels at a different pitch. Every pixel is biased via a polysilicon resistor, but only a fraction of pixels is read out.

Figure 6.5b is the result of the obtained spatial resolution when the device is scanned with a fine focus laser. Charge sharing between neighboring pixels through their interpixel capacitance improves the space resolution for a HAPS device with a pixel pitch of 100 μm and readout pitch of 200 μm by roughly a factor of 4 compared to the binary pixel pitch/$\sqrt{12}$ resolution. Values between 3 μm at the interleaved pixels (maximum charge sharing) and 10 μm at the readout pixel (minimal charge sharing) have been obtained (Fig. 6.5). This level of improvement can be extrapolated to small pixel pitches of

6.2 Limitations and Prospects of the Hybrid Pixel Technology

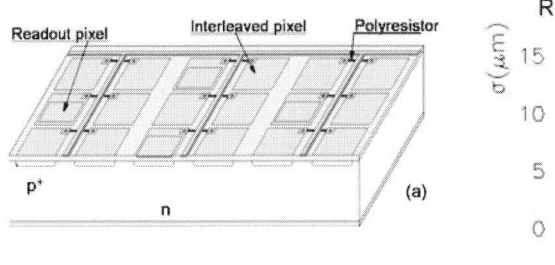
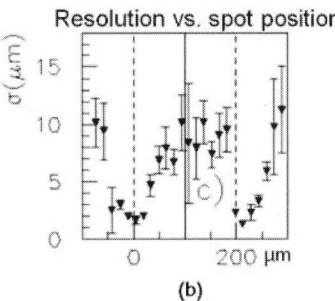

Fig. 6.5. (a) Layout of a HAPS sensor with different pixel and readout pitches. (b) Measured spatial resolutions by scanning with a fine focus laser [333]

20–25 µm provided a signal-to-noise ratio (S/N) >100 is maintained and the charge loss into neighboring pixels is less than 50% [333]. Needless to say that the interleaved pixel concept imposes lower constraints on the bumping process and reduces the total power consumption, when comparing to a device with amplification electronics for every pixel cell, of course on the expense of a degraded two-track resolution and the necessity for analog pulse height readout.

6.2.3 Active Edge Three-Dimensional Silicon Pixel Detectors

The features of doped silicon allow interesting geometries with respect to field shapes and charge collection. The so-called three-dimensional silicon detectors have been proposed [153] to overcome several limitations of conventional silicon pixel and also silicon strip detectors, in particular in high-radiation environments or applications which require a large active/inactive area ratio [287]. Examples are detectors which must operate close to particle beam lines and for which the radiation environment produces particle fluences in excess of 10^{15} particles/cm^2, with the additional complication that the irradiation is not homogeneous over the detector. The large active/inactive area ratio is of great advantage, e.g. for applications using Bragg diffraction with synchrotron light (see Sect. 5.2) where number and intensity of many Bragg spots need to be detected with high efficiency.

As shown in Fig. 6.6a, a three-dimensional structure is obtained by processing the n$^+$ and p$^+$ electrodes into the detector bulk rather than by conventional implantation on the surface by combining VLSI and MEMS (micro-electro-mechanical systems) technologies. Charge carriers are drifting inside the bulk parallel to the detector surface (Fig. 6.6b) over a typical drift distance of 50 µm. Another feature is the fact that the edge of the sensor can be a collection electrode itself [335], thus extending the active area of the sensor to within about 10 µm to the edge. The remaining distance results from the undepleted depth at this electrode. Edge electrodes also avoid

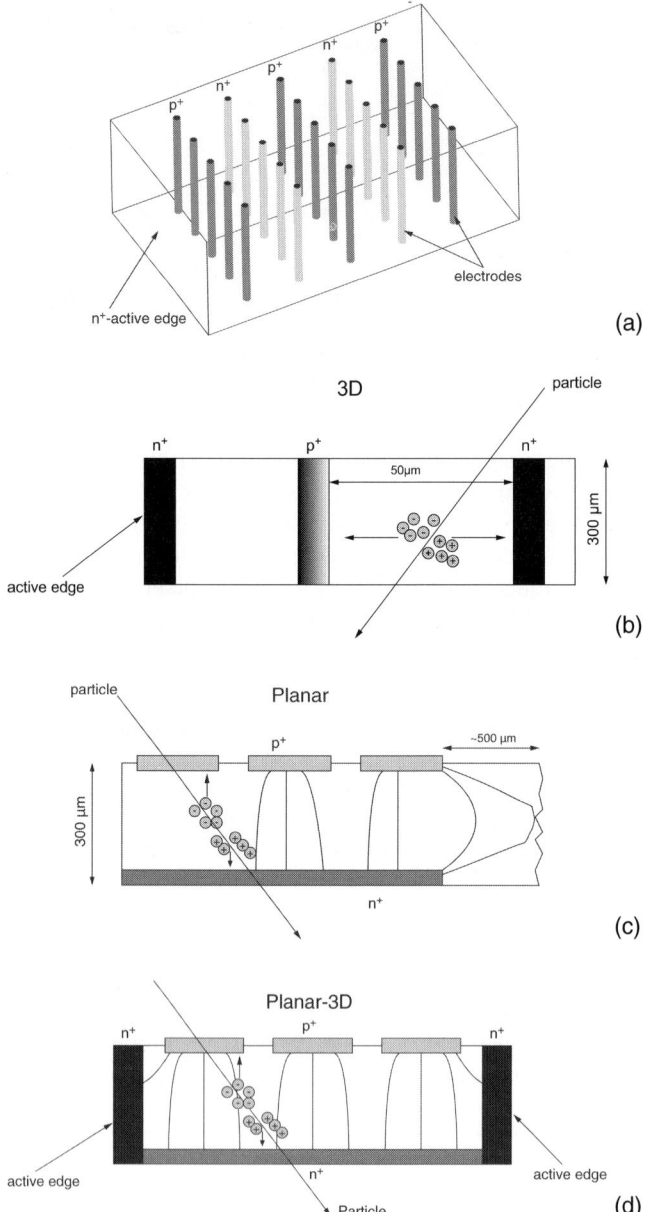

Fig. 6.6. (a) Schematic view of a three-dimensional silicon detector, (b) charge collection in a three-dimensional silicon structure and (c) in a conventional planar electrode detector, (d) a structure combining three-dimensional and planar features

inhomogeneous fields and surface leakage currents which usually occur due to chips and cracks at the sensor edges (Fig. 6.6c). The main advantages of three-dimensional silicon detectors are summarized in Table 6.1. They mostly come from the fact that the electrode distance is short (50 µm) in comparison to conventional planar devices at the same total charge, resulting in a fast (1–2-ns) collection time, low (<10 V) depletion voltage, and, in addition, a large active/inactive area ratio of the device. Note, however, that the three-dimensional electrodes inside the bulk volume constitute regions of dead area themselves.

Table 6.1. Comparison of characteristic properties of three-dimensional silicon detectors and standard planar detectors

	Three-dimensional detector	Planar detector
Charge collection path	50 µm	300 µm
Depletion voltage	<10 V	70 V
Charge collection time	1–2 ns	10–20 ns
Edge sensitivity	<10 µm	500 µm

The technical fabrication of three-dimensional silicon detectors is more complicated than for planar detectors and requires a bonded support wafer as well as reactive ion etching of the electrodes into the bulk. A compromise between three-dimensional and planar detectors, the so-called planar–three-dimensional detectors [336], maintaining the large active area, is shown in Fig. 6.6d. Planar technology is used for the charge collecting electrodes, but edge electrodes are included. The 3D structures are fabricated by diffusing the dopant from the deeply etched edge and then filling it with polysilicon. Prototype three-dimensional and planar–three-dimensional detectors using strip or pixel electronics have been fabricated. The three-dimensional detectors show encouraging results with respect to speed (3.5-ns rise time) and radiation hardness ($>10^{15}$ protons/cm^2) [337]. Figure 6.7 demonstrates the achievable energy resolution ($\sigma_E/E = 2\%$) observed in a measured ^{241}Am spectrum.

6.3 Monolithic and Semimonolithic Pixel Detectors

Because detector and readout electronics circuitry are in many fields of applications made of silicon as the base material, the idea of monolithic rather than hybrid devices is very appealing. They offer the possibility of very low input capacitances and hence very low noise figures, avoiding at the same time the costly and complex hybridization process. Several attempts were made in this direction in the 1980s [339–342]; however, the developments have so far been

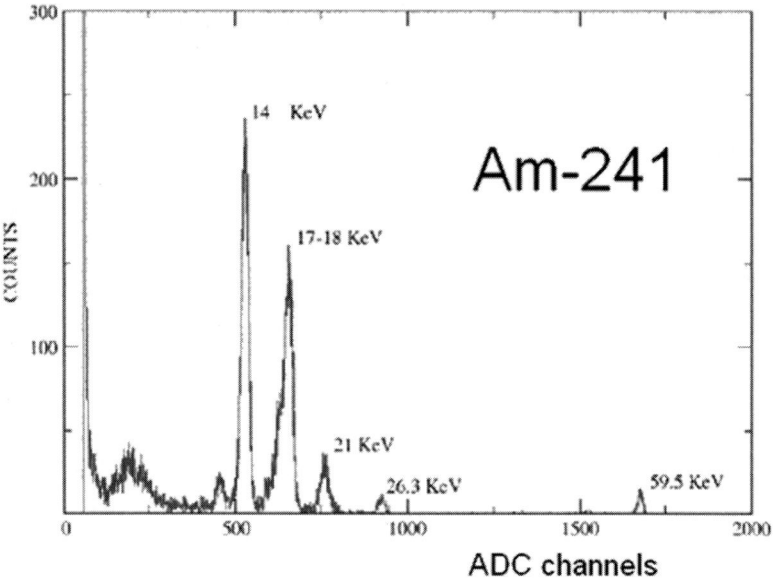

Fig. 6.7. Measured ^{241}Am spectrum measured with a three-dimensional silicon device with pixel electronics [338]

limited to small prototype devices. No real particle or radiation detector has emerged. Today, the situation regarding monolithic devices is still unchanged. Several developments look promising but still need to prove essential characteristics needed for high-energy particle detection, most notably the radiation tolerance. In this section some of these new concepts are introduced. The borderline between truly monolithic pixels, where sensor and amplification and readout circuitry are one entity, and semimonolithic devices, where some hybridization is still necessary, is not well defined. Hybrid pixels use electronic circuitry in every pixel cell with hundreds of transistors. In monolithic devices the maximum number implemented so far are 15 transistors [343], allowing only a simple logic implementation. Further signal treatment is deferred to the outer frame of a pixel matrix.

As to the ultimate goal to implement standard VLSI CMOS circuits on a high resistivity, fully charge collecting substrate material, the present compromises can be classified into:

- Nonstandard CMOS on high-resistivity bulk
- Standard CMOS with charge collection in an epitaxial silicon layer
- CMOS on SOI (nonstandard)
- Amorphous silicon on standard CMOS ASICs
- Amplification transistor implanted in high-resistivity bulk

6.3 Monolithic and Semimonolithic Pixel Detectors

The development of monolithic pixel devices has been challenged anew by the demands of future collider experiments, most notably by a linear e^+e^- collider [344], where very little material ($\ll 1\% X_0$) per layer, small pixel sizes ($\approx 20 \times 20\,\mu\mathrm{m}^2$), and high rate capability (80 hits/(mm^2 ms)) is required. This is due to the very intense bremsstrahlung of narrowly focused electron beams close to the interaction region which produce electron–positron pairs in vast numbers. Readout speeds of 50 MHz (row clocking rate) and full detector frame readout times of the order of 40 μs are required.

6.3.1 Monolithic Pixels on Bulk Silicon

The first successful development of a monolithic pixel detector prototype which could already be used as a particle beam telescope was made as early as 1992 [342, 345, 346]. As an historical achievement this development has already found its mentioning in Chap. 1. The device is based on a high-purity, high-resistivity p-type bulk PIN diode of which the junction is created by an n-type diffusion layer. On the top, an array of ohmic contacts to the substrate serves as collection electrodes. A schematic cross-sectional view is shown in Fig. 6.8. The circuitry on the top controls the readout of the pulse height information from each pixel cell to the off-shell electronics and consists of nine PMOS transistors. The diode junction is formed on the backside. The charge-collecting electrodes therefore sit in a region of minimal electric field

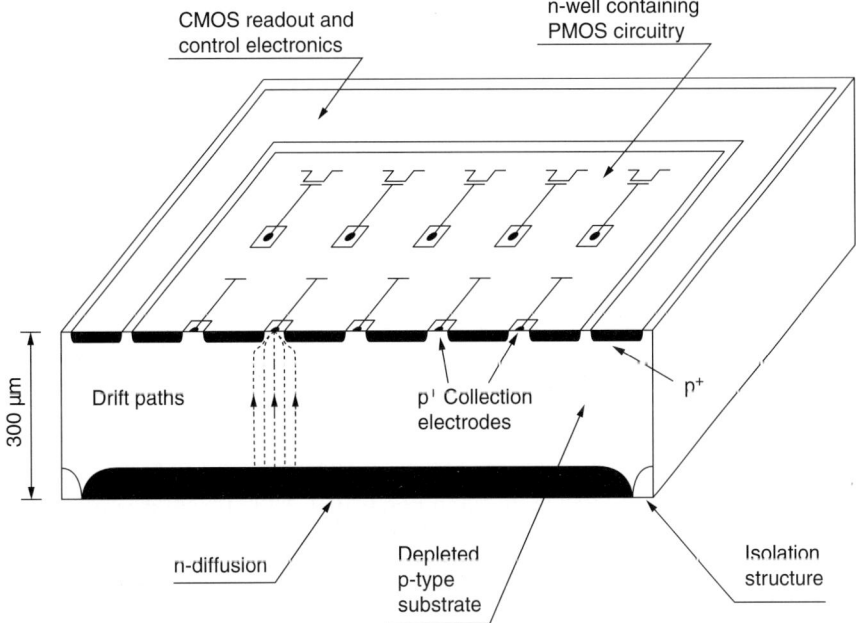

Fig. 6.8. Cross section of an early developed monolithic pixel detector [346]

when the device is operated in full depletion. Almost 90% of the sensor area is taken up by an n-well containing PMOS circuitry and at the same time serving as a shield between the bulk and the PMOS transistors. Note that no NMOS transistors can be implemented in the active area and that a dead area is introduced at the border of the device where the readout circuitry is placed.

An array of 10×30 pixels with 125×34 µm^2 pixel cells has been realized and characterized in a high-energy muon beam using a telescope arrangement of four-pixel arrays. Figure 6.9 shows an event display of the four planes passed by an ionizing high-energy muon. An S/N (single channel) of 55 was observed with a spatial resolution σ of 2.2 µm in the short 34-µm pixel direction.

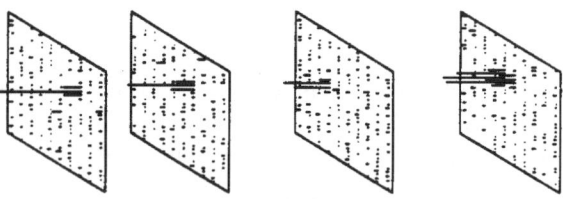

Fig. 6.9. Event display showing the passage of an ionizing track through four planes of a monolithic pixel detector [346]

While the achieved results were truly spectacular at this early stage of pixel developments, no large-scale detectors evolved from this approach mostly due to the use of nonstandard processing technologies.

6.3.2 Monolithic CMOS Pixels

Commercial CMOS technologies use low-resistivity silicon which is not suited for charge collection. However, in some technologies an epitaxial silicon layer of a few to 15-µm thickness can be used [347–350]. The generated charge is kept in the epitaxial layer by potential wells at the boundary and reaches an n-well collection diode by thermal diffusion. The signal charge is very small ($<1000\,e^-$) and its time development is inherently slower (≈ 100 ns) than that obtained in detectors with a highly resistive, depleted bulk. Very low noise electronics is required for this development. The principle of a monolithic active pixel sensor (MAPS) is shown in Fig. 6.10. CMOS pixels are potentially very cheap particle detectors, as standard processing technologies are used. Only NMOS transistors are allowed in the active area because of the n-well/p-epi collecting diode which does not permit other n-wells.

The fundamental difference between CMOS monolithic active pixel sensors and CMOS camera chips is that the former must have a 100% fill factor

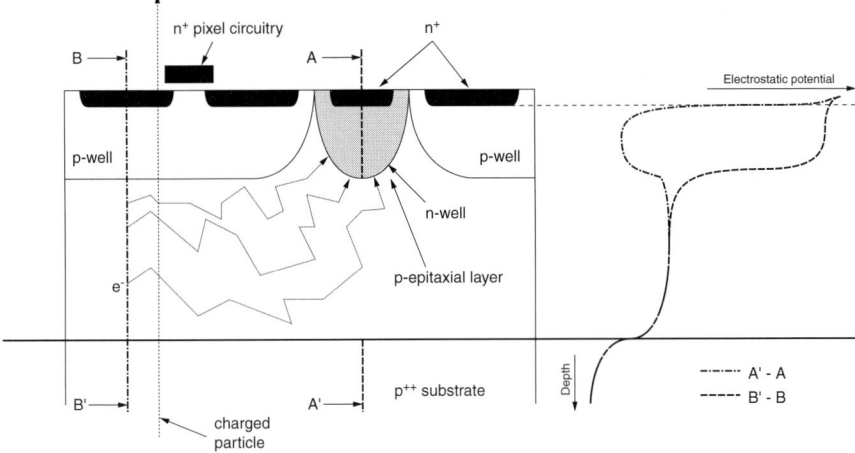

Fig. 6.10. Schematic cross section of a CMOS pixel detector cell (MAPS). The charge is collected in an n$^+$-well diode on the p-epitaxial layer [349]

for efficient particle detection. Standard commercial CMOS technologies are used which have a lightly doped epitaxial layer between the low-resistivity silicon bulk and the planar processing layer. The thickness of the epitaxial layer depends on the detail of the CMOS process technology. It can be at most 15 µm thick but not every CMOS process possesses an epitaxial layer. Different developments using similar approaches exist on both sides of the Atlantic [350–354]. Prototype detectors have been produced in 0.6, 0.35, and 0.25 µm CMOS technologies [343, 355].

The MAPS structure (Fig. 6.10) is depleted only directly under the n-well collection diode where full charge collection is obtained; the collection is incomplete in most parts of the epitaxial layer. Figure 6.11a shows the response to X-ray photons from a ^{55}Fe source at energies of 5.9 and 6.5 keV. The peaks correspond to photons absorbed in the small depleted region under the n-well. The main part of the spectrum, however, corresponds to the absorption of photons in the undepleted part of the epitaxial layer. A chip submission using a process without epitaxial layer, but with a low doped substrate of larger resistivity has also been tried (MIMOSA-4) [357]. The fabricated devices proved to function with high detection efficiency for minimum ionizing particles. While a complete abstention from the epitaxial layer is not a solution for CMOS pixels as particle detectors, this observation renders the uncertainty in and the need of an epitaxial layer with substantial thickness less constraining.

Matrix readout is performed using a standard three-transistor circuit (line select, source-follower stage, reset) commonly employed by CMOS matrix devices (see Fig. 6.11b), but can also include current amplification and current memory [343]. For an image two successive frames are

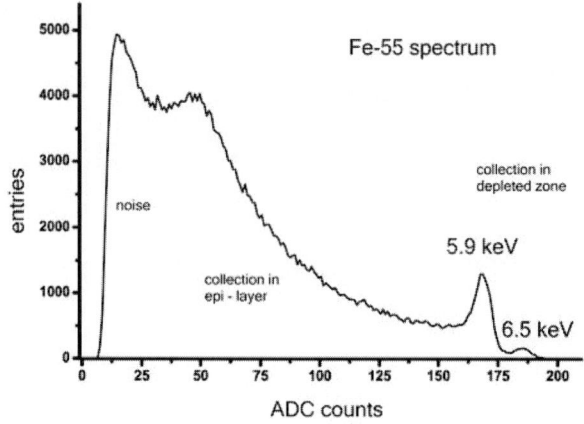

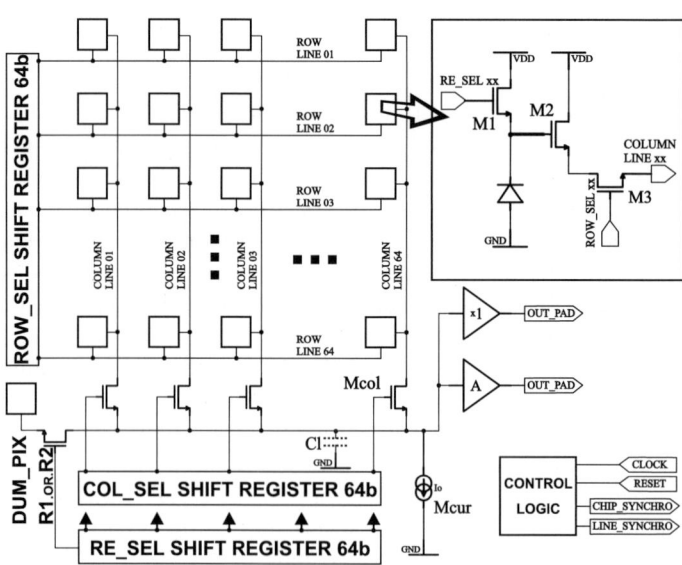

Fig. 6.11. (a) Response to a ^{55}Fe 5.9-keV X-ray source [356]. An interpretation is given in the text. (b) Matrix operation of MAPS monolithic pixels

subtracted from each other [343, 351] (correlated double sampling) which suppresses switching noise. From this the hit signal is obtained after a further pedestal and common mode noise frame subtraction. Noise figures of 10–30 e$^-$ and S/N > 20 have been achieved [343] with spatial resolutions below 5 µm.

The radiation hardness of CMOS devices is a crucial question for particle detection at high-intensity colliders. MAPS appear to sustain non-ionizing interaction damage up to $\approx 10^{12}$ n_{eq} [354, 358] and higher [359]. Damage effects of γ-rays and ionization from charged particles are still under investigation [357]. At a linear e^+e^- collider, radiation doses in the order of several kilograys, mostly from ionizing electron–positron pairs, are expected [344]. Irradiation of CMOS pixel devices with 10-keV X-rays to 2 kGy shows a degradation in S/N of about 15–20% [357], which appears to be close to being sufficient for this application. More recent studies indicate that ionizing doses of up to 10 kGy can be sustained if short integration times are chosen.

Apart from this the focus of further development lies in making larger area devices and in increasing the charge collection performance in the epitaxial layer. For the former, first results in reticle stitching, that is electrically connecting IC area over the reticle border, have been encouraging (Fig. 6.12a) [355]. Full CMOS stitching over the reticle border still has to be demonstrated. The poor charge collection by thermal diffusion is another area calling for ideas. Approaches using more than one n-well collecting diode [353], a photo-FET [343, 351, 352], or a photo-GATE [360] have been investigated. The photo-FET technique (Fig. 6.12b) has features similar to those of DEPFET pixels (see Sect. 6.3.5). A PMOS FET is implanted in the charge collecting n-well and signal charges affect its gate voltage, thus modulating the transistor current.

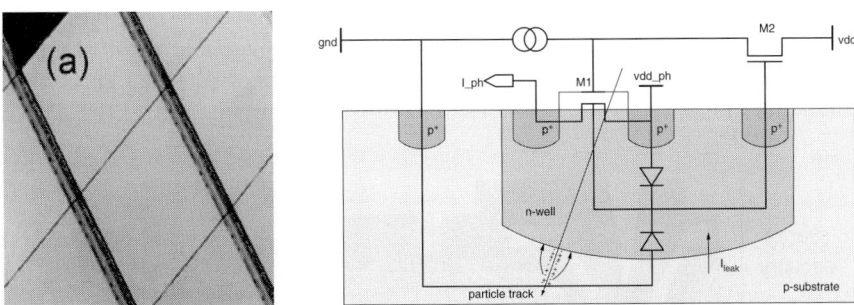

Fig. 6.12. (a) Reticules of a CMOS wafer. Seamless stitching over reticule boundaries is a possibility to obtain large area CMOS active pixels. (b) Schematic of a PMOS photo-FET to improve charge collection in MAPS

Above all, the advantages of a fully CMOS monolithic device relate to the adoption of standard VLSI technology and its resulting low-cost potential (potentially $\approx$ €25/cm^2). In turn the disadvantages also come largely from the dependence on commercial standards. The thickness of the epitaxial layer varies for different technologies. It becomes thinner for smaller processes and with it also the number of produced signal electrons decreases. Only a few processing technologies are suited. With the rapid change of commercial process technologies this is an issue of concern. Furthermore, only NMOS devices located in a p-well can be used in the active area. PMOS transistors would require an n-well which is reserved for the charge collection. The voltage signals are very small ($\approx$mV), of the same order as transistor threshold dispersions requiring very low noise VLSI design. Especially in high-energy physics experiments sophisticated electronics logic, e.g. for hit readout and zero suppression, including fast clock signals must be directly implemented at the chip level. Preventing cross coupling to the tiny analog signals is a challenge. The promising features of CMOS active pixels and the low cost potential of such devices have created a large R&D effort worldwide.

6.3.3 Monolithic SOI Pixels

In order to use high-resistivity bulk material for charge collection while maintaining CMOS processing of electronics circuits, the silicon-on-insulator (SOI) technology has been employed [361]. Figure 6.13 shows a cross-sectional view of an SOI monolithic pixel cell. By means of a thick buried oxide layer (BOX) in the bulk the active sensor volume is isolated from the electronics layers in which complete CMOS processing is now possible. At present, conventional bulk MOS technology on thick SOI substrate is still being used for prototype development [361]. Contact through the bulk oxide to the detector substrate is made by vias to p^+-implants. The detector is a conventional p^+-in-n, DC-coupled silicon detector.

The principal advantage of this pixel detector type is the possible use of high resistivity detector substrates with full charge collection in 200–300 μm. Also, in comparison to the CMOS active pixel sensors of Sect. 6.3.2, the SOI technique offers a less restricted use of full CMOS circuitry atop the active

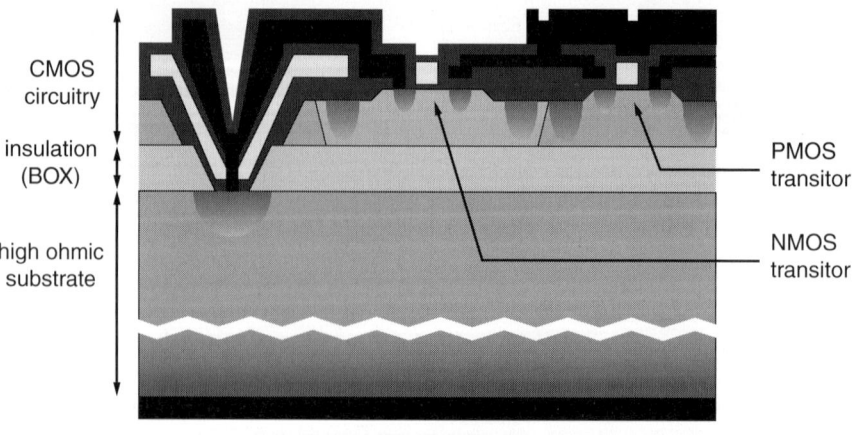

Fig. 6.13. Cross section through a SOI monolithic pixel cell. The buried oxide layer (BOX) is the insulation that separates the CMOS and sensor diode parts of the device

sensor area of the device. At present the technology is not a commercial standard, but would certainly become very attractive if industrial processing of bonded sensor- and CMOS-wafers became available.

6.3.4 Amorphous Silicon above CMOS Pixel Electronics

Hydrogenated amorphous-silicon (a-Si:H), where the H content is up to 20%, can be put as a film on top of CMOS ASIC electronics. a-Si:H has been studied as a sensor material long ago and has gained interest again [19, 362], also for particle detectors, with the advancement in low-noise, low-power electronics. Early developments of a-Si:H as detector material [363, 364] concluded that the signal for a minimum ionizing particle was too small to be detected. The charge signal in a film deposit of 10–30 μm thickness is 200–1,000 e which – due to the amorphous structure – is however spread only very little and remains confined in a sensitive, almost cubic pixel volume of less than 30 μm^3. The carrier mobility is very low ($\mu_e = 2$–5 cm^2/(V s), $\mu_h = 0.005$ cm^2/(V s)). Due to the large difference in the mobilities essentially only electrons contribute to the signal. The band gap of a-Si:H is increased to 1.8 eV with respect to crystalline silicon (1.1 eV). The challenge apart from developing a-Si:H material with sufficient charge collection lies in the development of low noise (ENC $< 5 \mathrm{e}^-$) CMOS electronics with very low power consumption per pixel. Figure 6.14 shows a cross section of a photodiode structure realized with amorphous silicon atop a CMOS circuitry [19]. From a puristic view it still a hybrid approach, but its main disadvantage, a complicated

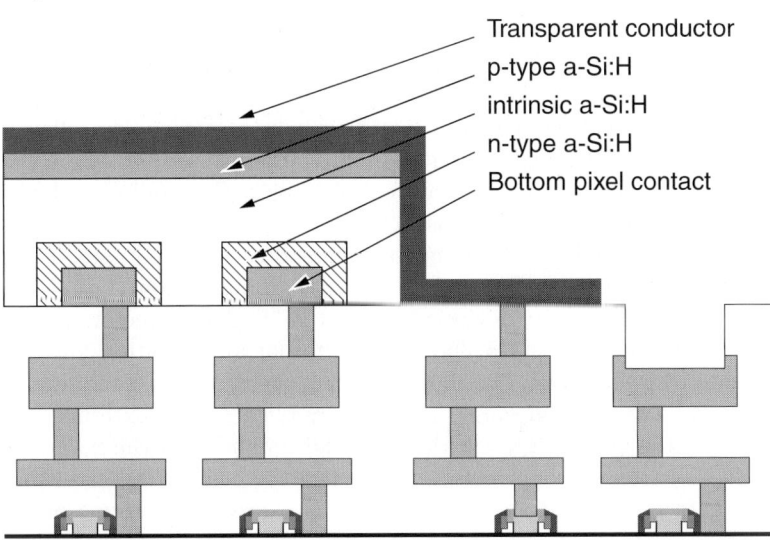

Fig. 6.14. Cross-sectional view of a structure using a-Si:H as a sensor deposited as a film atop a CMOS circuitry. The connection to the IC is made through vias [362]

interconnection process is absent. Basically any CMOS circuit and IC technology can be used and the dependence on industrial trends in IC technology is much reduced. The radiation hardness of a-Si:H detectors appears to be very high, in excess of $10^{15}\,\text{cm}^{-2}$ due to the defect tolerance, defect reversing ability of the amorphous structure, and the larger band gap. For high-Z applications polycrystalline HgI_2 constitutes a possible semiconductor film material. The potential advantages are small thickness, radiation hardness, and low cost. The development is still in its beginnings and – as for CMOS active pixel sensors – a real challenge to analog VLSI design.

6.3.5 DEPFET Pixels

The depleted field effect transistor (DEPFET) structure provides detection and amplification properties in every pixel jointly. The concept was proposed in 1987 [324]. It has been experimentally confirmed and successfully operated as single pixel structures [365] and as (64×64) pixel matrices [17, 326, 366–368].

The DEPFET detector and amplification structure is shown in Fig. 6.15. A field effect junction (JFET) or MOSFET is integrated onto a sideward-depleted detector substrate. Sideward depletion [369,370] provides a parabolic potential with a minimum for majority carriers (electrons in the case of n-type silicon) inside the detector volume, which can be moved by appropriate biasing toward the top surface closely ($\approx 1\,\mu$m) underneath the transistor structure. The transistor channel current can be steered by the voltage at the external gate and–important for the detector operation – also by the bulk potential. By means of a locally varying doping concentration (deep n-implant) and appropriate electrode potentials a local minimum for the charge carriers can be generated right underneath the transistor channel inside the substrate (internal gate). While (for n-type substrate material) holes created in the sensor are collected at the rear electrode, electrons are collected in the internal gate, thereby modulating the transistor current. The collected electrons are removed from the internal gate by a clear pulse applied to a dedicated contact outside the transistor region or by other clear mechanisms, e.g. from the external gate by punch-through to the internal gate. The very low input capacitance ($\approx$fF) and the in situ amplification (i.e. charge to current conversion) of the device make DEPFET pixel detectors attractive for low-noise operation [325, 368, 371]. External amplification enters only at the second level stage. The device is controlled from two terminals, the external and the internal gate. A voltage applied to the external gate as well as collected charges Q_{sig} in the potential minimum of the internal gate modulate the transistor current. The effect of the collected electrons in the internal gate on the output drain current is expressed in terms of $g_Q^{\text{int}} = \partial I_D / \partial Q_{\text{sig}}$ which is related to the transistor transconductance g_m^{ext} by the oxide capacitance. DEPFET structures with g_Q values of 300–$500\,\text{pA}/e^-$ have been fabricated [372].

6.3 Monolithic and Semimonolithic Pixel Detectors 279

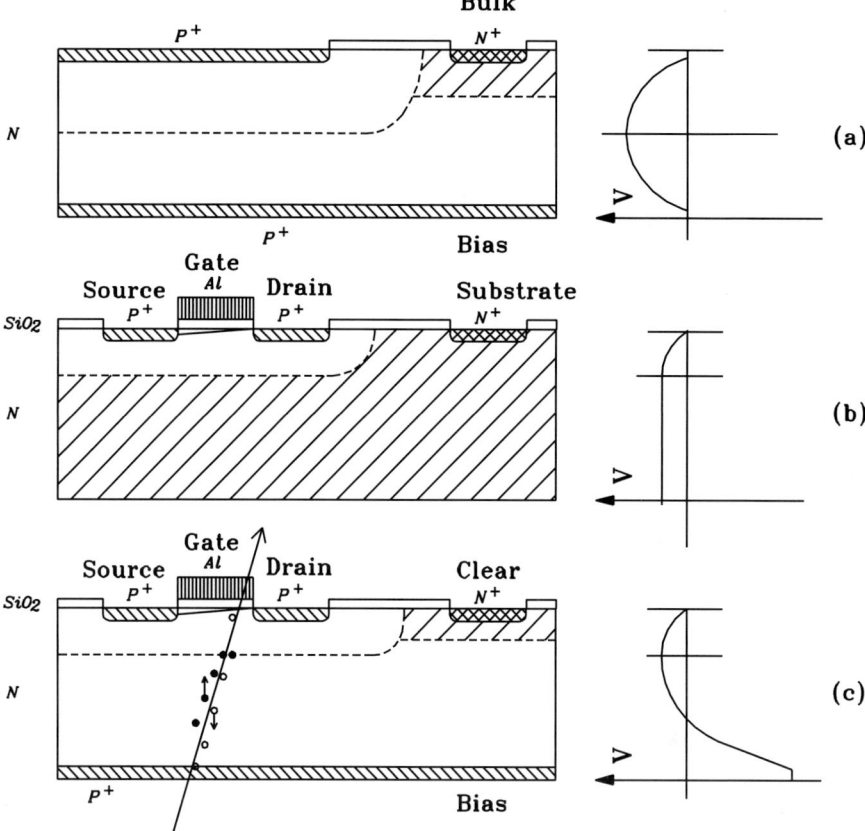

Fig. 6.15. The DEPFET detector and amplification structure (**c**) based on a sideward-depleted detector substrate material (**a**) with a planar field effect transistor (here MOS) embedded (**b**) in the pixel (from [4]). The potential shape is schematically drawn assuming ground potentials on the p$^+$-implants on top and bottom surfaces and negative voltage on the n$^+$ bulk contact

The noise performance of DEPFET structures measured at room temperature is shown in Fig. 6.16 by the response to a ^{55}Fe 6-keV X-ray source. The measured energy resolution is 134 eV for 6-keV X rays from an ^{55}Fe source. The low-energy tail is due to charge lost from the readout pixel by diffusion. The noise contribution is dominated by Fano noise ($\approx$14 e$^-$ at 6 keV and room temperature). The DEPFET structure itself contributes with ENC = 4.8±0.1 e$^-$ channel noise (insert of Fig. 6.16). With specially designed round single pixel structures, noise figures of 2.2 e$^-$ have been achieved.

Figure 6.17a shows the principle of operation of a complete DEPFET pixel matrix. Figure 6.17b is a photograph of the device with sequencing chips for gate control and clear, and the amplifying and scanning chip. The

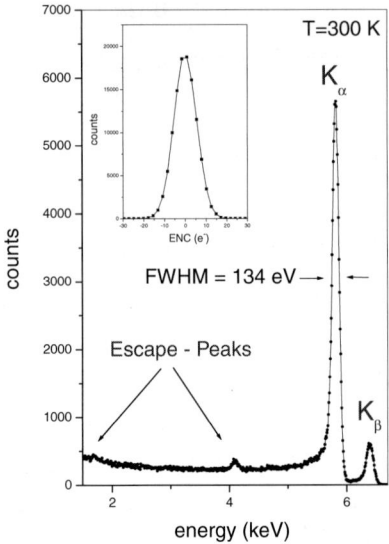

Fig. 6.16. Response of a DEPFET pixel structure to 6-keV X-rays from an ^{55}Fe source. The noise contribution (insert) to the signal peak is ENC = 4.8 ± 0.1 e$^-$

matrix is addressed row-wise by a gate-on voltage at the external gate and the DEPFET drain currents are read out column-wise and postamplified in the chip at the bottom.

The very good noise capabilities of DEPFET pixels are very important for low-energy X-ray astronomy and for autoradiography applications. For particle physics, where the signal charge is large in comparison, this feature can be used to design very thin detectors ($\approx$50 μm) with very low power consumption when operated as a row-wise-selected matrix [325, 373, 374]. Depending on the application, i.e. aiming at very low noise operation in spectroscopy or at a very fast readout in particle physics, the device can be operated in source follower readout mode or drain current readout mode, respectively [375]. What detectors for high-energy particles is concerned, the low noise feature can be exploited to make very thin sensors, still yielding a very good S/N performance. Thinning of DEPFET-like sensors to a thickness of 50 μm using a technology based on deep anisotropic etching has been successfully demonstrated [376].

The imaging performance of DEPFET pixels can be judged by Fig. 6.18 using a matrix of 64×64 pixels with 50-μm pixel diameter. With 50×50 μm^2 pixel cells spatial resolutions of

$$\sigma_{xy} = 4.3 \pm 0.8 \text{ μm} \quad \text{or} \quad 57 \pm 10 \text{ LP/mm} \quad \text{for} \quad 22 \text{ keV } \gamma$$
$$\sigma_{xy} = 6.7 \pm 0.7 \text{ μm} \quad \text{or} \quad (37 \pm 3) \text{ LP/mm} \quad \text{for} \quad 6 \text{ keV } \gamma$$

have been obtained in X-ray imaging [17].

6.3 Monolithic and Semimonolithic Pixel Detectors 281

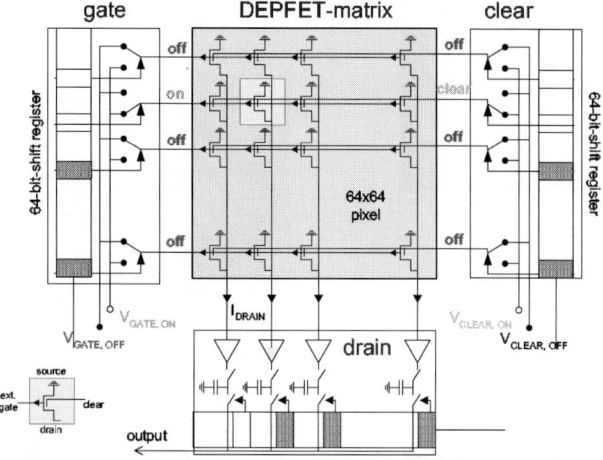

Fig. 6.17. Operation principle (*top*) and photograph (*bottom*) of a 64 × 128 DEPFET pixel system with pulsed clear. The DEPFET matrix sits in the center, CLEAR and gate-steering chips are on the sides, and the current amplifier chip is on the bottom

DEPFET pixels are being developed for very different application areas: vertex detection in particle physics [325, 373, 374], X-ray astronomy [291, 292, 377], and biomedical autoradiography [17, 325, 368, 378, 379]. They are also considered in Compton camera developments [301].

For the development of DEPFET pixels for a linear collider, high readout speeds and small pixel sizes are required. Matrix row rates of up to 50 MHz and frame rates for 512 × 4096 pixels of 25 kHz with readout at two sides are targeted. Using implanted (DEP)MOS transistors with linear instead of circular dimensions, sensors with small cell sizes of $20 \times 30\,\mu m^2$ have

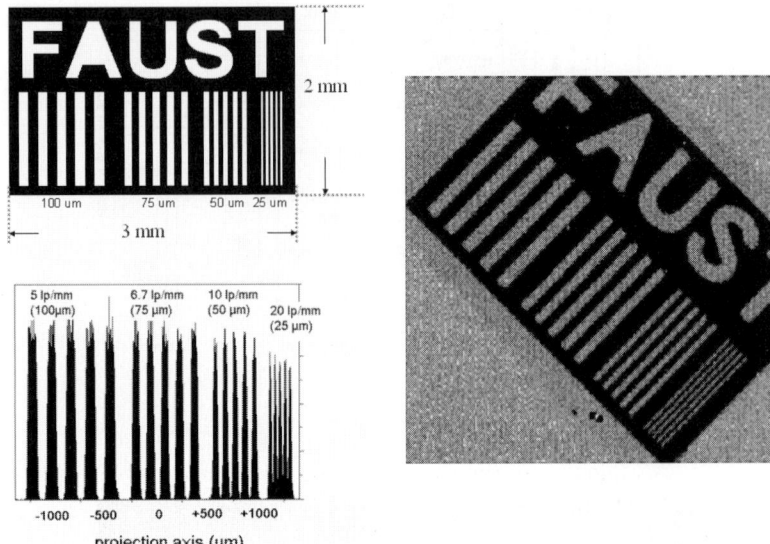

Fig. 6.18. Measurements of the imaging resolution of a DEPFET pixel matrix with 64 × 64 pixels and a cell size of 50-µm diameter. (**a**) Line contrast pattern images and (**b**) projection onto a plane vertical to the lines

been fabricated. Complete clearing of the internal gate, which is important for high-speed on-chip pedestal subtraction, has been demonstrated [325]. A matrix is read out using sequencer chips for row selection and clear, and a column readout chip based on current amplification and storage as shown in Fig. 6.17(bottom). With a power duty cycle of on/off = 1:200 the power consumption for a five-layer, 1.2-m^2 DEPFET pixel vertex detector is estimated to be as small as 5 W [325], rendering a very low mass detector without cooling pipes feasible. Radiation resistance of the system at a linear collider is required to several kilograys [344]. As expected for MOS devices, charge trapping in the interface (SiO_2 and Si_3N_4) leads to transistor threshold shifts which has been measured to amount to ~4 V for ionizing doses up to 10 kGy [380]. Threshold shifts in this voltage range can be conveniently compensated by a corresponding adjustment of the voltage levels delivered by the sequencer chips, provided that the shifts occur fairly homogeneously over the area of a detector.

Glossary

ADC	*Analog-to-Digital Converter* Electronic circuit that converts an analog voltage or current to a digital value
ALICE	*A Large Ion Collider Experiment* The Large Hadronic Collider detector dedicated to heavy ion studies
ASIC	*Application Specific Integrated Circuit* Integrated circuit with functionalities customized for a specific application
ATLAS	*A Toroidal Lhc ApparatuS* One of the large multi purpose experiments at the CERN LHC
BCB	*Benzocyclobutene* A dielectric polymer with very good film-forming properties
BTeV	*Beauty at TeVatron* The Tevatron detector dedicated to beauty particles studies Its construction was stopped in 2005
CCD	*Charged Coupled Device* Pixelated charge integrating device in which the charge is shifted from one pixel to the next to an output node
CDS	*Correlated Double Sampling* Readout technique in which double samples are taken and subtracted from each other
CERN	*European Organization for Nuclear Research* The world's largest particle physics laboratory located near Geneva, Switzerland
CFRP	*Carbon Fiber Reinforced Polymers* A composite made of carbon fiber molded with an epoxy resin and then cured
CME	*Linear Coefficient of Moisture Expansion* The fractional length change per unit mass variation due to the moisture desorption or absorption

CMOS	*Complementary Metal Oxid Semiconductor* Electronic circuit using both ("complementary") types of field effect transistors, NMOS and PMOS
CMS	*Compact Muon Solenoid* One of the large multi purpose experiments at the CERN LHC
CTE	*Linear Coefficient of Thermal Expansion* The fractional length change per degree of temperature increase
CVD	*Chemical wafer deposition* Process of depositing thin layers of dielectrics on a wafer from the gas phase
CZ process	*Czochralski process* Process of pulling a single crystal out of a melt. Named after its inventor Jan Czochralski (1885–1953)
DAC	*Digital-to-Analog Converter* Electronic circuit that converts a digital value to an analog voltage or current
DELPHI	*DEtector for Lepton, Photon and Hadron Interactions* CERN collider experiment dedicated to the study of the mediators of the weak interaction
DEPFET	*Depleted Field Effect Transistor* Pixel sensor with an implanted transistor in every pixel cell
Die	Unpackaged electronics chip
DOFZ material	*Diffusion Oxygenated Float Zone Material* Float zone silicon enriched with oxygen, by growing a thick oxide layer on the wafer surface and heating the wafer under inert atmosphere above 1000°C for more than 24 h. This material has shown to be more radiation-hard then standard float zone silicon
DSM	*Deep Submicron Technology* Semiconductor technologies with gate lengths $\lesssim 0.25\,\mu\text{m}$
ENC	*Equivalent Noise Charge* Root mean square noise at the output of a charge-sensitive amplifier normalized to the signal of a single electron input charge, i.e. the noise of an amplifier "in units of electrons"
EoC logic	*End of Column Logic* Part of the pixel front-end chip dedicated to data storage and input/output
ESRF	*European Synchrotron Radiation Facility* Synchrotron Radiation Laboratory in Grenoble, France

Eutectic	Mixture of two or more elements which has a lower melting point than any of its constituents
eV	*Electronvolt* Energy gained by an unbound electron falling through an electrostatic potential difference of 1 volt
FADC	*Flash Analog to Digital Converter* Fast ADC architecture. The analog input is simultaneously compared to many reference values
FE	*Front End* ASIC that reads out a matrix of pixel sensors
FET	*Field Effect Transistor* Electronic component in which the current between two terminals (source and drain) is steered by the potential of a third terminal (gate)
Fluence	Integral of a particle flux over a period of time
FNAL	*Fermi National Accelerator Laboratory* Particle physics laboratory near Chicago dedicated to the memory of the Italian physicist Enrico Fermi (1901–1954)
FZ process	*Float Zone Process* Process of producing a single crystal by partly melting a polycrystalline rod
HAPS	*Hybrid Active Pixel Sensors* Hybrid pixel technology using interleaved pixels with capacitive charge division
IC	*Integrated Circuit* Electronics circuit integrated on a chip
LEP	*Large Electron Positron Collider* Electron–Positron collider at CERN, Switzerland
LHC	*Large Hadronic Collider* Proton–proton collider at CERN, Switzerland
LSF	*Line Spread Function* Image of a line source object of unit intensity
LVDS	*Low Voltage Differential Signals* A standard in fast low power data transmission based on sending a low voltage ($\approx 350\,\text{mV}$) logic signal over a twisted pair cable
MAPS	*Monolithic Active Pixel Sensor* CMOS pixel detector

MCC	*Module Control Chip* ASIC that controls the input and output of data and performs event building in an ATLAS pixel module
MCM-D	*Multichip Module by Deposition* In this context: Module structure with additional conductive layers deposited on the sensor so that no extra printed circuit board is required
MEDIPIX	*Medical Pixel Chip* Pixel readout chip for medical imaging
MEMS	*Micro-Electro-Mechanical Systems* Small scale electromechanical technology
MIMOSA	*Minimum Ionizing Monolithic Sensor Array* CMOS pixel detector, special version of a MAPS detector
MIP	*Minimum Ionizing Particle* The energy deposited by an ionizing particle when passing through material depends, among other things, on its velocity. The energy loss increases for very slow and very fast particles and has a broad minimum. Particles with energy loss in this minimum are defined minimum ionizing particles.
MOSFET	*Metal-Oxide Semiconductor Field-Effect Transistor* Field effect transistor with isolated metal gate. Today, most FETs have polysilicon gates, but they are still often called MOSFET
MPEC	*Multipicture Element Counter* X-ray counting pixel chip
MTF	*Modulation Transfer Function* Ratio of image contrast to object contrast for a given spatial frequency
Multiple scattering	Change in direction of a charged particle track when traversing a medium primarily due to Coulomb scattering from nuclei
NIEL	*Nonionizing Energy Loss* Part of the absorbed energy which is not used for ionization but, e.g., for displacing lattice atoms. It was believed that radiation damage scales with the NIEL independently of the type of radiation an assumption which has not shown to be correct
OmegaD	*Pixel front-end chip used in experiment WA94 and WA97* One of the first front-end chip for reading out pixel matrices
PILATUS	*PixeL ApparaTUs for the SLS* Counting pixel detector for SLS synchrotron radiation beam lines at the Paul-Scherrer Institute, Switzerland

Pseudorapidity	Kinematical variable of a relativistic particle defined as $\eta = -\ln\tan(\theta/2)$, where θ is the particle zenith angle referenced to the direction of the crossing beams
PSF	*Point Spread Function* Image of a point source object of unit intensity
PSI	*Paul Scherrer Institute* Swiss multidisciplinary research center for natural sciences and technology, Villigen, Switzerland
PUC	*Pixel Unit Cell* Circuit block on the readout chip that is repeated for every pixel
RD19	CERN R&D experiment dedicated to the development of pixel detectors to be used at high luminosity colliders
SEU	*Single Event Upset* The flipping of a stored bit in an integrated circuit due to charge deposition by an ionizing particle
SLS	*Swiss Light Source* Swiss Synchrotron Radiation Facility at the Paul-Scherrer Institute, Switzerland
SOI	*Silicon on Insulator* Substrate type with a buried oxide layer to separate the "active" silicon from the substrate.
SPICE	*Simulation Program with IC Emphasis* Widely used software for the simulation of analog circuits
SSC	*Superconducting Super Collider* Particle accelerator planned for very high energy physics measurements in the United States. Its construction was stopped in 1994
Stoichiometry	Study of the proportion of elements that take part in chemical reactions
TBM	*Token Bit Manager* ASIC that controls the input and output of data of a CMS pixel module
Terabit	Unit of information storage corresponding to 10^{12} bits
Tevatron	Proton-Antiproton collider accelerator with the highest energy available today. It is located at FNAL, Batavia, Illinois, USA
Time walk	Time dispersion of a discriminated pulse due to the combined effect of the pulse heights spread and the single threshold discrimination

ToT	*Time over Threshold* Method to determine the amplitude of an analog signal. The signal is compared to a threshold and the duration of the output pulse is measured. ToT is often nonlinear
Trigger	Set of conditions that, once satisfied, select the event for further analysis
UBM	*Under Bump Metallization* Metal layer sequence under a solder bump
USB	*Universal Serial Bus* Popular serial interconnect to PCs
VLSI	*Very Large Scale Integration* Large scale integration technology of electronics circuits on silicon
W boson	Charged mediator of the weak nuclear interaction
WA92	CERN fixed target experiment dedicated to the study of particles containing heavy quarks
WA94	CERN fixed target experiment dedicated to the study of heavy ions collisions
WA97	CERN fixed target experiment dedicated to the study of heavy ions collisions
X_0	*Radiation length* Mean distance over which a high-energy electron looses all but $1/e$ of its energy by bremsstrahlung
XPAD	*X-ray Pixel Advanced Detector* Counting pixel detector for ESRF synchrotron radiation beam lines, Grenoble, France
Young's modulus	Measure of the stiffness of a given material (also called elasticity modulus)

References

1. R. Mellen, D. Buss (eds): *Charged-Coupled Devices: Technology and Applications* (IEEE Press, New York, 1997)
2. M.J. Howes, D.V. Morgan (eds): *Charged-Coupled Devices and Systems* (Wiley, New York, 1979)
3. J. Kemmer: Nucl. Instrum. Methods **42**, 499–502 (1980)
4. G. Lutz: *Semiconductor Radiation Detectors* (Springer, Berlin Heidelberg New York, 1999)
5. E.H.M. Heijne et al: RD19 status report. CERN/LHCC 97-059 (1997)
6. E.H.M. Heijne et al: Nucl. Instrum. Methods **178**, 331–343 (1980)
7. J.L. Blankenship, C.J. Borkowski: IRE Trans. Nucl. Sci. **7**(3), 190–195 (1960)
8. E.H.M. Heijne: Nucl. Instrum. Methods A **465**, 1–26 (2001)
9. M. Moll: Radiation damage in silicon particle detectors–microscopic defects and macroscopic properties. PhD Thesis, Universität Hamburg, Germany (1999). DESY-1999-040
10. E.H.M. Heijne et al: Nucl. Instrum. Methods A **349**, 138–155 (1994)
11. K.H. Becks et al: Nucl. Instrum. Methods A **418**, 15–21 (1998)
12. The ATLAS Collaboration: Technical design report of the ATLAS pixel detector. CERN/LHCC/98-13 (1998)
13. The CMS Collaboration: CMS tracker technical design report. CERN/LHCC/98-6 (1998)
14. The ALICE Collaboration: ALICE technical design report of the inner tracker system. CERN/LHCC/99-12 (1999)
15. G. Dipasquale, C. Schwarz, B. Mikulec, M. Campbell, J. Watt: Nucl. Instrum. Methods A **458**, 352–359 (2001)
16. P. Fischer et al: IEEE Trans. Nucl. Sci. **46**(4), 1070–1074 (1999)
17. J. Ulrici et al: Nucl. Instrum. Methods A **547**, 424–436 (2005)
18. G. Claus et al: Nucl. Instrum. Methods A **465**, 120–124 (2001)
19. P. Jarron et al: Nucl. Instrum. Methods A **518**, 366–372 (2004)
20. L.C.L. Yuan: Application of solid-state devices for high-energy particle detection. In: *Proc. Int. Conf. Instrumentation for High-Energy Physics*, Lawrence Rad Lab, Berkeley, Sept 1960, pp 177–181
21. G. Bellini et al: Nucl. Instrum. Methods A **196**, 351–360 (1982)
22. S. Benso et al: Nucl. Instrum. Methods A **201**, 329–333 (1983)
23. M. Adamovich et al: Nucl. Instrum. Methods A **379**, 252–270 (1996)
24. S. Gaalema: IEEE Trans. Nucl. Sci. **32**(1), 417–418 (1985)
25. J. Millaud, M. Wright, D. Nygren: A pixel unit cell targeting 16 ns resolution and radiation hardness in a column read-out particle vertex detector. Report LBL-32912 (1992)

26. M. Campbell et al: Nucl. Instrum. Methods A **290**, 149–157 (1990)
27. F. Anghinolfi et al: IEEE Trans. Nucl. Sci. **39**(4), 654–661 (1992)
28. M.G. Catanesi et al: Nucl. Phys. B (Proc. Suppl.) **32**, 260–268 (1993)
29. W. Snoeys et al: Nucl. Instrum. Methods A **465**, 176–189 (2001)
30. R. Dinapoli et al: Nucl. Instrum. Methods A **461**, 492–496 (2001)
31. N.S. Saks, M.G. Ancona, J.A. Modolo: IEEE Trans. Nucl. Sci. **31**(6), 1249–1255 (1984)
32. N.S. Saks, M.G. Ancona, J.A. Modolo: IEEE Trans. Nucl. Sci. **33**(6), 1185–1190 (1986)
33. DELPHI Collaboration: DELPHI 92-142 GEN 135 Internal Report (1992)
34. P. Chochula et al: Nucl. Instrum. Methods A **412**, 304–328 (1998)
35. T. Flick et al: Nucl. Phys. B (Proc. Suppl.) **125**, 85–89 (2003)
36. L.F. Miller: IBM, J. Res. Dev. **13**, 239–250 (1969)
37. E. Spenke: *Elektronische Halbleiter* (Springer, Berlin, 1965)
38. N.W. Ashcroft, N.D. Mermin: *Solid State Physics* (Saunders, Fort Worth, 1976)
39. C. Kittel: *Introduction to Solid State Physics* (Wiley, New York, 1976)
40. S.M. Sze: *Physics of Semiconductor Devices* 2nd edn. (Wiley, New York, 1981)
41. E.H. Nicollian, J.R. Brews: *MOS Physics and Technology* (Wiley, New York, 1982)
42. S.M. Sze: *Semiconductor Devices, Physics and Technology* (Wiley, New York, 1985)
43. B.J. Baliga: *Modern Power Devices* (Wiley, New York, 1987)
44. M.S. Tyagi: *Semiconductor Materials and Devices* (Wiley, New York, 1991)
45. G. Charpak, F. Sauli: Ann. Rev. Nucl. Part. Sci. **34**, 285–349 (1984)
46. The Particle Data Group: Eur. Phys. J. C **15**, 1–878 (2000)
47. J.D. Jackson: *Classical Electrodynamics*, 3rd edn (Wiley, New York, 1998)
48. J.F. Bak, A. Burenkov, J.B.B. Petersen, E. Uggerhøj, S.P. Møller: Nucl. Phys. B **288**, 681–716 (1987)
49. H. Bichsel: Rev. Mod. Phys. **60**, 663–699 (1988)
50. G.R. Lynch, O.I. Dahl: Nucl. Instrum. Methods B **58**, 6–10 (1991)
51. Y.S. Tsai: Rev. Mod. Phys. **46** (1974)
52. M.J. Berger et al: XCOM: Photon Cross Sections Database. URL: http://www.nist.gov/PhysRefData
53. R.D. Evans: *The Atomic Nucleus* (Krieger, New York, 1982)
54. U. Fano: Phys. Rev. **72**, 26–29 (1947)
55. R.C. Alig, S. Bloom, W.C. Struck: Phys. Rev. B **22**, 5565–5582 (1980)
56. R.C. Alig: Phys. Rev. B **27**, 968–977 (1983)
57. M. Krumrey, E. Tegeler, G. Ulm: Rev. Sci. Instrum. **60**, 2287–2290 (1989)
58. P. Lechner, R. Hartmann, H. Soltau, L. Strüder: Nucl. Instrum. Methods A **377**, 206–208 (1996)
59. F. Perotti, C. Fiorini: Nucl. Instrum. Methods A **423**, 356–363 (1999)
60. G.W. Fraser et al: Nucl. Instrum. Methods A **350**, 368–378 (1994)
61. A. Owens, G.W. Fraser, K.J. McCarthy: Nucl. Instrum. Methods A **491**, 437–443 (2002)
62. W. Shockley, W.T. Read: Phys. Rev. **87**, 835–842 (1952)
63. R.N. Hall: Phys. Rev. **87**, 387 (1952)
64. C. Jacoboni, C. Canali, G. Ottaviani, A. Alberigi Quaranta: *Solid-State Electron.* **20**, 77–89 (1977)

65. M.A. Green: J. Appl. Phys. **67**(6), 2944–2954 (1990)
66. I. Gorelev et al: Nucl. Instrum. Methods A **81**, 204–221 (2002)
67. A. Dorokhov et al: Nucl. Instrum. Methods A **530**, 71–76 (2004)
68. V. Bartsch et al: Nucl. Instrum. Methods A **497**, 389–396 (2003)
69. L. Andricek et al: Nucl. Instrum. Methods A **436**, 262–271 (1999)
70. B.D. Deal: IEEE Trans. Nucl. Sci. **27**, 606 (1980)
71. P. Azzi et al: Nucl. Instrum. Methods A **383**, 155–158 (1996)
72. L. Andricek et al: Nucl. Instrum. Methods A **409**, 184–193 (1998)
73. R. Ansari et al: Nucl. Instrum. Methods A **279**, 388–395 (1989)
74. E. Fretwurst et al: Nucl. Instrum. Methods A **288**, 1–12 (1990)
75. F. Hügging et al: Nucl. Instrum. Methods A **439**, 529–535 (2000)
76. B. Henrich, W. Bert, K. Gabathuler, R. Horisberger: Depth profile of signal charge collected in heavily irradiated silicon pixels. Technical Report CMS Note 1997/021, CERN, Geneva, Switzerland (1997). URL: http://cms-doc.cern.ch/documents/97/note97_021.pdf
77. M. Swartz et al: Type inversion in irradiated silicon: a half truth. e-print physics/0409049, Sept 2004
78. V. Chiochia et al: Simulation of heavily irradiated silicon pixel sensors and comparison with test beam measurements. Presented at the *IEEE NSS*, Rome, Italy, Oct 16–22 2004
79. C. Gemme: Measurement of the depletion voltage of irradiated sensors. URL: http://www.ge.infn.it/ATLAS/PixelWeek/home.html (2001)
80. E. Gatti, P.F. Manfredi: Riv. Nuovo Cimento **9**(1), 1–146 (1986)
81. V. Radeka: Ann. Rev. Nucl. Part. Sci. **38**, 217–277 (1988)
82. A. Cadeira, M. Estrada: IEEE Trans. Nucl. Sci. **44**, 63–66 (1997)
83. T. Rohe, F. Hügging, G. Lutz, R.H. Richter, R. Wunstorf: Nucl. Instrum. Methods A **409**, 224–228 (1998)
84. G. Gorfine, M. Hoeferkamp, G. Santistevan, S. Seidel: Nucl. Instrum. Methods A **465**, 70–76 (2001)
85. R. Kaufmann: Development of radiation hard pixel sensors for the CMS experiment. PhD Thesis, Universität Zürich, Switzerland (2001)
86. G. Bolla et al: Nucl. Instrum. Methods A **485**, 89–99 (2002)
87. S. Ramo: Proc. IRE **27**, 584–585 (1939)
88. E. Durand: *Electrostatique, Tome II* (Masson et Cie, Paris, 1966)
89. E. Belau et al: Nucl. Instrum. Methods **214**, 253–260 (1984)
90. V. Chiochia et al: Testbeam measurements taken at the CERN-SPS in 2003 and 2004. Unpublished manuscript, 2004
91. M. Edwards, G. Hall, S. Sotthibandhu: Nucl. Instrum. Methods A **310**, 283–286 (1991)
92. H.J. Ziock et al: IEEE Trans. Nucl. Sci. **38**(2), 269–276 (1991)
93. R. Wunstorf: Systematische Untersuchungen zur Strahlenresistenz von Silizium-Detektoren für die Verwendung in Hochenergiephysik-Experimenten. PhD Thesis, Universität Hamburg, Germany (1992)
94. D. Pitzl et al: Nucl. Instrum. Methods A **311**, 98–104 (1992)
95. F. Lemeilleur, M. Glaser, E.H.M. Heijne, P. Jarron, E. Occelli: IEEE Trans. Nucl. Sci. **39**(4), 551–557 (1992)
96. V.A.J. van Lint et al: *Mechanisms of Radiation Effects in Electronic Material* vol 1 (Wiley, Chichester, 1980)
97. L.S. Smirnov: *A Survey of Semiconductor Radiation Techniques* (Mir Publishers, Moskow, 1983)

98. The ROSE Collaboration: Notes on the fluence normalization based on the NIEL scaling hypothesis. Technical Note ROSE/TN/2000-02, CERN, Geneva, Switzerland (2000). URL: http://rd48.web.cern.ch
99. M. Moll, E. Fretwurst, M. Kuhnke, G. Lindström: Nucl. Instrum. Methods B **186**, 100–110 (2002)
100. G. Lindström et al: Nucl. Instrum. Methods A **466**, 308–326 (2001)
101. M. Moll, E. Fretwurst, G. Lindström: Nucl. Instrum. Methods A **426**, 87–93 (1999)
102. M. Moll et al: Nucl. Instrum. Methods A **439**, 282–292 (2000)
103. G. Lindström et al: Nucl. Instrum. Methods A **426**, 1–15 (1999)
104. H. Feick: Radiation tolerance of silicon particle detectors for high energy physics experiments. PhD Thesis, Universität Hamburg, Germany (1997). DESY F35D-97-08
105. R. Wunstorf, W.M. Bugg, J. Walter, F.W. Garber, D. Larson: Nucl. Instrum. Methods A **377**, 228–233 (1999)
106. I. Tsveybak: IEEE Trans. Nucl. Sci. **39**(6), 1720–1729 (1992)
107. Z. Li: Nucl. Instrum. Methods A **342**, 105–118 (1994)
108. The ROSE Collaboration: Third RD48 status report. Technical Report LHCC 2000-009, CERN, Geneva, Switzerland (2000). URL: http://rd48.web.cern.ch
109. G. Lindström et al: Nucl. Instrum. Methods A **465**, 60–69 (2001)
110. Z. Li, H.W. Kraner: Measurements of possible type inversion in silicon junction detectors by fast neutron irradiation. Technical Report BNL 46210, Brookhaven National Laboratory, USA (May 1991).
111. G. Lutz: Nucl. Instrum. Methods B **65**, 41–49 (1995)
112. R.H. Richter: Private communication, Munich, Germany (May 2001)
113. V. Eremin, Z. Li: IEEE Trans. Nucl. Sci. **41**(6), 1907–1912 (1994)
114. Z. Li, H.W. Kraner: Nucl. Phys. B (Proc. Suppl.), **32**, 398–409 (1993)
115. V. Eremin, E. Verbitskaya, Z. Li: Nucl. Instrum. Methods A **476**, 537–549 (2002)
116. G. Kramberger, V. Cindro, I. Mandić, M. Mikuž, M. Zavrtanik: Nucl. Instrum. Methods A **481**, 297–305 (2002)
117. J.M. McGarrity, F.R. McLean, T.R. Oldham, H.E. Boesch, Jr: Ionizing radiation effects in MOS-devices. In: *Semiconductor Silicon–Material Science and Technology*, ed by G. Harbecke and M. Schulz (Springer, Berlin Heidelberg New York, 1989)
118. C.T. Sah: IEEE Trans. Nucl. Sci. **23**(6), 1563–1568 (1976)
119. G.F. Derbenwick, B.L. Gergory: IEEE Trans. Nucl. Sci. **22**(6), 2151–2156 (1975)
120. J. Wüstenfeld: Characterization of ionization-induced surface effects for the optimization of silicon-detectors for particle physics applications. PhD Thesis, Universität Dortmund Germany (2001)
121. A. Longoni, M. Sampietro, L. Strüder: Nucl. Instrum. Methods A **288**, 35–43 (1990)
122. I. Ropotar et al: Nucl. Instrum. Methods A **439**, 536–546 (2000)
123. R.H. Richter et al: Nucl. Instrum. Methods A **377**, 412–421 (1996)
124. T. Oshugi et al: Nucl. Instrum. Methods **383**, 116–122 (1996)
125. S. Braibant et al: Nucl. Instrum. Methods A **485**, 343–361 (2002)
126. A. Bischoff et al: Nucl. Instrum. Methods A **326**, 27–37 (1993)
127. B.S. Avset, L. Evensen: Nucl. Instrum. Methods A **377**, 397–403 (1996)

128. L. Andricek et al: Nucl. Instrum. Methods A **439**, 427–441 (2000)
129. M. DaRold et al: IEEE Trans. Nucl. Sci. **44**(3), 721–727 (1997)
130. R.C. West (ed): *Handbook of Chemistry and Physics*, 54th edn (CRC Press, Cleveland, OH, 1973)
131. F. Antinori et al: Nucl. Instrum. Methods A **360**, 91–97 (1995)
132. A. Andreazza et al: Nucl. Instrum. Methods A **367**, 198–201 (1995)
133. S. Heising: Private communication. Villigen, Switzerland (Aug. 2002)
134. T.K. Nelson et al: Int. J. Mod. Phys. A **16**(Suppl. 1C),1091–1093 (2001)
135. Specifications for ATLAS silicon microstrip detectors for the ATLAS final design review. ATLAS SCT/Detector FDR/99-2. URL: http://allportp.home.cern.ch /allportp /jrc99.ps
136. CMS Collaboration: Addendum to the CMS tracker TDR. CERN-LHCC-2000–016 (2000)
137. J. Kemmer et al: Streifendetektor. Patentoffenlegungsschrift DE 19620081 A1 21.11.97, Munich, Germany (1997)
138. M.S. Alam et al: Nucl. Instrum. Methods A **456**, 217–232 (2001)
139. T. Rohe et al: IEEE Trans. Nucl. Sci. **51**(3), 1150–1157 (2004)
140. F. Ragusa et al: Nucl. Instrum. Methods A **447**, 184–193 (2000)
141. T. Rohe: Planung, Bau und Test des Sensor-Bausteins für einen hybriden Silizium-Pixel-Detektor zum Einsatz unter den extremen Strahlenbelastungen am LHC. PhD Thesis, Ludwig-Maximilian-Universität München, Germany (1999)
142. P.I. Hopman, J.P. Alexander, A.D. Foland, P.C. Kim, C.W. Ward: Nucl. Instrum. Methods A **383**, 98–103 (1996)
143. Y. Iwata et al: IEEE Trans. Nucl. Sci. **45**(3), 303–309 (1998)
144. G. Bolla et al: Nucl. Instrum. Methods A **461**, 182–184 (2001)
145. S. D'Auria et al: Nucl. Phys. (Proc. Suppl.) **78**, 639–644 (1999)
146. T. Rohe et al: Nucl. Instrum. Methods A **460**, 55–66 (2001)
147. D. Robinson et al: Nucl. Instrum. Methods A **426**, 28–33 (1999)
148. Y. Unno et al: Nucl. Instr. Methods A **541**, 40–46 (2005)
149. T. Nakayama et al: IEEE Trans. Nucl. Sci. **47**(6), 1885–1891 (2000)
150. M. De Palma et al: R&D proposal: Development of radiation hard semiconductor devices for very high luminosity colliders. CERN LHCC 2002-003/P6
151. K. Borer et al: RD39 status report. CERN LHCC 2002-004
152. A. Chilingarov, T. Sloan: Nucl. Instrum. Methods A **399**, 35–37 (1997)
153. S. Parker, C. Kenney, J. Segal: Nucl. Instrum. Methods A **395**, 328–343 (1997)
154. W. Adam et al: Nucl. Instrum. Methods A **447**, 244–250 (2000)
155. J. Kemmer: Nucl. Instrum. Methods **169**, 499–502 (1980)
156. K.A. Pickar: Ion implantation in silicon. In: *Applied Solid State Science*, vol 3, ed by R. Jung (Academic, New York, 1975)
157. R.H. Richter et al: Private communication, Munich, Germany (1999)
158. M. Rogalla: *Systematic Investigation of GaAs radiation detectors for HEP experiments* (Shaker, 1997)
159. J.S. Blakemore: *Gallium Arsenide* (American Institute of Physics, New York, 1987)
160. T. Kubicki et al: Nucl. Instrum. Methods A **345**, 468–473 (1994)
161. J.W. Chen et al: Nucl. Instrum. Methods A **365**, 273–284 (1995)
162. M. Lindner et al: Nucl. Instrum. Methods A **466**, 63–73 (2001)
163. C. Schwarz et al: Nucl. Instrum. Methods A **466**, 87–94 (2001)

164. M. Lindner et al: Nucl. Instrum. Methods A **405**, 53–59 (1998)
165. P. Siffert: Proc. SPIE **2305**, 98–109 (1994)
166. P.R. Bennett et al: IEEE Trans. Nucl. Sci. **45**, 417–420 (1998)
167. T. Takahashi et al: Nucl. Instrum. Methods A **436**, 111–119 (1999)
168. T. Takahashi et al: IEEE Trans. Nucl. Sci. **49**, 1297–1303 (2002)
169. S. Watanabe et al: IEEE Trans. Nucl. Sci. **49**, 1292–1296 (2002)
170. T. Takahashi et al: IEEE Trans. Nucl. Sci. **48**, 287–291 (2001)
171. M. Löcker et al: IEEE Trans. Nucl. Sci. **51**(4), 1717–1723 (2004)
172. M. Chmeissani et al: First experimental tests with a CdTe photon counting pixel detector hybridized with a Medipix2 readout chip. Presented at the IEEE Nuclear Science Symposium (NSS) and Medical Imaging Conference (MIC), Portland, OR, Oct 19–25, 2003
173. W.G. Eversole: Synthesis of diamond. U.S. Patent 3,030, 187 (1962)
174. A. Lettington, J.W. Steeds (ed): *Thin Film Diamond* (Chapman and Hall for the Royal Society, Great Britain, 1994)
175. P.K. Bachmann, D. Leers, H. Lydtin: Diamond Relat. Mater. **1**, 1 (1991)
176. R.S. Sussmann: Diamond Relat. Mater. 7 (1998)
177. H. Kagan: Recent advances in diamond detector development. In: *5th Int. Symp. on Development and Application of Semiconductor Tracking Detectors*, Hiroshima, Japan, June 2004
178. J. Isberg et al: *Science* **297** (5587), 1670–1672 (2002)
179. CERN RD42 Collaboration: Development of diamond tracking detectors for high luminosity experiments at the LHC. Technical Report LHCC 1997-003, CERN (1997)
180. D. Meier et al: Proton irradiation of CVD diamond detectors for high luminosity experiments at the LHC. In: *2nd Int. Conf. on Radiation Effects in Semiconductor Materials*, 1998. Preprint CERN-EP 1998-79
181. S. Han et al: Proton irradiation studies of CVD diamond detectors. CERN RD42 Internal Note (1996)
182. W. Dulinski: Electron irradiation of CVD diamond. CERN RD42 Internal Note (1995)
183. D. Meier: CVD diamond sensors for particle detection and tracking. PhD Thesis, University of Heidelberg, Germany (1999)
184. CERN RD42 Collaboration. Diamond as a particle detector. In: Proc. It. Phys. Society, Torino 1996, vol 52, p 105
185. J. Velthuis: Pumping effects in CVD diamond particle detectors. PhD Thesis, University of Twente, Enschede (1998). NIHKEF report 98/007
186. M. Ackers et al: IEEE Trans. Nucl. Sci. **46**, 2033–2038 (1999)
187. L. Blanquart et al: Nucl. Instrum. Methods A **456**, 217–231 (2001)
188. R. Wedenig et al: Nucl. Phys. B. (Proc. Suppl.), **78**, 497–504 (1999)
189. M. Keil et al: Nucl. Instrum. Methods A **501**, 153–159 (2003)
190. T. Lari et al: Nucl. Instrum. Methods A **537**, 581–593 (2005)
191. K.H. Becks et al: Nucl. Instrum. Methods A **386**, 11–17 (1997)
192. X. Llopart, M. Campbell, R. Dinapoli, D. San Segundo, E. Pernigotti: IEEE Trans. Nucl. Sci. **49**(5), 2279–2283 (2002)
193. H. Beker et al: Nucl. Instrum. Methods A **332**, 188–201 (1993)
194. D.C. Christian et. al: Nucl. Instrum. Methods A **435**, 144–152 (1999)
195. R. Baur et al: Nucl. Instrum. Methods A **465**, 159–165 (2001)
196. M. Campbell, E.H.M. Heijne, G. Meddeler, E. Pernigotti, W. Snoeys: IEEE Trans. Nucl. Sci. **45**(3), 751–753 (1998)

197. Ch. Brönnimann et al: Nucl. Instrum. Methods A **465**, 235–239 (2001)
198. P. Fischer, A. Helmich, M. Lindner, N. Wermes, L. Blanquart: IEEE Trans. Nucl. Sci. **47**(3), 881–884 (2000)
199. C. Grah et al: Nucl. Instrum. Methods A **465**, 211–218 (2001)
200. M. Campbell et al: Nucl. Instrum. Methods A **342**, 52–58 (1994)
201. E.H.M. Heijne et al: Nucl. Instrum. Methods A **383**, 55–63 (1996)
202. P. Fischer et al: Nucl. Instrum. Methods A **465**, 153–158 (2001)
203. M. Babera et al: Nucl. Instrum. Methods A **517**, 349–359 (2004)
204. W. Erdmann et al: The 0.25 µm front-end for the CMS pixel detector. In: *Proc. of 12th Int. Workshop on Vertex Detectors*, Low Wood, Lake Windermere, Cumbria, UK, Sept 14–19, 2003
205. W. Snoeys et al: Nucl. Instrum. Methods A **465**, 176–189 (2001)
206. M. Lindner, L. Blanquart, P. Fischer, H. Krüger, N. Wermes: Nucl. Instrum. Methods A **465**, 229–234 (2001)
207. P. Delpierre et. al: IEEE Trans. Nucl. Sci. **48**(4), 987–991 (2001)
208. F. Krummenacher: Nucl. Instrum. Methods A **305**, 527–532 (1991)
209. L. Blanquart, A. Mekkaoui, V. Bonzom, P. Delpierre: Nucl. Instrum. Methods A **395**, 313–317 (1997)
210. F. Hügging et al: Front end electronics and integration of ATLAS pixel modules. In: *Proc. 12th Int. Workshop on Vertex Detectors*, Low Wood, Lake Windermere, Cumbria, UK, Sept 14–19, 2003
211. R. Beccherle et al: Nucl. Instrum. Methods A **492**, 117–133 (2002)
212. E. Bartz: The token bit manager chip for the CMS pixel readout. Presented at *PIXEL 2002 (International Workshop on Semiconductor Pixel Detectors for Particles and X-rays)*, Carmel, CA, Sept 9–12, 2002
213. S. Gadomski et al: Nucl. Instrum. Methods A **320**, 217–227 (1992)
214. M. Artuso, J. Wang: Nucl. Instrum. Methods, A **465**, 115–119 (2001)
215. H.L. Hughes, J.M. Benedetto: IEEE Trans. Nucl. Sci. **50**(3), 500–521 (2003)
216. G. Anelli et al: IEEE Trans. Nucl. Sci. **46**(6), 1690–1696 (1999)
217. W. Snoeys et al: Nucl. Instrum. Methods A **439**, 349–360 (2000)
218. W.J. Snoeys, T.A. Palacios Gutierrez, G. Anelli: IEEE Trans. Nucl. Sci. **49**(4), 1829–1833 (2002)
219. M. Huhtinen, F. Faccio: Nucl. Instrum. Methods A **450**, 155–172 (2000)
220. F. Faccio et al: SEU effects in registers and in a dual-ported static RAM designed in a 0.25 µm CMOS technology for applications in the LHC. In: *Proc. 5th Workshop on Electronics for LHC Experiments*, Snow Mass, CO, Sept 20–25, 1999
221. D. Kotlinski: Nucl. Phys. B (Proc. Suppl.) **120**, 249–252 (2003)
222. G. Chiodini et al: Nucl. Instrum. Methods A **501**, 183–188 (2003)
223. T. Calin, M. Nicolaidis, R. Velazco: IEEE Trans. Nucl. Sci. **43**(6), 2874–2878 (1996)
224. X. Aragones, J.L. Gonzales, A. Rubio: Substrate coupling trends in future CMOS technologies. *Presented at PATMOS97*, Louvain-La-Neuve, 1997, pp 235–244
225. R. Dinapoli et al: Nucl. Instrum. Methods A **461**, 492–495 (2001)
226. C. Webb et al: ISSCC Digest of Technical Papers, February 1997, pp. 168–169
227. J. C. Chen, D. Sylvester, Chenming Hu: IEEE Trans. Semicond. Manufact. **11**(2), 204–210 (1998)
228. S.A. Kleinfelder et al: IEEE Trans. Nucl. Sci. **35**(1), 171–175 (1988)

229. E. Beauville et. al: Nucl. Instrum. Methods A **288**, 157–167 (1990)
230. S. Tedja, J.V. d. Spiegel, H.W. Williams: IEEE J. Solid State Circuits **30**(2), 110–118 (1995)
231. M. Manghisioni, L. Ratti, V. Re, V. Speziali: IEEE Trans. Nucl. Sci. **49**(4), 1783–1790 (2002)
232. G. De Geronimo, P. O'Connor, V. Radeka, B. Yu: Nucl. Instrum. Methods A **471**, 192–199 (2001)
233. G. De Geronimo, P. O'Connor: Nucl. Instrum. Methods A **421**, 322–333 (1999)
234. P.R. Gray, R.G. Meyer: *Analysis and Design of Analog Integrated Circuits* (Wiley, New York, 1993)
235. V. Vrba et al: Nucl. Instrum. Methods A **465**, 27–33 (2001)
236. G. Lutz et al: Nucl. Instrum. Methods A **263**, 163–173 (1988)
237. J.T. Walker, S. Parker, B. Hyams, S.L. Shapiro: Nucl. Instrum. Methods A **226**, 200–203 (1984)
238. A. Mekkaoui, J. Hoff: Nucl. Instrum. Methods A **465**, 166–175 (2001)
239. I. Perić: Design and realization of integrated circuits for the readout of pixel sensors in high-energy-physics and biomedical imaging. PhD Thesis, Universität Bonn, Germany (2004)
240. Z.Y. Chang, W.M.C. Sansen: *Low-Noise Wide-Band Amplifiers in Bipolar and CMOS Technologies* (Kluwer Academic, Boston, 1991)
241. G. Bertuccio, L. Fasoli, M. Sampietro: Nucl. Instrum. Methods A **409**, 286–290 (1998)
242. M. Manghisioni et al: IEEE Trans. Nucl. Sci. **51**(3), 980–986 (2004)
243. P. Delpierre, J.J. Jaeger: Nucl. Instrum. Methods A **305**, 627 (1991)
244. R. Horisberger: Nucl. Instrum. Methods A **465**, 148–152 (2001)
245. L. Blanquart et al: Nucl. Instrum. Methods A **439**, 403–412 (2000)
246. M. Wright et al: A column read-out pixel array prototype integrated circuit. Presented at the *Sixth Pisa Meeting on Advanced Detectors*, La Biodola, Elba, May 22–28, 1998
247. P. Fischer: Nucl. Instrum. Methods A **378**, 297–300 (1996)
248. D.W. Clark, L.J. Weng: IEEE Trans. Comput. **43**(5), 560–567 (1994)
249. P. Delpierre et al: IEEE Trans. Nucl. Sci. **49**(4), 1709–1711 (2002)
250. M. Caccia: Nucl. Instrum. Methods A **465**, 195–199 (2001)
251. L. Rossi: Nucl. Instrum. Methods A **501**, 239–244 (2003)
252. J.A. Appel: Overview of the BTeV pixel detector. FERMILAB-CONF-02/273-E (2002)
253. J.H. Laue (ed.): *Flip Chip Technologies* (McGraw-Hill New York, 1996)
254. E.K. Yung et al: IEEE Trans. Comp. Hybrids Manuf. Technol. **14**(3), 549 (1991)
255. O. Ehrmann, G. Engelmann, J. Simon, H. Reichl: A bumping technology for reduced pitch. In: *Proc. 2nd Int. TAB Symposium*, San Jose, 1990, pp 41–48
256. J. Wolf, G. Chmiel, H. Reichl: Lead/tin (95/5%) solder bumps for flip chip applications based on Ti:W(N)/Au/Cu underbump metallization. In: *Proc. 5th Int. TAB/Advanced Packaging Symposium ITAP*, San Jose, 1993, pp 141–152
257. J. Wolf, G. Chmiel, H. Reichl: Solderbumping–a comparison of different technologies. In: *Proc. 6th Int. TAB/Advanced Packaging Symposium*, San Jose, USA, 1994
258. J. Wolf: PbSn60 solder bumping by electroplating. In: *Pixel 2000 Conference*, Genova, Italy, June 2000. URL: http://www.ge.infn.it/Pix2000/slides.html

259. A.M. Fiorello: ATLAS bump bonding process. In: *Pixel 2000 Conference*, Genoa, Italy, June 2000. URL: http://www.ge.infn.it/Pix2000/slides.html
260. C. Brönnimann et al: Developement of an Indium Bump Bond Process for Silicon Pixel Detectors at PSI. In: Pixel 2005, Bonn, Germany, Sept. 2005. To be published in Nucl. Instrum. Methods A
261. P. Fischer et al: IEEE Trans. Nucl. Sci. **48**, 2401–2404 (2001)
262. R. Asgari, V. Romega-Thompson: Adv. Packag. **3**, 31–34 (2002)
263. S. Zimmermann et al: Nucl. Instrum. Methods A **465**, 224–228 (2001)
264. C. Gemme et al: Nucl. Instrum. Methods A **465**, 200–203 (2001)
265. G.L. Alimonti et al: Reworking of indium bump bonded pixel detectors. In: *Pixel 2002 Conference*, Carmel, CA, Sept 2002
266. P. Netchaeva et al: Nucl. Instrum. Methods A **465**, 204–210 (2001)
267. R. Dinapoli et al: The ALICE silicon pixel detector readout system: moving toward system integration. In: *Proc. 8th Workshop on Electronics for LHC Experiments*, Colmar, France, Sept 2002
268. P. Skubic et al: Nucl. Instrum. Methods A **465**, 219–223 (2001)
269. M. Olcese: Nucl. Instrum. Methods A **465**, 51–59 (2001)
270. C. Bayer et al: Development of fluorocarbon evaporative cooling recirculators and controls for the ATLAS inner silicon tracker. ATL-INDET-2000-024 (Dec 2000)
271. E. Anderssen et al: Fluorocarbon evaporative cooling developments for the ATLAS pixel and semiconductor tracking detectors. ATL-INDET-99-016 (Sept 1999)
272. G. Bonna et al: A radiation-hardened low-dropout voltage regulator for LHC and space applications. In: *Proc. 5th Workshop on Electronics for LHC Experiments*, Snow Mass, CO, Sept 20–25, 1998
273. R. Koss et al: Radiation-hard ASICs for optical data transmission in the ATLAS pixel detector. In:*13th Intl. Workshop on Room Temperature Semiconductor X-ray and Gamma-ray Detectors,* Portland, OR, Oct 17–25, 2003
274. P.K. Teng et al: Nucl. Instrum. Methods A **497**, 294–304 (2003)
275. K. Gill et al: Effect of neutron irradiation of MQW lasers to 10^{15}n/cm^2. CMS Note 1997/044 (May 1997)
276. C. Buttar et al: Estimating induced-activation of SCT barrel-modules in the ATLAS radiation environment. ATLAS-INDET-2002-013 (April 2002)
277. C. Gemme et al: Effect of accidental beam losses on the ATLAS pixel detector. ATL-COM-INDET-2005-002 (2005)
278. J.A. Appel: Overview of the BTeV pixel detector. FERMILAB-CONF-02/273-E (2002)
279. D. Barberis: Physics with 2nd generation pixel detectors. In: *Pixel 2002 Conference*, Carmel, CA, Sept 2002
280. C. Gemme: Nucl. Instrum. Methods A **501**, 87–92 (2003)
281. K.K. Gan et al: Nucl. Phys. B (Proc. Suppl.) **125**, 282–287 (2003)
282. The ATLAS Collaboration: Technical Design Report of the ATLAS Inner Detector, vol 1. CERN/LHCC/97-16 (1997)
283. The ATLAS Collaboration: Technical Design Report of the ATLAS Inner Detector, vol 2. CERN/LHCC/97-17 (1997)
284. D. Kotlinski: Nucl. Instrum. Methods A **465**, 46–50 (2001)
285. F. Meddi: Nucl. Instrum. Methods A **465**, 40–45 (2001)
286. C.R. Newsom: Nucl. Instrum. Methods A **465**, 34–39 (2001)

287. E.M. Westbrook: Pixels and proteins: better detectors for biological crystallography. In: IEEE Trans. Nucl. Sci. Portland, USA, Oct 2003
288. N. Wermes: IEEE Trans. Nucl. Sci. **51**(3, Pt 3), 1106–1115 (2004)
289. H. Graafsma: Detector needs at current and future synchrotron sources. In: *IEEE2003 Medical Imaging Conference*, Portland, USA, Oct 2003
290. N. Wermes: Nucl. Instrum. Methods A **541**, 150–165 (2005)
291. P. Holl et al: IEEE Trans. Nucl. Sci. **47**(4), 1421–1425 (2000)
292. L. Strüder et al: Fully depleted, backside illuminated, spectroscopic active *pixel* sensors from the infrared to X-rays. In: SPIE Int. Soc. Opt. Eng., Munich, July 2000, vol 4012, pp 200–217
293. P. Holl et al: Active pixel sensor for X-ray imaging spectrometers. In: SPIE Int. Soc. Opt. Eng., San Diego, March 2003, vol 4851, pp 770–778
294. T. Takahashi et al: IEEE Trans. Nucl. Sci. **48**(3), 287–291 (2001)
295. K. Nakazawa et al: IEEE Trans. Nucl. Sci. **51** (4, Pt 1), 1881–1886 (2004)
296. P. Holl et al: IEEE Trans. Nucl. Sci. **45**(3, Pt 1), 931–935 (1998)
297. L. Strüder et al: Imaging spectrometers for future X-ray missions. In: SPIE Int. Soc. Opt. Eng., San Diego, 2002, vol 4497, pp 41–49
298. B. Everett et al: Proc. IEEE **124**, 995–1000 (1977)
299. P.F. Bloser et al: New Astr. Rev. **46** (8–10), 611–616 (2002)
300. Y. Eisen, A. Shor, I. Mardor: IEEE Trans. Nucl. Sci. **51**(3, Pt 3), 1191–1199 (2004)
301. T. Conka-Nurdan et al: IEEE Trans. Nucl. Sci. **49**(3, Pt 1), 817–821 (2002)
302. T. Takahashi et al: Proc. SPIE **5501**, 229–240 (2004)
303. J.R.J. Baker: *Autoradiography: A Comprehensive Overview* (Oxford University Press, Oxford, 1989)
304. A.W. Rogers: *Techniques of Autoradiography,* 3rd edn (Elsevier, Amsterdam, 1979)
305. M. Overdick et al: Nucl. Instrum. Methods A **392**, 173–177 (1997)
306. M. Overdick: Digital autoradiography using silicon strip detectors. PhD Thesis, Universität Bonn, Germany (1998) BONN-IR-98-05
307. J.C. Dainty, R. Shaw: *Image Science* (Academic, London, 1974)
308. N. Jung et al: SPIE **3336**, 396–407 (1998)
309. P.R. Granfors et al: SPIE **4320**, 77–86 (2001)
310. M. Choquette et al: SPIE **4320**, 501–508 (2001)
311. F. Busse et al: SPIE **4682**, 819–827 (2002)
312. O. Tousignant et al: SPIE **4682**, 503–510 (2002)
313. MEDIPIX Collaboration, S.R. Amendolia et al: Nucl. Instrum. Methods A **509**, 283–289 (2003)
314. F. Edling et al: Performance of a pixel readout chip with two counters for X-ray imaging. In: *IEEE2002 Nuclear Science Symposium,* Lyon, France, Nov 2002, vol 1, pp 29–32
315. P. Fischer, A. Helmich, M. Lindner, N. Wermes, L. Blanquart: IEEE Trans. Nucl. Sci. **47**(3), 881–884 (2000)
316. M. Lindner, L. Blanquart, P. Fischer, H. Krüger, N. Wermes: Nucl. Instrum. Methods A **465**, 229–234 (2000)
317. M. Löcker et al: IEEE Trans. Nucl. Sci. **51**(4, Pt 1), 1717–1723 (2004)
318. G. Mettivier, M.C. Montesi, P. Russo: Nucl. Instrum. Methods A **516**, 554–563 (2004)
319. P. Delpierre et al: IEEE Trans. Nucl. Sci. **48**(4), 987–991 (2001)

320. P. Delpierre et al: IEEE Trans. Nucl. Sci. **49**(4), 1709–1711 (2002)
321. E. Eikenberry et al: Nucl. Instrum. Methods A **501**, 260–268 (2003)
322. C. Brönnimann et al: J. Synchrotron Radiat. **7**, 301 (2000)
323. B. Schmitt et al: Nucl. Instrum. Methods A **518**, 436–439 (2004)
324. J. Kemmer, G. Lutz: Nucl. Instrum. Methods A **253**, 356–377 (1987)
325. N. Wermes et al: IEEE Trans. Nucl. Sci. **51**(3, Pt 3), 1121–1128 (2004)
326. W. Neeser et al: IEEE Trans. Nucl. Sci. **47**(3), 1246–1249 (2000)
327. O. Ehrmann: Private communication, Berlin, 2004
328. M.N.T. Volpert, M. Fendler, F. Marion, L. Mathieu, J.-M. Debono, et al: Ultra fine pitch hybridization of large imaging detectors. In: *2003 IEEE Nuclear Science Symposium*, Portland, USA, Oct 2003
329. J. Wolf et al: High density pixel detector module using flip chip and thin film technology. In: SPIE Conference, Denver, USA, 2000, vol 4217, pp 553–556
330. P. Gerlach, C. Linder, K.H. Becks: Nucl. Instrum. Methods A **473**, 102–106 (2001)
331. E.H.M. Heijne, L. Hubbeling: Nucl. Instrum. Methods **178**, 331–341 (1980)
332. J. Kemmer, G. Lutz: Nucl. Instrum. Methods A **273**, 588–598 (1988)
333. W. Kucewicz et al: Acta Phys. Pol. B **30**, 2075–2083 (1999)
334. M. Amati et al: Nucl. Instrum. Methods A **511**, 265–270 (2003)
335. C. Kenney, S. Parker, J. Segal, C. Storment: IEEE Trans. Nucl. Sci. **48**(4), 1224–1236 (1999)
336. C.J. Kenney, S. Parker, E. Walckiers: IEEE Trans. Nucl. Sci. **48**(6), 2405–2409 (2001)
337. S. Parker, C.J. Kenney: IEEE Trans. Nucl. Sci. **48**(5), 1629–1638 (2001)
338. C.J. Kenney et al: IEEE Trans. Nucl. Sci. **48**(2), 189–193 (2001)
339. W.R.T. ten Kate, C.L.M. Van der Klauw: Nucl. Instrum. Methods A **228**, 105–109 (1984)
340. S. Holland: Nucl. Instrum. Methods A **275**, 537–541 (1989)
341. G. Vanstraelen, I. Debusschere, C. Claeys, G. Declerck: Nucl. Instrum. Methods A **275**, 574–579 (1989)
342. W. Snoeys, J. Plummer, S. Parker, C.J. Kenney: IEEE Trans. Nucl. Sci. **39**, 1263–1269 (1992)
343. W. Dulinski et al: CMOS monolithic active pixel sensors for high resolution particle tracking and ionizing radiation imaging. In: *Proc. Frontier Detectors for Frontier Physics 2003*, Elba, May 2003
344. R.-D. Heuer, R. Settles T. Behnke, S. Bertolucci (eds): TESLA Technical Design Report. DESY-01-011, vol IV (2001)
345. W. Snoeys et al: Nucl. Instrum. Methods A **326**, 144–149 (1993)
346. C.J. Kenney et al: Nucl. Instrum. Methods A **342**, 59–77 (1994)
347. B. Dierickx, G. Meynants, D. Scheffer: Near 100% fill factor CMOS active pixel. In *Proc. IEEE CDD&AIS Workshop*, Brugge, Belgium, June 1997
348. G. Meynants, B. Dierickx, D. Scheffer: Proc. SPIE–Int. Soc. Opt. Eng. (USA), **3410**, 68–76 (1998)
349. R. Turchetta et al: Nucl. Instrum. Methods A **458**, 677–689 (2001)
350. G. Claus et al: Nucl. Instrum. Methods A **473**, 83–85 (2001)
351. G. Deptuch, W. Dulinski, Y. Gornushkin, C. Hu-Guo, I. Valin: Nucl. Instrum. Methods A **512**, 299–309 (2003)
352. G. Deptuch et al: Monolithic active pixel sensor with in-pixel double sampling and column-level discrimination. In: *IEEE Nuclear Science Symposium*, Portland, USA, Oct 2003, vol 1, pp 551–555

353. R. Turchetta: Monolithic active pixel sensors (MAPS) for particle physics and space science. In: *Proc. VERTEX 2003*, Low Wood, Lake Windermere, UK, Sept 2003
354. H. Matis et al: IEEE Trans. Nucl. Sci. **50**, 1020–1025 (2003)
355. A. Gay: High resolution CMOS sensors for a vertex detector at the linear collider. In: *Proc. Vertex 2003 Conference*, Lake Windermere, UK, Sept 2003
356. L. Jungermann: Space-qualified electronics for the AMS02 experiment and medical radiation imaging. PhD Thesis, Universität Karlsruhe, Germany (2005) Report IEKP-KA/2005-6
357. W. Dulinski et al: IEEE Trans. Nucl. Sci. **51**(4, Pt 1), 1613–1618 (2004)
358. Y. Gornushkin et al: Nucl. Instrum. Methods A **513**, 291–295 (2003)
359. J.J. Velthuis et al: IEEE Trans. Nucl. Sci. (2005), in press
360. S. Kleinfelder et al: Novel integrated CMOS pixel structures for vertex detectors. In: *IEEE Nuclear Science Symposium 2003*, Portland, USA, Oct 2003, vol 1, pp 335–339
361. J. Marczewski et al: IEEE Trans. Nucl. Sci. **51**(3): 1025–1028 (2004)
362. J.A. Theil et al: a-Si:H photodiode technology for advanced CMOS active pixel sensor imagers. In: *19th Int. Conf. on Amorphous Materials and Semiconductors*, Nice, Aug 2001
363. J.S. Drewery et al: Nucl. Instrum. Methods A **310**, 165–170 (1991)
364. B. Equer, J.B. Chevrier: Mater. Res. Soc. 1045–1055 (1992)
365. J. Kemmer et al: Nucl. Instrum. Methods A **288**, 92–98 (1990)
366. P. Klein et al: Nucl. Instrum. Methods A **392**, 254–259 (1997)
367. P. Fischer et al: Nucl. Instrum. Methods A **451**, 651–656 (2000)
368. J. Ulrici et al: Nucl. Instrum. Methods A **465**, 247–252 (2001)
369. E. Gatti, P. Rehak: Nucl. Instrum. Methods A **225**, 608–614 (1984)
370. P. Rehak et al: Nucl. Instrum. Methods A **235**, 224–234 (1985)
371. G. Cesura et al: Nucl. Instrum. Methods A **377**, 521–528 (1996)
372. R. Richter et al: Nucl. Instrum. Methods A **511**, 250–256 (2003)
373. P. Fischer et al: A DEPFET based pixel vertex detector at TESLA. DESY Linear Collider Note, LC-DET-2002-004 (2002)
374. M. Trimpl et al: Nucl. Instrum. Methods A **511**, 257–264 (2003)
375. P. Fischer, W. Neeser, M. Trimpl, J. Ulrici, N. Wermes: Nucl. Instrum. Methods A **512**, 318–325 (2003)
376. L. Andricek, G. Lutz, M. Reiche, R.H. Richter: IEEE Trans. Nucl. Sci. **51**(3, pt 3), 1117–1120 (2004)
377. P. Holl et al: Proc. SPIE–Int. Soc. Opt. Eng. (USA), **4851**, 770–778 (2003)
378. W. Neeser et al: Nucl. Instrum. Methods A **477**, 129–136 (2002)
379. P. Klein et al: Nucl. Instrum. Methods A **454**, 152–157 (2000)
380. L. Andricek: Radiation hardness of DEPFET detectors. Presented at the *2005 International Linear Collider Workshop*, Stanford, March 18–25, 2005

Index

δ-electrons 31
η-function 63

accelerator 3, 13, 21
acceptor 28
accumulation 47, 80
   counting 261
   integration 261
ALICE 221, 225, 241
aluminum 118
aluminum spikes 118
amorphous silicon 277
annealing
   beneficial 74
   leakage current 70
   reverse 75
   short term 74
   surface damage 81
ASIC 13, 233
ATLAS 15, 224, 225, 228, 232
attenuation length 248
autoradiography 246, 280
   biomedical 256, 281
   tritium 248, 256

Bethe–Bloch formula 29
Bragg
   diffraction 267
   peak 32
   spot 254
breakdown voltage 43
bricked pixel pattern 66
BTeV 225, 243, 246
built-in voltage 40
bulk damage 68
bump bonding 203, 242, 244
bumping 262
   Au stud 210

C4 process 204
   indium 206, 207
   indium reflow 210
   reworking 213
   solder(PbSn) 206
   UBM 204

capacitance
   backplane 57
   input 269
   interpixel 58, 266
   pixel 57
carbon–carbon 235
carriers
   majority carriers 28
   minority carriers 28
CdTe 11, 248, 262
CdZnTe 248
CERN 6, 21
CFRP 241
channeling 117
charge amplifier
   basics 136
   cascode 162
   optimal collector current 185
charge sharing 65, 266
charge trapping 77
charmed particles 13
chemical vapor deposition 113
CME 216, 220
CMOS 18, 270
   technology 272
CMS 221, 225, 237
collection electrode 267, 271
collider 261
   e^+e^- 262
   hadronic 201, 202, 225
   linear 281

Compton camera 281
Compton scattering 34
contrast 249, 252, 261
counting pixels 251, 254, 256
cross talk 59
  in sensor 149
  on chip 157
crystallography 246, 254
  protein 250, 254
CTE 216, 220
current
  dark 43, 51
  leakage 70
current-to-voltage characteristic 52
CV-measurement 54
Czochralski process 111

damage
  constant 70
  rate 70
  stable 74
DELPHI 21, 23
DEPFET pixel 12, 257, 278
depletion 47
  deep 48
  depth 42
  full 272
  zone 2, 40
dew point 222
diamond 11
diffraction 254
diffusion 9, 37, 116
  layer 271
  oxygenation 116
discriminator 141
  differential 168
DOFZ 116
donor 28
donor removal 74
doping 28
  by diffusion 116
  by implantation 116
  effective 72
double junction 77
drift 37, 267
  distance 267
dynamic range 252

edge electrode 267

electrical breakdown 52
electron accumulation 80
electroplating 264
ENC 147
End of Column logic 229, 233
energy
  resolution 256
  windowing 252
epitaxial layer 273
equivalent noise charge 147
  in CR–RC-shaper 183
  without shaper 175
etching 115
evaporative cooling system 221

fabrication of sensors 110
Fano factor 35
Faraday cage 223
FE, front end 14, 15, 233
feedback
  capacitor 137, 163
  circuit 138, 163
Fermi
  energy 27
  function 26
  intrinsic level 28
  level 28
Fermilab 225, 244
flat-panel imager 251
flex hybrid 264
flip-chip 207, 209
float zone process 112

GaAs 11, 248, 262
generation of charge carriers 35

Hamburg model 72
HgI_2 248
Hybrid active pixel sensors 266
hybrid pixel detector 1
hybridization 248, 262, 269

imaging 261
  in astronomy 246
  in biomedicine 247, 261
  in crystallography 250
  in radiology 261
  X-ray 246
impact parameter 236, 241

implantation   116
inversion   48
   space charge sign   72
   strong   48
   type   72
IV-curve   52

Joule effect   222
junction   6, 8, 40

kapton   222

Landau distribution   31
latency   189
latent heat   221
leakage current   133
   compensation   163
LEP, Large Electron Positron collider   21
LHC, Large Hadron Collider   18, 215, 223, 225, 226
lifetime   3
liquid cooling system   220, 221
Lorentz
   angle   9, 39, 65
   boost   18
   factor   40
   force   9, 235, 238
LSF, line spread function   248

MAPS   13, 272, 273
mask alignment   262
MCC, module control chip   15, 233, 234
MCM-D   216, 263, 264
metallization   118
microdischarge   52
microstrip detectors   4, 6, 14, 26
MIMOSA   273
minimum ionizing particle   31
mobility   38, 277
   Hall   40
   low field   38
module controller chips   143
monolithic pixels   271
MOS structure   46
MOS structure, shift of flat band voltage   80
MOSFET   17

MQW, multiquantum well   223
MTF, modulation transfer function   249
multiple scattering   33, 241

net doping   72
NIEL   69
n-material   28
noise   279
   equivalent noise charge   175
   Fano   257, 279
   in bipolar amplifier   184
   in FET amplifier   183
   measurement   148
   requirement   147
   with semi-gaussian shaper   178, 182
   without shaper   171

OmegaD   18–20
open p-stop   105
oxidation
   dry   113
   wet   113
oxide charge   47, 80
oxygen enriched material   75

pad detector   50
pair production   34
parallax   247
photoelectric effect   34
photolithography   114
p-material   29
pitch   262, 265
pixel
   geometry   145
pixel chip
   geometry   130
   photograph   134
   table   132
pixel detector
   counting   251
   hybrid   23, 262
pixel unit cell   135
planar
   device   269
   technology   110
pn-junction   40
polyimide   206
polysilicon resistor   266

power
  density 262
  duty cycle 282
process sequence 118
PSF, point spread function 248
PSI, Paul Scherrer Institute 240
punch through 48

radiation damage 68
radiation hardness
  design for 154
  DICE cell 157
  single event upset 156
radiation length 242, 243, 245
radiography 252
radiology 246
readout architecture 187
  BTeV 196
  column drain 194
  conveyor belt 191
  counter 197
  DELPHI 188
  time stamp 192
  timer 190
recombination of charge carriers 35
residual 63
resistivity 7
resolution
  analog 63
  binary 61
  spatial 61, 261
reticle 275
reverse annealing 224

saturation of oxide charge 80
Schockley–Read–Hall process 36
segregation 113
semiconductor
  doped 26
  intrinsic 26
sensor
  equivalent circuit 133
SEU, single event upset 156, 229
shaper 140, 178
  pulse shapes 178
signal formation 59
small pixel effect 61
SOI, silicon on insulator 270, 276
space charge region 40

sparse CMOS 265
SSC, Superconducting Super Collider 18
stitching 275
surface barrier 44
surface damage 80

TBM, Token Bit Manager 239
Tevatron 215, 225, 244, 246
thermal
  conductivity 219
  runaway 78
  conductivity 220
  oxidation 113
thinning 214
threshold
  trim 141, 170
  variations 146
threshold dispersion 252
time walk 141, 227
timing 261
tracker 236
transconductance 278
trapping 77
trapping time 77
trigger 4, 17, 189, 226, 233
Tritium 256
type inversion 218, 229, 238

VCSEL, vertical cavity surface emitting lasers 223
vertex detector 262
VLSI 275
voltage
  breakdown 43, 52
  flat band 46, 80
  full depletion 53
  punch through 49
  threshold 48

W boson 21
weighting field 59

X-ray 5, 13, 265
  astronomy 280

yield 264
Young's modulus 220

zero suppression 188

Printing and Binding: Strauss GmbH, Mörlenbach